Soil Biodiversity in Amazonian and Other Brazilian Ecosystems

Soil Biodiversity in Amazonian and Other Brazilian Ecosystems

Edited by

F.M.S. Moreira and J.O. Siqueira

Department of Soil Science
Federal University of Lavras
Brazil

and

L. Brussaard

Department of Soil Quality
Wageningen University
The Netherlands

CABI Publishing

CABI Publishing is a division of CAB International

CABI Publishing
CAB International
Wallingford
Oxfordshire OX10 8DE
UK
Tel: +44 (0)1491 832111
Fax: +44 (0)1491 833508
E-mail: cabi@cabi.org
Website: www.cabi-publishing.org

CABI Publishing
875 Massachusetts Avenue
7th Floor
Cambridge, MA 02139
USA
Tel: +1 617 395 4056
Fax: +1 617 354 6875
E-mail: cabi-nao@cabi.org

A catalogue record for this book is available from the British Library, London, UK.

Library of Congress Cataloging-in-Publication Data

Soil biodiversity in Amazonian and other Brazilian ecosystems/edited by F.M.S. Moreira and J.O. Siqueira and L. Brussaard.
p. cm.
Includes bibliographical references and index.
ISBN-13: 978-1-84593-032-5 (alk. paper)
ISBN-10: 1-84593-032-0 (alk. paper)
1. Soil invertebrates--Ecology--Brazil. 2. Soil microbiology--Brazil. 3. Biological diversity--Brazil. I. Moreira, F.M.S. (Fattima M.S.) II. Siqueira, J.O. (José Oswaldo) III. Brussaard, L. (Lijbert) IV. Title.
QL365.45.B6S65 2005
577.5'7'0981—dc22

2005015579

ISBN-10: 1-84593-032-0
ISBN-13: 978-1-84593-032-5

Typeset by SPI Publisher Services, Pondicherry, India.
Printed and bound in the UK by Cromwell Press, Trowbridge.

Contents

The colour plate section can be found following p. 22.

Contributors

Acioli, A.N.S., *PPG Entomologia – INPA, Caixa Postal 478, 69011-970, Manaus, AM, Brazil.*

Barros, E., *Instituto Nacional de Pesquisas d Amazonia–Agronomia, Av. Andrê Araújo 2936, Manaus-AM, 69083-000 Brazil.*

Bernardi, A.C.C., *EMBRAPA Pecuaria Sudeste, Rodovia Washington Luiz, km 234, Fazenda Canchim, Caixa Postal 339, CEP 13560-970, Sao Carlos, SP, Brazil.*

Brown, G.G., *EMBRAPA Soja, Rod. Carlos Joao Strass acesso Orlando Amaral, CP 231, Londrina, PR 86001-970, Brazil.*

Brussaard, L., *Department of Soil Quality, Wageningen University, PO Box. 8005, 6700 E.C. Wageningen, The Netherlands.*

Cares, J.E., *Universidade de Brasilia, Instituto de Ciencias Biologicas, Departamento de Fitopatologia, Caixa Postal 4457, CEP 70, 904-970 Brazil.*

Coelho, M.R., *EMBRAPA Solos, Rua Jardin Botanico 1024, CEP 2246-000, Rio de Janeiro, Brazil.*

Constantino, R., *Department of Zoology, University of Brasilia, 70910-900 Brasilia, DF Brazil.*

de Abreu, L.M., *Departamento de Fitopatologia, Universidade Federal de Lavras, 37200-000 Lavras MG, Brazil.*

de Morais, J.W., *Instituto Nacional de Pesquisas da Amazonia (INPA), Coordenacao de Pesquisas em Entomologia (CPEn), CP 478, 69011-970 Manaus, AM, Brazil.*

dos Santos, H.G., *EMBRAPA Solos, Rua Jardin Botanico 1024, CEP 2246-000, Rio de Janeiro, Brazil.*

Fidalgo, E.C.C., *EMBRAPA Solos-Rua Jardin Botanico 1024-CEP 2246-000, Rio de Janeiro, Brazil.*

Franklin, E., *Instituto Nacional de Pesquisas da Amazonia (INPA), Coordenacao de Pesquisas em Entomologia (CPEn), CP 478, 69011-970 Manaus, AM, Brazil.*

Huang, S.P., *Universidade de Brasilia, Instituto de Ciencias Biologicas, Departamento de Fitopatologia, Caixa Postal 4457, CEP 70, 904-970 Brazil.*

James, S.W., *Kansas University Natural History Museum and Biodiversity Research Centre, Lawrence, Kansas, 66045, USA.*

Lavelle, P., *Institut de Recherche pour le Développement, UMR 137 BIOSOL, 32 Avenue Henri Varagnat, 93143 Bondy Cedex, France.*

Machado, P.L.O.A., *EMBRAPA Solos, Rua Jardin Botanico, 1024-CEP 2246-000, Rio de Janeiro, Brazil.*

Manzatto, C.V., *EMBRAPA Solos, Rua Jardin Botanico, 1024-CEP 2246-000, Rio de Janeiro, Brazil.*

Mathieu, J., *Institut de Recherche pour le Développement, UMR 137 BIOSOL, 32 Avenue Henri Varagnat, 93143 Bondy Cedex, France.*
Mendonça-Santos, M.L., *EMBRAPA Solos, Rua Jardin Botanico 1024, CEP 2246-000, Rio de Janeiro, Brazil.*
Moreira, F.M.S., *Departamento de Ciencia do Solo, Universidade Federal de Lavras, Caixa Postal 3037, Lavras, MG, CEP 37 200-000, Brazil.*
Nascimento, A.R.L., *Instituto Nacional de Pesquisas de Amazonia–Agronomia, Av. Andrê Araújo 2936, Manaus–AM, 69083-000 Brazil.*
Pfenning, L.H., *Departamento de Fitopatologia, Universidade Federal de Lavras, 37200-000 Lavras MG, Brazil.*
Siqueira, J.O., *Departamento de Ciencia do Solo, Universidade Federal de Lavras (UFLA), Caixa Postal 3037, Lavras, MG, CEP 37200-000, Brazil.*
Sturmer, S.L., *Departamento de Ciencias Naturais (DCN), Universidade Regional de Blumenau (FURB), Caixa P. 1507, 89010-971 Blumenau, SC Brazil.*
Tapia-Coral, S., *Instituto Nacional de Pesquisas de Amazonia - Agronomia, Av. Andrê Araújo 2936, Manaus–AM, 69083-000 Brazil.*
Vasconcelos, H.L., *Institute of Biology, Federal University of Uberlandia (UFU), CP 593, 38400-902 Uberlandia, MG, Brazil.*

Foreword

Over the last decade humans have become increasingly concerned about the impact that they have on the environment. These concerns, originating locally in our own backyards, have become global and are enshrined in international conventions to combat desertification, climate change and the loss of biological diversity. This book targets the third of these concerns but has high significance for the other two. It addresses the state of biological diversity in the Amazon. Although not a new subject at first glance, this book is not so much concerned with rainforest trees, mammals or birds as with the diversity of life hidden from our general consciousness below the ground, i.e. in the soil.

Why should we be concerned about this element of diversity? There are various reasons why we value and seek to preserve biological diversity. For many it is a question of belief and ethics – why should the human species assume that they have more right to live than any other species? For others it is more aesthetic, because they love and get pleasure from the beauty and romance of the living world. For many, however, their concern is based on the conviction that our own survival as a species is inextricably dependent on the maintenance of efficiently functioning ecosystems, a concern that easily translates into the need to maintain biological richness. For most of us, perhaps the concern is a mixture of all of the above.

How does soil biodiversity fit into these concerns? Few of us will instinctively mourn the disappearance of a microorganism or a microscopic worm. Perhaps it will occasionally register if a previously common fungus, beautiful and tasty, is missing from our autumnal environment, or if the earthworms disappear from our garden. In general, however, any change in the diversity of soil is likely to go unremarked by all but the specialist. But we should be concerned – because many of the natural processes that provide for our food and comfort are dependent on these hidden organisms.

A multitude of soil organisms – bacteria, fungi and soil animals – are the primary agents of decomposition and drivers of nutrient cycling and thence food and fibre production. They are major contributors to greenhouse gas emissions, so any imbalance in their activities affects our climate. They regulate the dynamics of soil organic matter and thus the storage of carbon in the soil, which can counter greenhouse gas emissions. They modify soil physical structure and thence regulate the availability of water to plants as well as the susceptibility of soil to erosion. Soil microorganisms have been the source of many important medicines, including most of the early antibiotics. The inventory of functional importance can go on. But despite this functional significance the biota of soil remains substantially hidden – to scientific understanding as well as to the common gaze. There is now, however, a major attempt to gain the understanding that will enable us to better manage this crucial resource and conserve and protect these beautiful beings. This book has been written as part of this endeavour.

Among the wonderful varieties of life on land there are perhaps two types of ecosystems that have most captured the human imagination: the tropical savannah because that is where we originated as a species and the tropical rainforest because it is the richest and the most dark and mysterious and beautiful of environments. And among tropical rainforests the Amazon is the greatest in actual extent and looms largest in our interest and concern. We are thus fortunate to have this book as an addition to the scientific literature charting the multiple faces of the Amazon, adding indeed a dimension hitherto missing.

High levels of diversity demand a wide range of expertise to describe them. This is evident in the book, which has contributions from 27 authors, every one an expert in his or her own part of the diversity puzzle. The book is about diversity but is by no means just a catalogue: the functional importance of the soil biota is explicitly or implicitly addressed at all turns, in particular with respect to the practices of agriculture and food production. The Amazon is not a pristine forest but a home to humans and has been the source of livelihood for thousands of years.

Human impact on nature dates substantially from the origins of our history as agriculturalists. The human footprint is evident on the organisms below the ground just as it is above, so the book opens by considering agricultural practices in the Amazon, and in particular the management of soil, before moving on to consider the abundance and diversity of the soil organisms. The key to the maintenance of a mutually sustainable relationship between humans and forests rests on the way in which we manage the resources that the forest gifts. The biological diversity below ground is part of that resource, and is susceptible to mismanagement in the same way as is the forest of which it is a part. It remains an act of faith that better understanding of our biological resources will lead to improvements in their management. This book is an important contribution to the validation of that belief.

Mike Swift
Former Director
Tropical Soil Biology & Fertility Institute
of CIAT
Nairobi
Kenya

1 Soil Organisms in Tropical Ecosystems: a Key Role for Brazil in the Global Quest for the Conservation and Sustainable Use of Biodiversity

F.M.S. Moreira,[1] J.O. Siqueira[1] and L. Brussard[2]

[1]*Departamento de Ciência do Solo, Universidade Federal de Lavras, Caixa Postal 37,Lavras, MG, CEP 37 200-000, Brazil, e-mail: fmoreira@ufla.br, siqueira@ufla.br;* [2]*Department of Soil Quality, Wageningen University, P.O. Box. 8005, 6700 EC Wageningen, The Netherlands, e-mail: lijbert.brussaard@wur.nl*

Setting the Scene

The Convention on Biological Diversity (CBD), which resulted from the United Nations World Conference on Environment and Development held in Rio de Janeiro in 1992, was an expression of the worldwide concern that the alarming rate at which we are losing species would somehow affect human life. Brazil was the first country to sign the convention and has installed a number of rules and regulations (Box. 1.1) to follow up on the numerous recommendations made by the so-called Conferences of the Parties (COP) held by the signatories to the convention every other year.

In the minds of the public at large, biodiversity is predominantly associated with visible plants and animals, admired for their beauty or their size. However, most biodiversity probably resides in the soil and is hardly visible to the naked eye, but extremely important when it comes to the continuous supply of goods (in agriculture and medicine) and maintenance ecosystem services.

The processes soil organisms carry out, such as organic matter (dead plants, animals and microorganisms) decomposition, nutrient cycling, biological control of pests and diseases, purification of water and the breakdown of organic residues and toxic substances, among many others, are of vital importance to the survival of all beings on this planet. 'Functional redundancy' of species results in the 'resilience' (ability to recover) of these processes when an adverse condition happens, because biodiversity reflects different adaptations to an ever-changing environment. If the environmental conditions turn adverse for some species, other species adapted to the new environment replace the previous ones in their function(s). Microbial plasticity and adaptation is such that even extremely adverse conditions such as temperatures above 100°C, high salinity (174 g/l NaCl), high contents of harmful heavy metals and nuclear wastes that make the survival of most species rather difficult are optimal for some microbial species (e.g. *Thermus aquaticus*, *Acidianus infernos*, *Halobacterium*

 Soil Biodiversity in Amazonian and Other Brazilian Ecosystems (eds F.M.S. Moreira *et al.*)

Box. 1.1. Brazilian policies related to the Convention on Biological Diversity.

The Ministry of Environment has five secretariats (http://www.mma.gov.br/), one of which is the secretariat of 'Biodiversity and Forests', created in 1999 (Decree no. 2972, 26 February 1999), which also covers soil biodiversity. Its main objectives are the proposing of policies and rules, the definition of strategies and the implementation of projects and programmes related to the following themes:

1. Shared management of sustainable use of natural resources.
2. Knowledge, conservation and sustainable use of biodiversity.
3. Access to genetic resources.
4. Reforestation and recovery of deforested areas.
5. Sustainable use of ichthyofauna and fishing resources.
6. Management of the national system of conservation units.
7. Sustainable use of forests, including preservation and control of forest slash and burning.

After the publication of the first National Report to the Convention on Biological Diversity by the Ministry of Environment-MMA (1998), results of the projects have been compiled in a series of publications named 'Biodiversidade' (Biodiversity), of which six volumes have been already released.

Main laws and decrees submitted to and approved by the Brazilian environmental legislation and relevant to biodiversity were:

- Order no. 55, 14 March 1990 – Rule on the collection of scientific material by foreigners.
- Project of Senate law no. 306/95 – Concerning legal instruments to control access to genetic resources and other measures.
- Law of environmental crimes no. 9605/98.
- Decree no. 2519 (16 March 1998) – Promulgate the Convention on Biological Diversity signed in Rio de Janeiro on 5 June 1992.
- Decree 3179/99, Art. 14 – Penalties in collecting zoological material for scientific purposes without special license delivered by a competent authority.
- Law 9.985/00 – National system of Nature Conservation Units: *in situ*, 36 categories.
- Decree no. 3.420, 20 April 2000 – National Programme of Forests.
- Provisional act 2.186–16/01 – Access to genetic resources, protection of and access to traditional knowledge associated with the sharing of benefits and the access to and transfer of technology for its conservation and utilization.
- Presidential decree no. 4339/02 – National Policy for Biodiversity.
- Presidential decree no. 4703/03 – National Commission on Biodiversity (including representatives of indigenous people and the Brazilian Society for the Progress of Science).

saccharovorum, *Micrococcus radiophilus*). Thus, virtually all places on Earth have inhabitants and their activities can also modify their environments, turning them into suitable places for other species. For instance, the bare rock surface colonized by bacteria, algae and fungi for millions of years becomes the soil in which we produce our food nowadays. Or to give a short-term example, some microbial strains of *Burkholderia* sp. are able to remove toxic wastes from the environment (sea, soil, lakes, etc).

Air, water, soil and biodiversity constituting the global environment and 'infrastructure' provide the raw materials, or natural capital, for all goods and services that have added value to humankind. They used to be considered free goods, but societies that regulate, by law or convention, the use of natural resources for production purposes and the disposal of by-products are becoming the rule not the exception. These aspects have become a matter of controversy between those exploiting natural resources, such as hydroelectric power, land mining and agriculture, and the conservationists when it comes to goals and means of biodiversity conservation. So we increasingly see price tags for the rights to exploit land or water resources at the national level and the use of transferable market rights in the global arena, e.g. fishing and milk production quotas in the European Union and greenhouse gas emission rights worldwide. But the market is not working well towards saving natural capital and avoiding its overexploitation and degradation.

Biodiversity is a case in point. Wild land biodiversity, to begin with, is under severe pressure as a consequence of human population growth, intensive agriculture with high inputs of pesticides, fertilizers and fossil fuels, and expansion, leading to destruction and fragmentation of biodiversity-rich areas. The trade-offs for rights to exploit natural resources are subject to market forces in a number of cases, whereas the resource itself, i.e. the raw material of biodiversity in its natural habitat, is not assigned a value. The capital represented by this resource can be effectively sold (out). This signals a basic flaw, at least in the industrialized world: our social arrangements are interest- not capital-oriented. As such, natural capital will be lost before future generations can profit from possibilities of interest that are yet to be discovered.

In the case of agrobiodiversity the market works in the sense that crop varieties and landraces can be patented by breeders. But who are they? Increasingly, they are breeding/seed companies who sell their products to farmers willing to do so in the face of large gains in the short term. Fair enough, but in the process we see the diversity of varieties and landraces decrease. The breeders used to be innovative farmers, who capitalized on their wit and craftsmanship, working with nature in a coevolutionary way, where human knowledge (technology) and natural evolution intrinsically interact to increase diversity. Diversity also pays off, albeit locally, and with less short-term revenue, but also with less risk, meaning more sustainability. Control of diseases, weeds and pests and maintaining soil fertility by agrochemicals works in the short term, but can create problems in the long run, if improperly managed.

Hence, wild land biodiversity and agrobiodiversity are subject to the same process of genetic erosion. This fact emphasizes that we need an approach where scientists from different disciplinary backgrounds work together and with other 'knowledge-bearers', including traditional farmers, policymakers and politicians, towards a truly integrative scientific approach to deal with biodiversity. This should address biodiversity as a natural capital that humans can save without losing interest, but rather by gaining gradual interest. We believe that Brazil, where the controversy between conservationists and agricultural producers is sometimes very tense, can set an example in visionary policy to the world for reconciling the conflict, building on the international prestige it acquired in 1992.

As scientists we need to describe and reveal biodiversity, in order to explain the role it plays in the working of nature, and to assist policymakers and other stakeholders to make informed choices about its conservation and sustainable use. As already mentioned, soil biota is a major component of terrestrial ecosystems, but as yet we know appallingly little of the biodiversity in our soils. Here again, however, Brazilian scientists have been instrumental in launching an international project, cofunded by the Global Environmental Facility (GEF) on the 'Conservation and Sustainable Management of Below-Ground Biodiversity' (CSM-BGBD) (Box. 1.2). For most of the soil organisms studied, the 'state of the art' in Brazil before the beginning of the project had to be known. This marks the relevance of the present book, which was written by researchers participating in the project and other collaborating experts. Preliminary results of the project already indicate a great contribution to the knowledge of below-ground biodiversity as well as a huge contribution to official collections of soil organisms.

Another important milestone happened in 2002, when Brazil was again one of the driving forces in convincing the COP to add a technical paragraph on soil biodiversity to the convention and to launch an 'International Initiative for the Conservation

Box. 1.2. Conservation and Sustainable Management of Below-Ground Biodiversity (Project GF/2715-02).

The initial step for this project was a workshop funded by the United Nations Environment Programme (UNEP) and convened by the Tropical Soil Biology and Fertility Programme (TSBF), which was held at the International Crops Research Institute for the Semi-Arid Tropics (ICRISAT) Centre, Hyderabad, India, in January 1995. Forty-four scientists from 15 countries attended the workshop with the purpose of exploring the implications of soil biodiversity loss for small-scale agricultural systems in the tropics and for evaluating the potential for improving sustainable agricultural production by management of soil biota. As a result of this meeting a special issue of Applied Soil Ecology (no. 6, 1997) was published and the full report was published by TSBF (1996). A further step was in 1997 when Brazilian scientists of different areas, including below-ground biodiversity, started working together in the project 'Alternatives to Slash and Burn in Brazil' from which a summary report and a book were published later on (Lewis *et al.*, 2002, Bignell *et al.*, 2005).

The project 'Conservation and Sustainable Management of Below-Ground Biodiversity' was submitted for funding to GEF in November 1998 after endorsement of the focal points from the participating countries. In the case of Brazil, it was in agreement with the demands of the new legislation and political programme as shown above. The contract with UNEP was signed in August 2002. The project is carried out in seven countries: Brazil, Ivory Coast, India, Indonesia, Kenya, Mexico and Uganda. Its objective is to enhance awareness, knowledge and understanding of below-ground biodiversity (BGBD) important to sustainable agricultural production in tropical landscapes by the demonstration of methods for conservation and sustainable management. The project will explore the hypothesis that by appropriate management of above- and below-ground biota, optimal conservation of biodiversity for national and global benefits can be achieved in mosaics of land uses at differing intensities of management and furthermore result in simultaneous gains in sustainable agricultural production. The primary outcomes of the project are:

1. Internationally accepted standard methods for characterization and evaluation of BGBD, including a set of indicators for BGBD loss.
2a. Inventory and evaluation of BGBD in benchmark sites representing a range of globally significant ecosystems and land uses.
2b. A global information exchange network for BGBD.
3. Sustainable and replicable management practices for BGBD conservation identified and implemented in pilot demonstration sites in representative tropical forest landscapes in seven countries.
4. Recommendations of alternative land use practices and an advisory support system for policies that will enhance the conservation of BGBD.
5. Improved capacity of all relevant institutions and stakeholders to implement conservation management of BGBD in a sustainable and efficient manner.

Selected groups of below-ground organisms studied by the seven countries comprise:

- leguminosae nodulating bacteria;
- arbuscular mycorrhizal and ectomycorrhizal fungi;
- pathogenic and antagonist fungi;
- nematodes;
- mesofauna;
- macrofauna, including earthworms, ants, beetles and termites;
- pests.

In Brazil, except for ectomycorrhizal fungi, all these functional groups are studied (see further information about participants and activities at: http://www.biosbrasil.ufla.br, http://www.bgbd.net/, http://www.ciat.cgiar.org/tsbf_institute/csm_bgbd.htm#partners).

and Sustainable Use of Soil Biodiversity' (Box. 1.3). Hence, it is not a coincidence that this initiative was launched under the section 'Agricultural Biodiversity'.

So, where do we stand right now as far as soil biodiversity is concerned? The organisms on Planet Earth comprise at least five kingdoms: Animalia, Plantae, Fungi, Protoctista and Bacteria (synonyms: Prokaryota, Procariota, Monera) (Margulis and Schwartz, 1998). The greatest controversy regarding this classification is related to the Prokaryota, recognized as being sufficiently polyphyletic to comprise the two kingdoms Archaebacteria (Archaea) and Eubacteria (Bacteria) (Cavalier-Smith, 1993) or, along with Eucarya, constituting the three domains in life (Woese *et al.*, 1990) widely accepted by bacteriologists and used by important databases such as the National Center for Biotechnology Information (NCBI) (http://www.ncbi.nlm.nih.gov).

Methods for assessment of soil biodiversity are described briefly or referenced in this book. Macroscopic organisms are usually classified based on morphological characteristics. Their assessment is mainly limited by sampling size, which is related to spatial and temporal heterogeneity. In the case of microscopic organisms (mainly Archaea, Bacteria and Fungi) the main well-known limitation is that about 99% of

Box. 1.3. Progress towards an International Initiative for the Conservation and Sustainable Use of Soil Biodiversity.

In Decision VI/5 (CBD, 2002), the COP of the Convention on Biological Diversity (CBD) decided to establish the 'International Initiative for the Conservation and Sustainable Use of Soil Biodiversity as a cross-cutting initiative within the Programme of work on Agricultural Biological Diversity', and invited the 'FAO and other relevant organizations, to facilitate and coordinate this initiative' (see further information and activities of FAO and partners at http://www.fao.org/ag/AGL/agll/soilbiod/).

As an initial collaborative activity, an international technical workshop on the Biological Management of Soil Ecosystems for Sustainable Agriculture was jointly organized by FAO and EMBRAPA-Soybean, in Londrina, Brazil, in June 2002, in order to discuss the concepts and practices of integrated soil management, share successful experiences of soil biological management and identify priorities for action under the Soil Biodiversity Initiative (SBI). The discussions among renowned experts from some 20 countries and several organizations led to the formulation of a set of principles and two main aims for the SBI in regard to expanding cooperation and coordinated action worldwide among interested partners:

First, to raise awareness of the importance of soil biodiversity, a seriously neglected but vital aspect of land resources management and sustainable agricultural systems, including improved understanding of the key roles of functional groups and of the impacts of different land uses and management practices.

Second, to improve management of soil biodiversity and promote ownership and adaptation by farmers of integrated soil biological management practices as an integral part of their agricultural and sustainable livelihood strategies.

Three strategic areas were identified for collaborative action by partners and countries during this technical meeting for which proposed activities were elaborated:

- Increasing recognition of the essential services provided by soil biodiversity across all production systems and its relation to sustainable land management.
- Capacity building to promote integrated approaches and coordinated activities for the sustainable use of soil biodiversity and enhancement of agroecosystem functions, including assessment and monitoring, adaptive management and targeted research and development.
- Developing partnerships and cooperative processes through mainstreaming and coordinated actions among partners to actively promote the conservation, restoration and sustainable use of soil biodiversity and enhanced contribution of beneficial soil organisms to the sustained productivity of agroecosystems.

For further information see the full workshop report published by FAO (2003) as World Soil Resources Report no. 101 (http://www.fao.org/ag/AGL/agll/soilbiod/docs.stm) and the Embrapa Soybean Documents no. 182 (Brown *et al.*, 2002).

these organisms are unculturable. Morphological characteristics, especially in the case of Prokaryota, are not suitable for their classification, but after the great breakthrough of the 1980s, the molecular techniques developed were demonstrated to be suitable for the assessment of unculturable organisms. NCBI, USA, was established in 1988 as a national resource for molecular biology information and a public database. Sequences of 2460 'species' of Archaea and 48,088 species of Bacteria were available in the NCBI molecular database on 5 January 2005. When unculturable organisms were excluded from the database these figures decreased to 844 and 35,747 species, showing how helpful molecular techniques are to reveal this formerly unknown biodiversity. Figures in this database are increasing exponentially day by day, so many more microorganisms, both culturable and unculturable, will have their genetic make-up revealed. Also, molecular techniques are beginning to be applied to macroorganisms revealing more reliable phylogenetic relationships as a useful tool for classification. In spite of this great advance, the numbers of described species are far beyond those presented in Table 1.1.

Brazil has the greatest biological diversity among the large nations of the planet (National Report to the Biological Convention on Biological Diversity, 1998), and this is at least partially related to a high diversity of soils (Mendonça-Santos *et al.*, Chapter 2, this volume) and ecosystems (Plate 1). The most important ecosystems are the Amazon forest, the Atlantic forest and 'cerrado'. Also, the seasonally flooded areas in the central west part of the country called 'pantanal' and the caatinga in the north-east cover significant portions of Brazil. Each of these vegetation types has unique botanical and edaphic characteristics, contributing to the high biodiversity of the country. The first National Report to the Convention on Biological Diversity (1998) reported 55,000 plant species (22% of total species on the planet), 524 mammalian species (131 endemic), 517 amphibian species (294 endemic), 1622 bird species (191 endemic), 468 reptile species (172 endemic), 3000 fish species and between 10 million and 15 million insect species in Brazil. The number of species of the smaller animals, like insects, could only be estimated. Furthermore, in the 283-page report only one page was devoted to microorganisms, without any precise information about their diversity, and only very few of all the organisms listed in the report qualified as soil organisms. This indicates an urgent need for the assessment of the biodiversity of these groups.

Soil is teeming with life (Tiedje *et al.*, 2001). According to Young and Crawford (2004), 1 g of fertile soil contains 10^{12} bacteria, 10^4 protozoa, 10^4 nematodes, 25 km of fungi and countless other species, i.e. more organisms than the number of human beings that have ever lived on the planet. But the current rate of deforestation, topsoil loss and land degradation is far greater than the rate of soil formation, causing loss of habitat and constituting a major threat to soil biodiversity. This situation sets the scene for the present book.

Contents of the Book

The book sets out with a description by Mendonça-Santos *et al.* (Chapter 2, this volume) of the major soils and land uses in Brazil, which are inextricably related to soil biodiversity. The Brazilian Amazon territory extends to approximately 5,000,000 km^2 over nine states. This area has been subject to major changes as a consequence of different occupation cycles where mining, civil construction, agriculture and cattle raising activities have intensified in the last two decades. Latossols and Argissols make up approximately 62% of the Amazon surface and Plintossols and Alissols cover another 16%. By and large, these soils are highly weathered and of low inherent fertility. Deforested soils are characterized by high Al saturation, low nutrient availability and low organic matter and cation-exchange capacity (CEC) and this increases soil leaching capacity. These are the major constraints for sustainable crop

Table 1.1. Number of described species in the main taxonomic categories of plants and of soil biota, considering those phyla with highest species numbers.

Taxonomic categories[a] (total number of extant Phyla) (examples of soil organisms/common names)	Number of described species
Domain Eucarya	
Kingdom Plantae (12 phyla)	255,000
Phylum Bryophyta (mosses)	10,000
Phylum Hepatophyta (liverworts)	6,000
Phylum Filicinophyta (ferns)	12,000
Phylum Anthophyta (angiosperms)	235,000
Monocotyledons	65,000
Dicotyledons	130,000
Kingdom Animalia (37 phyla)	10 million
Phylum Tardigrada[b]	750
Phylum Mollusca (snail)[b]	50,000
Phylum Annelida (earthworms, enchytraeids, leeches)[b]	15,500
Class Polychaeta	9,000
Class Oligochaeta	6,000
Class Hirudinea	500
Phylum Crustacea (>6 classes)[b]	45,000
Class Malacostraca (Isopoda – wood mites (10,000) and Decapoda – shrimp, prawn, crab, lobster, krill)	25,000
Phylum Mandibulata (Arthropoda)	
Class Hexapoda (Insecta)	750,000
Order Coleoptera (beetles)	350,000
Order Isoptera (termites)	2,800
Order Hymenoptera	
Family Formicidae (ants)	11,826
Order Collembola (springtails)	7,500
Order Diplura	659
Class Myriapoda	15,162
Order Diplopoda (millipeds)	10,000
Order Chilopoda (centipeds)	2,500
Class Symphlyla	200
Class Pauropoda	700
Phylum Chelicerata (3 classes)[b]	75,000
Class Arachnida (11 orders)	93,455
Order Palpigrada (micro whipscorpions)	80
Order Acari (mites)	45,000
Order Pseudoscorpionida (pseudoscorpions)	3,235
Order Aranae (spiders)	38,884
Order Scorpionida (scorpions)	1,100
Phylum Gastrotricha (gastrotriqueos)[b]	400
Phylum Acanthcephalla (worms)[b]	1,000
Phylum Rotifera[b]	2,000
Phylum Nemertina (worms)[b]	900
Phylum Nematoda (nematodes)	15,000
Phylum Plathyheminthes (worms)[b]	20,000
Kingdom Protoctista (30 phyla)	
Phylum Rhizopoda (amoebae – protozoa and moulds)	Large number not determined
Phylum Dinomastigota (dinoflagellates)[b]	4,000
Phylum Ciliophora (ciliates–protozoa)	10,000
Phylum Discomitochondria (flagellated and zooflagellated protozoa)	800
Phylum Diatomacea[b]	10,000

Continued

Table 1.1. Number of described species in the main taxonomic categories of plants and of soil biota, considering those with highest species numbers. – cont'd

Taxonomic categories[a] (total number of extant phyla) (examples of soil organisms/common names)	Number of described species
Phylum Oomycota (oomycetes)	Hundreds of species
Phylum Rhodophyta (red algae)	4,100
Phylum Chlorophyta (green algae)	16,000
Phylum Chytridiomycota	1,000
Kingdom Fungi (four phyla)[c]	60,000
Phylum Zycomycota	1,100
Phylum Basidiomycota	22,250
Phylum Ascomycota	30,000
Domain Archaea (four phyla)[bd]	844
Domain Bacteria (52 phyla)[bd]	35,747

[a]Taxonomic categories from the highest to the lowest level: domain, kingdom, phylum, class, order, family, genus, species. Prokaryote (domains Archaea and Bacteria) classification according to Woese *et al.* (1990) and Eucarya kingdoms classified according to Margulis and Schwartz (1998).
[b]Includes soil and aquatic organisms.
[c]Phylum Glomeromycota included.
[d]Includes unclassified and unspecified species.
Sources: Bellinger *et al.* (1996–2005), Margulis and Schwartz (1998), Platinick (2000), Rappé and Giovannoni (2003), NCBI (2005), chapters of this book.

production in the Amazon region. Once primary forest is cleared, soil organic matter is rapidly oxidized, leading to a depletion of C and nutrients, and the soil environment changes substantially. In addition to the reduction of nutrients, soil physical characteristics also change: soil temperature increases and water retention decreases.

Barros *et al.* (Chapter 3, this volume) give evidence of the enormous impact such changes have on the community structure of the soil fauna, irrespective of regional differences due to geology and climate.

Soil microbiology and microbial ecology has been a much more common area of science than soil zoology and soil animal ecology, probably due to the widely recognized importance of microorganisms for biochemical processes in soil. The recent upsurge in the use of molecular techniques has been a major boost to the taxonomy of microorganisms, which in itself is spurred by the search for organisms and, for that matter, genes of economic importance in the food and medicinal industries. Fortunately, it has been more commonly accepted over the last 20 years that the soil fauna comprises 'ecosystem engineers', i.e. organisms that affect the availability of resources to other species through physical changes in their habitat. Earthworms, termites and ants move around enormous amounts of soil and create structures above and below the ground that may last for decades. So, while soil microorganisms are pivotal for biochemical transformations, soil fauna are pertinent to biophysical transformations.

Unfortunately, this realization has so far not resulted in an equally intensified taxonomic effort as with microorganisms. Yet, if we add the high biodiversity and abundance of these engineers, it is clear that understanding the functioning of ecosystems is impossible without thorough knowledge of the taxonomy of these groups. Here we have a real challenge. For example, James and Brown (Chapter 4, this volume) estimate that at the rate of 50 earthworm species descriptions per man-year, 60 man-years of full-time taxonomy will be needed to describe the estimated remainder of 3000 undescribed species worldwide. Given the estimated species richness in Brazil, most of this manpower will have to become available in this country. Although the number of people working on soil fauna in Brazil has greatly increased over the past 10 years, this will not happen unless taxonomic training and capacity building is given the highest priority by funding agencies. It is of course necessary that we know how to sample the soil fauna in the first place and it is very appropriate that the authors of Chapters 3–8 give this aspect due attention.

Although earthworms are among the most visible soil fauna, this does not mean that their diversity and ecology are well known. Yet, according to James and Brown (Chapter 4, this volume), the earthworm biodiversity of Brazil will likely be the highest of all large nations of the world, for biogeographical and climatic reasons. Many species will be endemic, caused by contractions and expansions of tropical vegetation during (de)glaciation periods in the northern hemisphere. The limited evidence available also suggests that soil type is an important habitat-defining factor. Although earthworms are rightly associated with soil, many of the habitats where they are found can be temporarily aquatic, which has given rise to behavioural adaptations, like climbing up trees or horizontal seasonal migration. Still other species live in truly aquatic environments. In terms of species conservation, it is a matter of concern that only a few native species are found in disturbed habitats like secondary forest and agricultural land, where the earthworm fauna is dominated by peregrine species and exotics. Unfortunately, the importance of native earthworms for plant growth is still unknown, because experiments to assess such effects have been done with exotic or peregrine earthworms only.

As Constantino and Acioli point out in Chapter 5, termite faunas differ markedly between the major ecosystems of Brazil: the Amazon rainforest (harbouring almost half of the neotropical termite fauna), cerrados and the Atlantic forest. Very little is known yet about soil type as a habitat-defining factor. The limited evidence available suggests that termites are highly sensitive to habitat fragmentation and disturbance, particularly the humus feeders. Although termite taxonomy is well developed in Brazil relative to other countries in Latin America, it still is a major impediment to the study of termite ecology and distribution, and to their management and control for beneficial and detrimental activities in urban and agricultural areas.

Ants are among the most species-rich terrestrial invertebrates. Of the estimated 20,000 species, Vasconcelos (Chapter 6, this volume) reckons that approximately 25% occur in the neotropics. It is not known how many of these will be found in the Amazon. Considerable differences in ant diversity exist between habitats such as *várzea* and *terra firme* forest and between forest and savannah. Even within a seemingly homogeneous habitat, ant diversity differs over short distances, partly related to differences in topography, but also to natural disturbances such as flooding, wildfires and treefall gaps. Transformation of the forest to plantation or pasture results in drastic reductions of ant diversity, along with an equally drastic change in community composition. Logging effects appear to be much less strong. The good news is that recovery of the ant fauna during reforestation is more or less complete after 25 years (but depending on previous land use), if colonization can take place from nearby undisturbed forest. This is much quicker than recovery of the forest itself.

The soil mesofauna, measuring between 0.2 and 2 mm in body width, largely comprise springtails (Collembola), mites (Acari) and the smaller worms (Oligochaeta). Although they are generally considered to be of minor importance to soil metabolism and element transformation, this is not justified in soils where the ecosystem engineers are not abundant or lacking. Even where earthworms, termites and ants are abundant, the mesofauna can be considered ecosystem engineers, albeit at less conspicuous spatial scales. In the absence of earlier work, Franklin and Morais (Chapter 7, this volume) had to dwell largely on recent studies carried out by them and co-workers. Their strong inclination to experimental work and the impressive amount of research done in a short period of time make their research particularly valuable in reaching the conclusion that the species richness and abundance of the soil mesofauna, and their clear response to natural and human disturbances, make them suitable as indicators of environmental change. Once again, only if abundant resources become available to train taxonomists will we be able to exploit this result to its full potential.

The microfauna (body width less than 0.2 mm) comprises several taxonomic groups, but the only group receiving considerable attention is the nematodes. Different feeding modes (which can be relatively easily deduced from the morphology of their mouthparts) and different life history strategies make soil nematodes well suited for reflecting environmental changes in their community structure and composition. In studying both rainforest and cerrados and both natural and agricultural habitats, Huang and Cares (Chapter 8, this volume) were able to show that nematode diversity is closely related to vegetation diversity (both in natural and in agricultural systems), with, somewhat counterintuitively, plant parasites the most important functional group in native vegetation versus bacterial feeders in agricultural systems. Nematode abundance is generally greater in agricultural systems. Soil type clearly has an influence on abundance as well, but the nature of this phenomenon is little understood.

In Chapter 9, Pfenning and Abreu discuss soil microfungi. This group of soil organisms comprises zygomycetes, ascomycetes with fruiting bodies smaller than 2 mm and conidial states of ascomycetes (formerly fungi imperfecti or deuteromycetes). They represent an important functional group of soil heterotrophs that are responsible for organic matter decomposition and biogeochemical processes in soil ecosystems. Several microfungi are also plant pathogens, antagonists and insect pathogens. In the soil and rizosphere environments, they interact with the whole microbial community, including other fungi, bacteria and fauna components. Several studies on the occurrence and diversity of soil microfungi have been conducted in the tropics. They exhibit high incidence and diversity, especially in the litter layer in forest ecosystem. Soil cultivation may cause sudden shifts in the fungal community. In general, fungal communities comprise many cosmopolitan species, but there are reports on the occurrence of rare species. Studies reported by Pfenning and Abreu in the eastern Amazon show that the proportion of dominant species is lower in forest stands than in cultivated sites and that introduced crops in forest-cleared areas resulted in increased numbers of plant-parasitic microfungi such as *Fusarium* sp. In addition to their functional role in the soil, soil microfungi are an important component of soil biomass, and therefore represent a crucial component of the ecosystem that can be highly affected by the land use. The study of this group is limited by techniques, because of the lack of culture media mimicking soil conditions. Hence, data based on cultured species may not represent the soil community as it is in the real world. As reviewed by the authors, molecular techniques have been successfully applied to a variety of studies on soil microfungi.

Another important group of soil organisms is the arbuscular mycorrhizal fungi (AMF), reviewed by Stürmer and Siqueira in Chapter 10. In spite of their ubiquitous occurrence throughout the world, a complete inventory of AMF has not been conducted. The AMF are obligate symbionts that originated 353–462 million years ago and AM associations are characterized by typical two-way interactions in which plant communities affect fungal occurrence and different fungal assemblages may have different effects on host plants. Because of the close plant–fungal relationship, AMF represent an important link between soil and plant. The AMF major host effect is enhancement of absorption of nutrients from the soil, but their most consistent effects are observed on P and micronutrients such as Zn and Cu. In addition, they favour plant–water relationships, improve soil aggregate formation and stability and can reduce plant damage caused by soil-borne plant pathogens. The systematics of this group of fungi has been rather problematic because of our inability to grow them in defined media in the laboratory. It is based upon phenotypic characters of soil-collected spores and their classification has experienced major changes after the advent of the molecular phylogenetic analysis based on SSU rRNA sequences. The AMF were recently removed from a phylogenetic group in the Zygomycota and placed in a newly erected monophyletic group as Glomeromycota. They do not form a very rich group, with only 160 formally described species in five families and seven genera in existence. These fungi have been well studied in Brazil. A total of 79 AMF species have being reported in 28 surveys. The distribution of AMF species is highly affected by soil characteristics and vegetation type, but prediction of the distribution of a given species or population make-up is difficult. Although these studies were concentrated mostly in the south-east, they indicate that Brazilian ecosystems are an important source of AMF diversity, deserving more attention in terms of germplasm conservation policy. The ecology of AMF in Amazon ecosystems has been overlooked, but it is expected that land use will have a great impact on the occurrence and species diversity of these fungi. They may have a great potential for use in Amazonian agriculture where soils are severely P-deficient and phosphate fertilizers are unavailable. Because different land use systems affect AMF populations, selecting efficient isolates for field tests may contribute to the exploitation of biodiversity by local communities in the Amazon. The authors stress the need to strengthen the research on AMF in Brazil, which is considered a major centre of biodiversity of these fungi.

Biological nitrogen fixation (BNF) is one of the most important functions of the soil–plant system for the maintenance of life on Earth. The enzymatic machinery capable of reducing N_2 to NH_3 is restricted to some bacterial species from which part of them can establish a symbiotic relationship with leguminous plants, the leguminosae nodulating bacteria (LNB). The LNB have high economic value because of their efficiency in supplying atmospheric N_2 to terrestrial ecosystems. Their role offers an opportunity for improving agricultural productivity in an environmentally sound way. This is a well-developed technology for grain legumes (e.g. soybean) in which bacterial inoculants are used as seed treatment to replace chemical fertilizers. N_2 fixed in soybean in Brazil alone is approximately 2×10^6 mg N per year and this represents a global saving of US$ 2.0 billion for Brazil's economy, in addition to the ecological benefits of reducing the amount of reactive N in the environment.

There has been tremendous progress in surveying nodulating legume species in Brazil. As reviewed in Chapter 11, the vast majority of Mimosoideae and Papilionoideae species do nodulate, whereas only 24% of the Caesalpinioideae are compatible with the LNB. The name rhizobia has been used for a long time as a collective name for LNB. It originates from Rhizobiaceae, which were known to include all LNB, but with the discovery of LNB in other phylogenetic branches of Prokaryotes, this name became inappropriate. The taxonomy of the LNB has experienced great advances, in spite of the fact that most studies are restricted to isolates from a few host species. Currently, 47 species belonging to 11 genera have been formally described, but very few of them were described based on isolations from tropical ecosystems. Current figures indicate that the nodulating ability of around 11,200 leguminous species around the world is unknown. Considering the great diversity of leguminous plants in the tropics, it is evident that the diversity of LNB is still poorly understood and deserves more intensive investigation. Aspects of the evolution of this symbiosis, diversity of LNB in Brazil, efficiency and application of LNB isolates are discussed in Chapter 11.

The Way Forward

Although this book is mainly about the Brazilian Amazon in its natural state, most authors already make references to the other major ecosystems of Brazil and to agricultural systems derived from nature. Hence, this book adequately sets the scene for expanding on exactly these aspects in the GEF project on CSM-BGBD (Box. 1.2). The groups of organisms selected for research in the project are only a subset of the total biodiversity in soil (see Table 1.1). In the face of uncertainty regarding the functionality of many species, these groups may just reflect the taxa about which we know the most. Yet, they are believed to constitute or contain the species (assemblages) that are, qualitatively and quantitatively, the most important in terms of ecosystem processes, agricultural goods and environmental services, as recently reviewed by Wall (2004). Because the CSM-BGBD project builds further on soil biodiversity inventories, standardization of methods of assessment and the creation of a global database, there is currently no other major project that should be better able to produce the scientifically sound soil biodiversity data needed to deliver what the world so urgently needs (e.g. van Noordwijk *et al.*, 2004), i.e. recommendations of alternative land use practices and an advisory support system for policies that will enhance the conservation of below-ground biodiversity, as well as an improved capacity of all relevant institutions and stakeholders to implement conservation management of BGBD in a sustainable and efficient manner. The stakes are high.

References

Bellinger, P.F., Chrostiansen, K.A. and Janssens, F. (1996–2005) *Checklist of the Collembola of the World.* Available at: http://www.collembola.org

Bignell, D.E., Tondoh, J., Dibog, L., Huang, S., Moreira, F.M.S., Pereira, E.G., Nwaga, D., Pashanasi, B., Susilo, F. and Swift, M. (2005) Belowground biodiversity assessment: the ASB functional approach. In: Palm, A.A., Vosti, S.A., Sanchez, P.A., Ericksen, P.J. and Ruo, A.S.R. (eds) *Slash and Burn: The Search for Alternatives.* Columbia University Press, New York.

Brown, G.G., Hungria, M., Oliveira, L.J., Bunning, S. and Montañez, A. (eds) (2002) *Programme, Abstracts and Related Documents of the International Technical Workshop on Biological Management of Soil Ecosystems for Sustainable Agriculture.* Embrapa Soja, Série Documentos 182, Londrina, Brazil, 256 pp.

Cavalier-Smith, T. (1993) Kingdom Protozoa and its 18 Phyla. *Microbiological Reviews* 57, 953–994.

FAO (2003) *Biological Management for Soil Ecosystems for Sustainable Agriculture. World Soil Resources Report* 101. FAO, Rome, 102 pp.

Lewis, J., Vosti, S., Witcover, J., Ericksen, P.J., Guevara, R. and Tomish, T. (2002) *Alternatives to Slash and Burn in Brazil: Summary Report and Synthesis of Phase II.* ASB/ICRAF, Nairobi, Kenya, 93 pp.

Margulis, L. and Schwartz, K.V. (1998) *Five Kingdoms: An Illustrated Guide to the Phyla of Life on Earth.* W.H. Freeman, New York, 497 pp.

National Report to the Biological Convention on Biological Diversity (1998) Ministry of Environment, Ministério do Meio Ambiente, Brazil, 283 pp.

NCBI – National Center for Biotechnology Information. Available at: http://www.ncbi.nlm.nih.gov

Platinick, N.I. (2000) The world spider catalog. American Museum of Natural History. Available at: http://research.amnh.org/entomology/spiders/catalog/COUNTS.html

Rappé, S.J. and Giovannoni, S. (2003) The uncultured microbial majority. *Annual Review of Microbiology* 57, 369–394.

Tiedje, J.M., Cho, J.C., Murria, A., Treves, D., Xia, A. and Zhou, J. (2001) Soil teeming with life: new frontiers for soil science. In: Rees, R.M., Ball, B.C., Campbell, C.D. and Watson, C.A. (eds) *Sustainable Management of Soil Organic Matter.* CAB International, Wallingford, UK, pp. 393–412.

van Noordwijk, M., Cadisch, G. and Ong, C.K. (eds) (2004) *Below-Ground Interactions in Tropical Agroecosystems.* CAB International, Wallingford, UK, 440 pp.

Wall, D.H. (ed.) (2004) *Sustaining Biodiversity and Ecosystem Services in Soils and Sediments.* Island Press, Washington, DC, 275 pp.

Woese, C.R., Kandler, O. and Wheelis, M.L. (1990) Towards a natural system of organisms: proposal for the domains Archaea, Bacteria and Eucarya. *Proceedings of the National Academy of Sciences USA* 87, 4576–4579.

Young, I.M. and Crawford, J.W. (2004) Interactions and self-organization in the soil–microbe complex. *Science* 304, 1634–1637.

2 Soil and Land Use in the Brazilian Amazon

M.L. Mendonça-Santos,[1] H.G. dos Santos,[1] M.R. Coelho,[1] A.C.C. Bernardi,[2] P.L.O.A. Machado;[1] C.V. Manzatto[1] and E.C.C. Fidalgo[1]

[1]*EMBRAPA Solos, Rua Jardim Botânico, 1024, CEP 22460-000, Rio de Janeiro, RJ, Brazil, e-mail: loumendonca@cnps.embrapa.br, humberto@cnps.embrapa.br, mrcoelho@cnps.embrapa.br, pedro@cnps.embrapa.br, manzatto@cnps.embrapa.br, efidalgo@cnps.embrapa.br;* [2]*EMBRAPA Pecuária Sudeste, Rodovia Washington Luiz, km 234, Fazenda Canchim, Caixa Postal 339, CEP 13560-970, São Carlos, SP, Brazil, alberto@cppse.embrapa.br*

Introduction

The Legal Amazon accounts for 60% of the National Territory, with approximately 5,000,000 km^2 corresponding to the political and geographical unit on which most of the planning and development programmes have been based. It is located between the latitudes 5°N and 16°S and the longitudes 44°W and 74°W and consists totally or partially of the following states: Acre, Amapá, Amazonas, Maranhão, Mato Grosso, Pará, Rondônia, Roraima and Tocantins (Rodrigues, 1996; Fearnside, 2002).

The large extension of Legal Amazon results in a great diversity of environments, characterizing different ecosystems mostly composed of different types of equatorial and tropical forests, savannahs and tropical grasslands (BRASIL, 2002a,b). In addition to the large diversity of ecosystems, the interactions among climatic, geological, geomorphological and biological factors result in a large diversity of soil types.

A brief analysis of the three subregions of this area – east, west and south Amazon – shows different spatial changes mostly due to historical occupation, flooding and their effects on the spatial dynamics of the Amazon. Hence, major changes in each of these subregions were due to mining activities and construction of railways, highways, and large agriculture- and pasture-based cattle raising. As a consequence of different occupation cycles, there have been various pressures on the vegetation cover, contributing to intensive forest clearing in each region.

In spite of the widespread concepts about the low natural fertility and the high aluminium saturation in most of the Amazon soils, an increasing number of agricultural establishments and the enlargement of the existing space have been observed, mainly in Rondônia and Pará states. This phenomenon represents part of the so-called development arc (otherwise known as 'the deforestation arc') of the Amazon.

In these areas, the conversion of forest into pastures represents the legal instrument to obtain property rights by large farmers and land owners. On the other hand, for small producers, pastures are the immediate alternative for valuation of the land, even degraded, after cyclic use with annual crops. Another factor that exerts pressure over the forest is the growing demand for timber in the home market, for furniture manufacturing and firewood used to dry grains in the local areas. The largest producers of timber are still the states of Pará and Mato Grosso, followed by Rondônia, mainly to supply the markets in south-east Brazil (37.4% of the production), while the foreign market absorbs 14% (Egler, 2001).

The introduction of modern agriculture in Amazonia is a historical novelty in an area that has always lived on extractive activities of natural resources. The symbolic crop of the new model is soybean that, accompanied by rice and maize, in the mid-1990s had just advanced over the savannah-bordering areas of Legal Amazon. It already occupied new and significant areas by the end of the 1990s, making the state of Mato Grosso one of the main producers of grains and fibres in the country.

Therefore, the diversity of Amazonia as well as the changing processes in the course of time suggest the importance of characterizing the soil and its use in the context of the Amazon area, through the accomplishment of diagnostic and detailed environmental studies of sensitive variables to the changes and impacts caused by the dynamics of land use, such as soil microbial biodiversity. In this sense, this chapter characterizes the Brazilian soils, emphasizing the main classes of soils of the Amazon area and accomplishing a scenario of potentialities and limitations. Thereafter, in order to understand the land use patterns, a brief history of the occupation of Amazon soils as a whole is reported, as well as the technologies and available tools for the accomplishment of those studies.

Brazilian Soil: General Classes

Soil is defined as a natural body resulting from the interactions of climate, organisms, relief and parent material acting together with varying intensities during a certain period of time. These soil-forming factors define the nature of the soils and their distribution and settings in the landscape. In a profile of the landscape they represent individuals. In a landscape they constitute a continuum having a set of physical, chemical, mineralogical and biological attributes. This concept is linked to the evolution of the soil and the patterns of distribution in the landscape, showing where and why certain types of soils occur as such; they constitute geographical bodies equivalent to the 'pedons' and 'polypedons' (Knox, 1965).

In an utilitarian approach soil is considered a collection of natural tridimensional bodies resulting from the interactions of soil-forming factors. Soils consist of solid, liquid and gaseous phases, formed by mineral and living and dead organic matter and occupying most of the surface mantle of the continental extensions of the Earth.

The continental dimension of the Brazilian territory is the cause for a great diversity of soil types, corresponding directly to the intensity of interactions of relief types, climate, parent materials, vegetation and associated organisms that contribute to the most diversified ecosystems. This diversity is responsible for the nature of the country conditioning its potentialities and constraints for the use of the geographical space and, largely, to the regional differences concerning the several occupation, economic development and cultural patterns of the territory.

On the basis of the Brazil Soil Map (Plate 2) and the current Brazilian Soil Classification System (EMBRAPA, 1999) 13 major representative soil classes of the Brazilian territory can be distinguished (Table 2.1). These classes are subdivided into different soil types according to morphological characteristics and chemical, physical and mineralogical properties separating them into more homogeneous units. The set of attributes that defines and distinguishes the most varied types of Brazilian soils is taxonomically organized and systematized in the Brazilian Soil Classification System (EMBRAPA, 1999).

The soil classes are briefly characterized below and some of the most expressive attributes that define and differentiate them, as well as some aspects of their extension,

Table 2.1. Extent and percent distribution of Brazilian soils.

	Brazil		% Region				
Soil classes	Absolute values (km²)	% Total area	North	North-east	Centre-west	South-east	South
Alissolos	371,874.48	4.4	8.7	0.0	0.0	0.0	6.3
Argissolos	1,713,853.49	20.0	24.4	17.2	13.8	20.7	14.8
Cambissolos	232,139.19	2.7	1.1	2.1	1.6	8.6	9.3
Chernossolos	42,363.93	0.5	0.0	1.0	0.3	0.2	3.9
Espodossolos	133,204.88	1.6	3.1	0.4	0.3	0.4	0.0
Gleissolos	311,445.26	3.7	6.4	0.8	2.8	0.5	0.4
Latossolos	3,317,590.34	38.7	33.9	31.0	52.8	56.3	25.0
Luvissolos	225,594.90	2.6	2.7	7.6	0.0	0.0	0.0
Neossolos	1,246,898.89	14.6	8.5	27.5	16.4	9.4	23.2
Nitossolos	119,731.33	1.4	0.3	0.1	1.2	2.6	11.5
Planossolos	155,152.13	1.8	0.2	6.6	1.7	0.2	3.0
Plintossolos	508,539.37	6.0	7.6	4.7	8.8	0.0	0.0
Vertissolos	169,015.27	2.0	3.2	1.0	0.4	1.2	2.6
Water bodies	160,532.30	1.9	3.2	0.4	0.3	1.2	2.6
Total	8,547,403.50	100.0	100.0	100.0	100.0	100.0	100.0

Source: Coelho *et al.* (2002).

geographical distribution (Table 2.1) and correlation with other classification systems (Table 2.2), such as Soil Taxonomy (Soil Survey Staff, 1999) and the World Soil Reference Base (WRB) (FAO, 1998), are generalized.

Latossolos

These soils are highly weathered due to strong alterations of parent materials or to their genesis related to preweathered sediments (Oliveira *et al.*, 1992). They are characterized by a clay fraction dominated by minerals in the last weathering stage such as: (i) clay silicates of low activity (kaolinite); and (ii) iron and aluminium oxides (haematite, goethite and gibbsite). The sand fraction predominantly consists of highly weathering-resistant minerals. The Latossolos show uniformity in their morphological, physical, chemical and mineralogical properties, with little horizon differentiation. They have a variable texture from loam to fine clay and may also present a sandy texture in surface horizons. They are commonly very deep and porous, friable and permeable, showing low increments in the clay content in depth and are generally of low natural fertility. They are most representative of all the soils in Brazil, occupying approximately 40% of the total area of the country (Table 2.1) and distributed in the entire National territory (Plate 2). There are several types of Latossolos whose differentiation can be made by: (i) colour; (ii) natural fertility; (iii) the content of iron oxides; (iv) the existence of intermediate attributes with other soil classes; (v) the type of surface horizon; (vi) soil texture; and (vii) soil mineralogy, among other attributes.

Argissolos

The Argissolos have a significant increase in the clay content at depth, although this characteristic can be absent in certain soils. Argissolos with no textural gradient require the use of other morphological attributes for their complete identification, such as the degree of structural development, clay content and the presence of clay skins. In

Table 2.2. Approximate correlation between the Brazilian System of Soil Classification (SiBCS) (EMBRAPA, 1999), Soil Taxonomy (Soil Survey Staff, 1999) and the Soil World Reference Base (WRB) (FAO, 1998).

SiBCS	Soil taxonomy	WRB
Alissolos	Ultisols	Acrisols, alisols
Argissolos	Ultisols, alfisols	Acrisols, lixisols
Cambissolos	Inceptisols	Cambisols
Chernossolos	Mollisols	Leptosols, kastanozems, greyzems, chernozems, phaeozems
Espodossolos	Spodosols	Podzols
Gleissolos	Inceptisols, ultisols, mollisols, alfisols, entisols	Fluvisols, gleysols
Latossolos	Oxisols	Ferralsols
Luvissolos	Alfisols	Luvisols
Neossolos	Entisols	Fluvisols, leptosols, regosols, arenosols
Nitossolos	Ultisols, alfisols	Nitisols
Planossolos	Alfisols, ultisols, mollisols, aridisols	Planosols
Plintossolos	Oxisols, ultisols, inceptisols, entisols, alfisols	Sexquisols
Vertissolos	Vertisols	Vertisols

Source: adapted from Palmieri *et al.* (2003).

general, they are well structured and deep, mainly reddish and yellowish, varying in texture from sand to clay in the surface horizons and from loam to fine clay in the subsurface horizons. Their fertility is variable, predominantly with relatively low nutrient contents, low clay activity and mineralogy, mostly kaolinitic. The Argissolos account for approximately 20% of the total area of the country. In terms of geographical extension they are close to Latossolos and distributed practically all over Brazil (Plate 2).

Alissolos

They comprise all soils with low natural fertility, high contents of extractable aluminium ($Al^{3+} \geq 4$ $cmol_c$/kg soil) and clay activity equal to or higher than 20 $cmol_c$/kg clay. In some soils of this class a significant increase of clay content at depth may be observed, while in other soils this increase may be less pronounced. In general, they are well structured and are distributed in the subtropical area of Brazil, especially in the southernmost states (Paraná, Santa Catarina and Rio Grande do Sul). However, the largest extensions are in western Amazon (Plate 2), predominantly under tropical and equatorial conditions (Oliveira *et al.*, 1992).

Cambissolos

Due to the heterogeneity of the parent material, relief forms and climatic conditions, the characteristics of these soils vary significantly in the different areas of Brazil. However, a common characteristic is the incipient stage of development, with the surface horizon generally showing fragments of rocks mixed with the soil mass, presence of primary easily weatherable minerals, besides low (or nil) clay increments at depth. They are also distributed all over Brazilian territory, predominantly in quite dissected reliefs, although they may occur on old fluvial terraces in flat relief. Extensive Cambissolo areas are exceptionally found in the eastern part of the plateaux of the southern states (Rio Grande do Sul, Santa Catarina and Paraná), where they present high contents of organic matter and extractable aluminium. Other expressive occurrences are those related to the Serra do Mar, a mountain chain extending from the state of Rio Grande do Sul to Espírito Santo, Serra da Mantiqueira and areas in Minas Gerais state (Oliveira *et al.*, 1992).

Chernossolos

They consist of soils with high clay activity in the subsurface horizons and surface horizon of chernozemic type (thick, dark, well structured, rich in organic matter and with high content of exchangeable cations). These soils are very well structured, usually not deep (< 100 cm), with or without increasing clay content at depth. Chernossolos are dark, not very coloured, moderately acidic to strongly alkaline. Therefore, they show high natural fertility as a result of the presence of clay minerals such as smectite and/or vermiculite in significant proportions. The largest areas of these soils are found in the states of Rio Grande do Sul and Bahia.

Gleissolos

These soils are strongly influenced by water table movement. Hence, they are permanently or seasonally flooded, except if artificially drained. They are generally found in recent sediments near water streams and channels, in alluvial deposits subject to hydromorphic conditions as well as in areas of flat relief of fluvial, lacustrine or marine terraces. They are characterized by strong gleysation, easily identified by the grey colour that starts at a depth of 50 cm. They may also develop bluish and/or greenish colours. They are found in all humid areas of the Brazilian territory, where the groundwater level is high most of the year. Significant occurrences, however, are related to the Amazon lowlands along the Araguaia River in the states of Goiás and Tocantins, along the river Paraíba do Sul in the states of São Paulo and Rio de Janeiro and surrounding Lagoa dos Patos, Mirim and Mangueira in Rio Grande do Sul State (Oliveira *et al.*, 1992).

Luvissolos

These are moderately acidic to alkaline soils with high natural fertility, generally with low or no content of extractable aluminium and significant and variable amounts of 2:1 clay minerals responsible for the high ion-exchange capacity (soils with high clay activity and high base saturation) in the subsurface horizons. They are commonly not deep (< 100 cm), red or yellow in colour, with well-developed structures, showing occasionally significant increases in the clay content at depth. The semiarid zone of north-east Brazil is the region that shows the highest occurrence of these soils. However, Luvissolos with high contents of exchangeable basic cations and extractable aluminium are found in the state of Acre (north-west of Brazil).

Neossolos

These soils are weakly developed and generally not very thick due to the low expression of the processes responsible for their formation, which did not lead to expressive alterations of the parent material. They are very variable in the landscape and are strongly influenced by the parent material. There are four great types of Neossolos with the following characteristics:

1. Neossolos Litólicos – shallow soils with thickness lower than 50 cm and generally a narrow layer of earthy material over rock in different alteration stages.
2. Neossolos Regolíticos – deeper soils with thickness higher than 50 cm and showing easily weatherable minerals in the sand fraction (mineral nutrients reservoir) or fragments of partially weathered rock that originated in these soils.
3. Neossolos Quartzarênicos – generally deep soils essentially sandy over the whole

profile and with no or low nutrient supplies (no or low content of weatherable minerals). **4.** Neossolos Flúvicos – soils derived from alluvial sediments. They usually present a darkened surface horizon overlying stratified layers without strong pedogenetic relationships with them. The irregular distribution of organic carbon content at depth is another important characteristic of these soils.

Neossolos Litólicos, in general, are associated with steep reliefs and rock outcrops. In soil maps they are present as narrow and long strips reflecting the crests and more unstable parts of the landscape (Resende *et al.*, 1988). There is no regular distribution by region and they are distributed all over Brazil. Neossolos Regolíticos are also common in Brazil as a whole and closely associated with Neossolos Litólicos in the landscape. However, extensive areas occur in semiarid north-eastern Brazil, where they are deeper. Neossolos Quartzarênicos are located in low-level topography and the largest occurrences are in the states of São Paulo, Mato Grosso do Sul, Mato Grosso, west and north Bahia, south Pará, south and north Maranhão, in Piauí and Pernambuco. Neossolos Flúvicos rarely occupy extensive and contiguous areas because they are restricted to the borders of the waterways, lakes and coastal plains, where they occupy small extensions of the lowlands (Oliveira *et al.*, 1992).

Espodossolos

They are mainly sandy soils with significant accumulation of illuvial organic matter at depth, which is associated with aluminium complexes containing iron compounds in some cases. Although Espodossolos with significant contents of exchangeable cations have been found mainly associated with the presence of shells in the profile, they are generally nutrient-poor soils and normally show relatively high contents of extractable aluminium in comparison with other exchangeable basic cations. Their occurrences are not continuous in the landscape and they are distributed in the coastal plains of Brazil, especially in the states of Rio de Janeiro, Bahia, Sergipe, Alagoas and Rio Grande do Sul as well as in western Amazon, where they occur over extensive areas.

Nitossolos

These are clayey textured soils presenting low or no clay increase at depth. They are usually very deep (> 200 cm), well-drained and well-developed subsurface horizons in terms of soil structure and clayskins (shiny peds). Predominant colours are reddish or brownish; they are moderately acidic with low clay activity and variable chemical fertility (with low or high base saturation). They eventually show high contents of extractable aluminium and the largest continuous areas are found in the southern states of Brazil. However, extensive areas are also found in the basaltic plateau in the states of São Paulo and Rio Grande do Sul.

Planossolos

These are poorly drained soils, generally with sandy textured surface horizon and abruptly contrasting with the underlying compacted soils and extremely hardened under dry conditions. Normally the well-structured and very slowly permeable B horizon shows relevant clay accumulation, sometimes responsible for a perched water table. They are found in areas of gently undulating relief used mainly for irrigated rice in the state of Rio Grande do Sul as well as for cattle raising pastures in the north-eastern states of Brazil (Resende *et al.*, 1988).

Plintossolos

Their most outstanding characteristic is the remarkable presence of plinthite in the soil profile and is generally associated with red mottling, both originating from iron segregation. The plinthite basically consists of a mixture of different clay types such as kaolinite and iron oxides. Plinthite is poor in organic carbon and rich in iron, sometimes associated with aluminium, with the sand fraction dominated by quartz. Plinthite is easily identified in the profile by the large contrast with the soil matrix either by its colour or by its consistency. The Plintossolos may contain continuous and hardened ferruginous layers as well as petroplinthite. Petroplinthite consists of nodular materials or iron concretions originating from irreversible hardening of the plinthite after successive cycles of moistening and drying. Plintossolos are frequently acidic with low stocks of nutrients and variable texture. They are commonly found in gently undulating reliefs, in depressed areas, alluvial plains and in the lower part of the backslope that favours slow water percolation. The largest extensions are found in the Amazon basin (upper Amazon river in Brazil), the state of Amapa, the island of Marajó, lowlands of the state of Maranhão, north of the state of Piauí, the south-eastern state of Tocantins and the north-eastern state of Goiás, the Pantanal region in the state of Mato Grosso and the island of Bananal in the state of Tocantins (Oliveira *et al.*, 1992).

Vertissolos

These are soils that present pronounced changes of volume with increasing water contents, morphologically expressed by the presence of deep cracks in the dry periods, grooved aggregate surfaces (slickensides) and wedge-shaped structures, slanted in relation to a horizontal direction. They present grey and black colours and sometimes yellowish or reddish colours with small variation in the clay content at depth. They have a clayey texture and high chemical fertility despite showing problems related to physical properties, i.e. soil swelling and contraction. The largest extensions of Vertissolos are located in the semi-arid zone in north-eastern Brazil, in the Pantanal region of the state of Mato Grosso do Sul, in the Campanha Gaúcha and in Recôncavo Bahiano (Oliveira *et al.*, 1992).

Amazon Soils

Several ecosystems are identified in the Legal Amazon consisting of different types of tropical and equatorial forests and savannahs (Vieira and Santos, 1987; EMBRAPA, 1992). The soils as a component of the natural resource complex also vary considerably. General aspects and some peculiarities of the Amazon soils, emphasizing their main characteristics, potentialities and constraints to agricultural use, are presented below.

Soil types and their characteristics

Low natural fertility and high exchangeable aluminium saturation are the most common aspects of the Amazon soils as shown in soil surveys carried out by EMBRAPA (1976, 1978, 1980a,b, 1981a,b, 1982a,b, 1983a,b, 1986) and RADAMBRASIL (BRASIL, 1975, 1977a,b, 1978a–c) in that region for the last 40 years. However, small areas of fertile soils occur in the Amazon region as reported by some more detailed soil surveys conducted along the highways and in selected agriculture and settlement areas.

Kaolinite is the predominant mineral of the clay fraction in Amazon soils with, however, low cation-exchange capacity (CEC) and base saturation. Hence, the soils have low stocks of nutrients, which is a limiting factor for the productivity and sustainability of agricultural and agroforestry production systems. Some continuous areas that can be represented in small-scale maps (Plate 3) located in the north-west of Brazil (state of Acre) contain soils with high natural fertility (Cambissolos, Argissolos and Luvissolos), although in steep slopes and mainly conditioned by the type of parent material.

A singularity in the Amazon region is the existence of anthropogenic dark earth (Plate 4), the so-called Terra Preta de Indio (Indian Black Earth, TPI in Portuguese) distributed in isolated patches of different sizes (5–500 ha) (Falesi *et al.*, 1972; Kern *et al.*, 2003). TPI is the name given to these soils built either intentionally or unintentionally by prehistoric indigenous populations that inhabited the margins of the rivers of the Amazon basin (Woods, 2003). These special places served as dwelling sites in the prehistoric past, where the populations deposited residues of vegetable origin (leaves and shafts of several palm trees, cassava peels, seeds, etc.) and of animal origin (bones, blood, fat, excrement, shells of all kinds, etc.), besides a great amount of ashes and residues of bonfires (Kern, 1996). They are characterized by high levels of promptly absorbed nutrients for plants as well as high contents of organic matter and favourable physical conditions for growing cultivated plants. In addition, they have a highly contrasting intensive biological activity in relation to the surrounding kaolinitic, more weathered soils with low organic matter content (Madari *et al.*, 2003).

However, the acidic and more weathered Latossolos (Plate 5) and Argissolos (Plate 6) with low natural fertility make up approximately 62% of Legal Amazon total area (Table 2.3 and Plate 3). In spite of their low fertility, which can be easily corrected by the use of fertilizers, Latossolos have good physical conditions and topography very favourable for mechanized agriculture.

Expressive areas of Plintossolos (Plate 7) and Alissolos (Plate 8) are distributed in approximately 16% of the Legal Amazon region (Table 2.3). Associated with the low nutrient stock, the presence of plinthite and petroplinthite in Plintossolos and the high contents of extractable aluminium in Alissolos (>4 $cmol_c$/kg soil) make it more difficult or even impossible to implement agriculture in parts of these areas.

Other classes of soils in the Amazon domains are the Gleissolos, Neossolos, Espodossolos, Cambissolos, Nitossolos, Planossolos and Vertissolos (Table 2.3), accounting for approximately 22% of the Legal Amazon.

Potentialities and constraints of the Amazon soils

The differences in soil types need to be understood in terms of nutrient availability and other related attributes such as pH, organic matter and CEC, and soil texture (Moran and Brondizio, 1998). Besides the difference in increasing clay content with depth, Latossolos and Argissolos present some common properties or characteristics that are strongly related to the availability of soil nutrients to plants and to the conditions of plant growth (Sanchez, 1976). Demattê (1988) reported that chemical

Table 2.3. Extent and percent distribution of Amazonian soils.

Soil classes	Area (km^2)	Relative area (%)
Latossolos	1,900,996.38	37.53
Argissolos	1,229,606.27	24.28
Plintossolos	457,262.39	9.03
Alissolos	337,578.44	6.67
Gleissolos	299,192.67	5.91
Neossolos quartzarênicos	255,942.64	5.05
Neossolos litólicos	227,035.60	4.48
Espodossolos	126,075.69	2.49
Cambissolos	44,432.38	0.88
Neossolos flúvicos	33,220.28	0.66
Nitossolos	16,361.73	0.32
Planossolos	11,997.87	0.24
Vertissolos	390.59	0.01
Water bodies	124,764.52	2.50
Total	5,064,857.45	100.00

properties related to soil fertility are more constraining than physical properties.

In order to provide an overview of the soil fertility in the Amazon region, the main constraints in acid tropical soils are presented. A summary of the extent of soil-related limitations, both physical and chemical, in the acid infertile soils of the Amazon region was given by Cochrane and Sanchez (1982) and is presented in Table 2.4. Deficiency of P was shown to be the most severe chemical limitation to crop growth. The list of major constraints is completed by the toxicity of Al, deficiency of K, high P fixation and low CEC. Other physical hindrances are shown but they are of minor relevance. Many authors confirmed that crop growth in Amazon soils is limited by P, Ca and Mg, rather than by N (Cuevas and Medina, 1986, 1988; Vitousek and Matson, 1988).

Phosphorus sorption and deficiency

Acid tropical soils normally contain a limited P reserve and often have a high sorption capacity (Rodrigues, 1996; Novais and Smyth, 1999). According to Sanchez and Uehara (1980), there are two main processes responsible for P fixation in acid soils: (i) precipitation by exchangeable Al; and (ii) adsorption on the surface of sesquioxides. Phosphorus fixation tends to be high in acid soils where the Fe and Al oxyhydroxides are ubiquitous. The reversibility of P sorption is important since desorption is often a limiting factor in the uptake of phosphorus by crops. Hence, P is considered to be the most limiting nutrient in Amazon soils and frequently found only as a trace (below 1 mg/kg of soil). Phosphorus deficiencies limit annual crop production in 90% of Amazon upland soils (Sanchez, 1976; Cochrane and Sanchez, 1982). Later, Corrêa and Reichardt (1995) observed that pasture establishment and growth was limited by phosphorus deficiency.

Aluminium toxicity and subsoil acidity

The epipedon of acid tropical soils and their typically kaolinitic subsoil is generally dominated by exchangeable Al (Rodrigues, 1996). The high amounts of Al, and sometimes Mn, and the low contents of Ca, Mg, and other nutrients frequently account for the low productivity of crops grown on these acid soils. High concentrations of aluminium inhibit root development and tend to limit absorption of other nutrients, especially of Ca and Mg that are closely related to root growth and plant development (Lathwell and Grove, 1986).

Cation-exchange capacity (CEC) and basic cation deficiencies

The magnitude of CEC results from the nature of the mineral and organic colloids and of the pH of the soil. The clay fraction

Table 2.4. Summary of main constraints in the Amazon Basin under native vegetation.

Soil constraint	Million hectares	Per cent of Amazon
Phosphorus deficiency	436	90
Aluminium toxicity	353	73
Drought stress	254	53
Low potassium reserves	242	50
Poor drainage and flood hazard	116	24
High phosphorus fixation	77	16
Low cation-exchange capacity	64	13
High erodibility	39	8
Steep slopes (> 30%)	30	6
Laterite hazard if subsoil exposed	21	4

Source: Cochrane and Sanchez (1982).

of Latossolos and Argissolos is usually dominated by sesquioxides, gibbisite, kaolinite and intergrade minerals. These compounds have low intrinsic quantity of negative charges and, therefore, most of the CEC of these soils depends on organic matter (see below) and on the soil solution pH. As a consequence, such soils exhibit a strong relationship between charge and pH. In some cases the soils may show net positive charge at low pH, which affects the availability of some nutrients (Sanchez, 1976). CEC is responsible for the equilibrium of ions in the solid/liquid interface in soils. So the usually low values of CEC combined with low pH lead to leaching of K, Ca and Mg.

Low concentrations of K, Ca and Mg and the low CEC associated with high Al contents are serious fertility constraints in acid tropical soils. Evaluation of these parameters in subsurface layers (below 0.3 m) should be undertaken. Liming is a low-cost and effective way to neutralize soil acidity. Liming reduces Al and Mn toxicity, improves P, Ca and Mg availability, increases CEC, promotes N_2 fixation and improves soil structure. Overall, liming improves soil capacity to supply needed nutrients and the ability of plants to absorb nutrients and water due to better root growth. Also an increase in exchangeable bases and pH can stimulate decomposition and mineralization of organic matter by creating a more favourable environment for microbial populations (Sanchez, 1976).

Importance of soil organic matter

Commonly the most important function of organic matter in soil is a reserve of nitrogen and other nutrients required by plants (Craswell and Lefroy, 2001). Nevertheless, soil organic matter (SOM) also plays an extremely important role in tropical soils, since it affects soil properties such as electrical charge and nutrient supply (Sanchez, 1976). The main factor responsible for negative charges, and therefore for CEC, is SOM, which contributes 60–80% of total soil CEC (Raij, 1969). The organic matter content is affected by vegetation type, as well as the parent materials from geological formations and increases with soil clay content and rainfall (Tognon *et al.*, 1998). Alfaia (1988) confirmed the role of organic matter showing increases of soil CEC by the deposition of organic matter on the soil surface, and the direct relation between soil properties with soil organic carbon content and pH. However, changing original forest to cropland leads to an increase in the decomposition rate of SOM (Vitorello *et al.*, 1989).

SOM can be increased by addition of crop residues, cover crops, green manure crops, compost, animal manure, by reduced or no-tillage and by avoiding residue burning. Enhanced SOM increases soil aggregation, water-holding capacity and P availability and reduces P fixation, toxicity of Al and Mn, and nutrient leaching by enhancing exchangeable Ca, Mg and K (Baligar and Fageria, 1997). SOM also provides a source of nutrients, as was shown by Pereira *et al.* (2000) who evaluated changes in chemical properties of a Xanthic Hapludox managed under pasture, using two rotational systems with *Brachiaria brizantha* and *Panicum maximum*. The organic material incorporated into the soil through vegetable and animal residues influences the chemical characteristics, increasing the levels of Ca, Mg, K, P, N, C, OM and pH, and decreasing the Al levels.

Soil fertility and forest

Although highly weathered and leached of nutrients, the Amazon soils often support dense evergreen rainforests that have evolved an array of efficient nutrient conservation mechanisms to cope with the paucity of soil nutrients. Deforestation results in replacement of the primary forest species, which are efficient in cycling nutrients, by crops or pioneer and second-growth species, which are less efficient. The nutrient cycles are also modified and the nutrient-conserving mechanisms are lost. In the Legal Amazon, the nutrient supply of the vegetation depends strongly on the humus-enriched topsoil and, notably, on the nutrients in the biomass of the rain-

Plate 1.

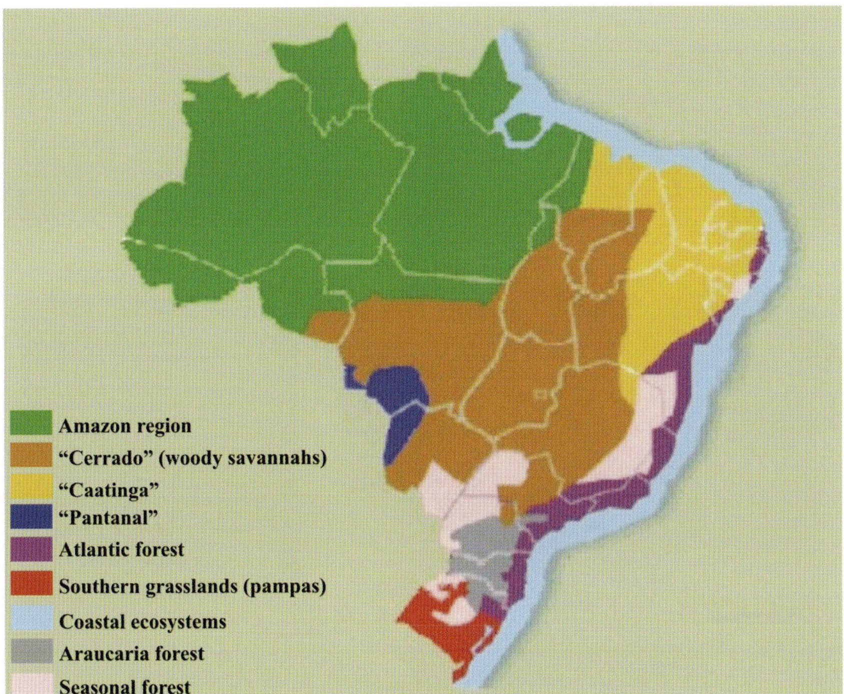

Plate 2.

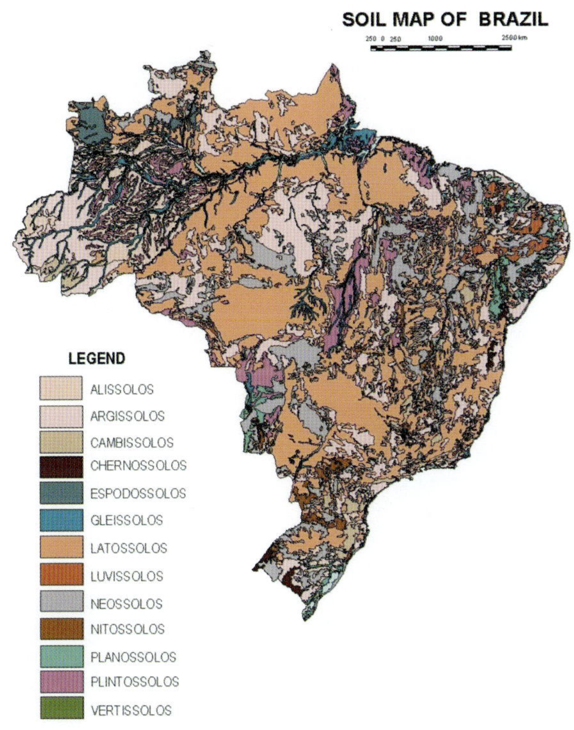

Plate 1. Brazilian ecosystems (see Chapter 1).
(http://www.ibge.gov.br/home/presidencia/noticias/noticia_visualiza.php?id_noticia=169&id_pagina=1)

Plate 2. Brazilian soil map (see Chapter 2). (Source: adapted from EMBRAPA, 1981.)

Plate 3.

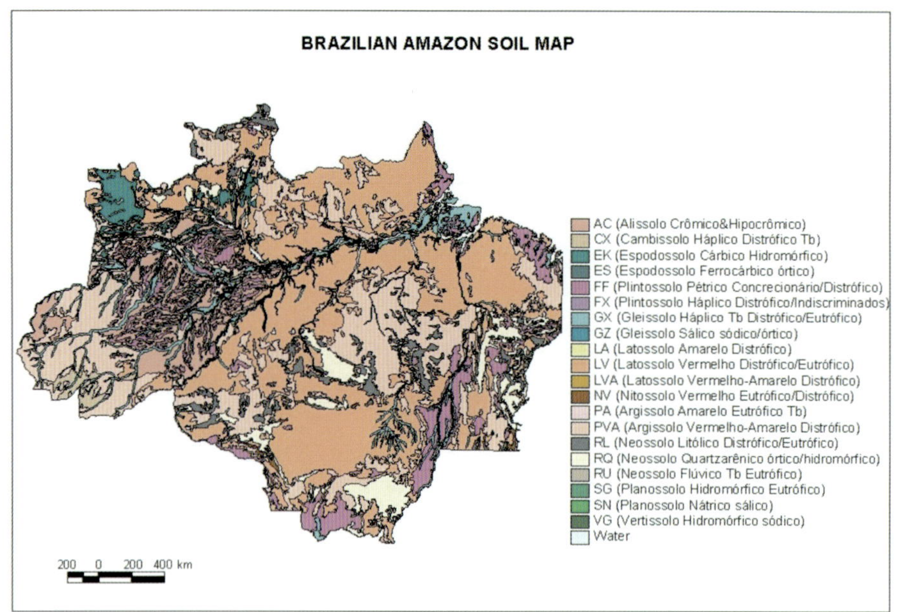

Plate 4.

Plate 5.

Plate 3. Soils of the Legal Amazon (see Chapter 2). (Source: adapted from EMBRAPA, 1981.)
Plate 4. Photo showing the Indian Black Earth profile near Manaus, Amazonas State, in the left bank of Solimões River. Note the expressive presence of pottery fragments within the soil profile (see Chapter 2). (Source: Beata Emoke Madari, EMBRAPA Solos soil researcher.)
Plate 5. Profile of Latossolo Vermelho-Amarelo loamy texture used for pasture in the Apuí County, Southern Amazonas State (see Chapter 2). (Source: Tony Jarbas Ferreira Cunha, EMBRAPA Solos soil researcher.)

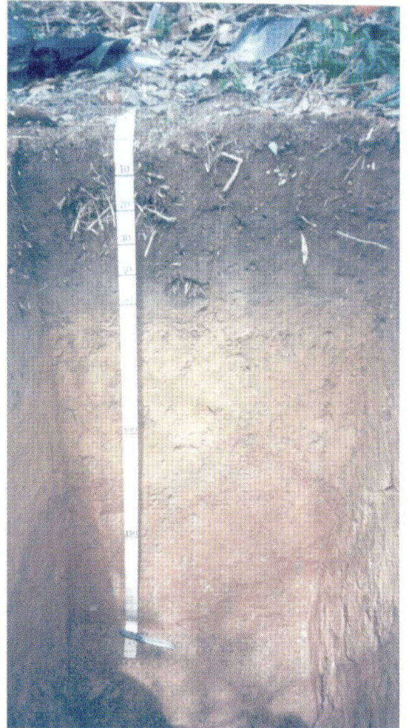
Plate 6.

Plate 7.

Plate 8.

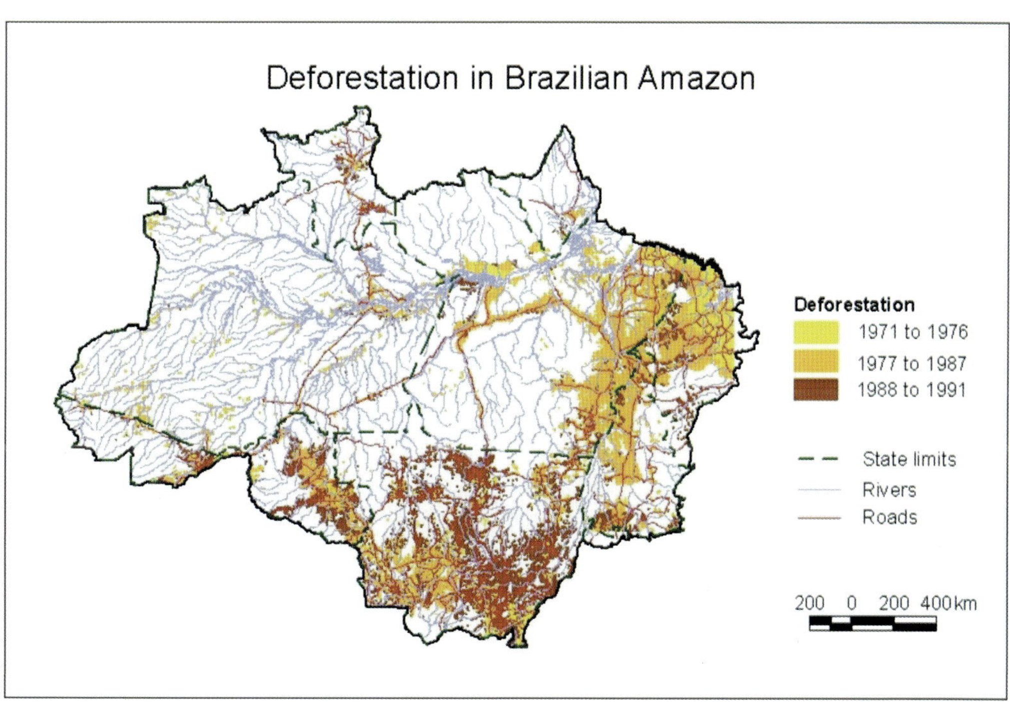

Plate 9.

Plate 6. Profile of Argissolo Amarelo loamy/clay texture, plinthic, under native forest, in Humaitá County, Southern Amazonas State (see Chapter 2). (Source: Tony Jarbas Ferreira Cunha, EMBRAPA Solos soil researcher.)
Plate 7. Profile of Plitossolo clay texture with expressive petroplinthite nodules within the soil, in Paragominas County, Pará State (see Chapter 2). (Source: Marcelo Nunes Camargo, EMBRAPA Solos Records.)
Plate 8. Profile of Alissolo clay texture with plinthite in depth used with pasture in Humaitá County, Southern Amazonas State (see Chapter 2). (Source: Nilson Rendeiro Pereira, EMBRAPA Solos soil researcher.)
Plate 9. Deforestation in Brazilian Amazon (see Chapter 2). (Source: BRASIL, 2002b.)

forest vegetation, which are effectively recycled from decomposing organic materials by the dense, superficial roots and their mycorrhiza (Stark and Jordan, 1978; Herrera and Jordan, 1981; Cuevas and Medina, 1988).

Soil management and nutrient dynamics

Cultivation of acid soils in the Amazon is preceded by cutting and removing the economically important trees and burning the remaining aerial biomass (Martins *et al.*, 1991). Alterations to both climatic and ecological patterns due to forest clearing and burning have been reported (Watson *et al.*, 2000). These land-clearing methods often lead to an immediate effect on the initial levels of nutrients in soils and, consequently, affect nutrient dynamic patterns (Martins *et al.*, 1991).

The traditional system of slash-and-burn clearing is part of the shifting cultivation system employed by Amazonian farmers. The practice is controversial and pressure is rising to seek alternatives to burning. In shifting cultivation, an important function of the secondary vegetation is the accumulation of nutrients in the aerial parts of the plants and the fast liberation of these nutrients by burning as a means to improve soil fertility. Burning also provides: (i) increases in soil pH due to the ash alkalinity; (ii) improved access for sowing; and (iii) reduction of weeds as well as pests and diseases. The improvement of soil fertility depends on the quantity of ash, which for its part depends on the burned biomass and the age of the secondary vegetation (Kato *et al.*, 1999). Attempts to eliminate burning of woody fallow vegetation are limited by difficulties in handling huge amounts of biomass without the use of heavy equipment (Seubert *et al.*, 1977). The disadvantages of burning are losses due to volatilization of nitrogen and sulphur as well as smaller quantities of phosphorus and potassium. Hölscher *et al.* (1997) estimated such losses as amounting to 96%, 76%, 47% and 48%, respectively, of these nutrients in the aboveground material. Nutrients released by burning may also be rapidly leached, and as a consequence, multiple nutrient deficiencies develop early in annual crop rotations (Cravo and Smyth, 1997). Many studies report the favourable effects of burning on soil chemical properties initially following forest clearing, and the nutrients released after burning mature forest usually support 2 or 3 years of no-input annual cropping before fields are abandoned to fallow (Sanchez *et al.*, 1983; Ewel, 1986; Serrão *et al.*, 1996). McGrath *et al.* (2001) demonstrated that these soil changes resulting from the slash-and-burn conversion of forest to agroforest may persist at least 6 years after agroforest establishment. However, due to recent concerns related to global climate change and mitigation of greenhouse gases by agricultural systems, alternatives to the use of slash-and-burn are being investigated.

The losses with fire can be eliminated by preparing fields without the use of fire, offering the hope of more efficient nutrient cycling and improved sustainability (Luna-Orea and Wagger, 1996). Nevertheless, a mechanized clearing would remove the vegetation along with part of topsoil (Seubert *et al.*, 1977).

A sustainable alternative is just mulching or incorporation of slashed vegetation. The organic material serves as a carbon-rich substrate that is decomposed to SOM by microbial organisms, thereby initially immobilizing a large fraction of the available soil nutrients (Braakhekke *et al.*, 1993). When left on the surface, residues are subject to rapid drying and decompose slowly, resulting in slow rates of mobilization. Residues mixed with soil often remain moist and decomposers have easier access to soil nutrients so that decomposition is much more rapid than for residues left on the soil surface (Sanchez *et al.*, 1989; Myers *et al.*, 1994; Woomer *et al.*, 1994).

Because of the low chemical fertility of the soils, the correction of high acidity and initial nutrient deficiencies as well as the replacement of nutrient exports in the harvested biomass will normally also be necessary in permanent agricultural systems (Szott and Kass, 1993). Then a continuous monitoring of the declining nutrient availability may be used as a guideline for establishing well-advised fertilization programmes for sustained productivity (Sanchez *et al.*, 1983).

Depending on the position in the landscape, soils in the Legal Amazon may also be grouped into *terra firme* (non-flooded) and *várzea* (floodplain) soils. Considering that the total area of *terra firme* is 4,469,215.8 km^2 (estimate based on data given in Table 2.3), Latossolos and Argissolos cover 70% of the non-flooded area. These soils are normally under both perennial and annual crop cultivation, pasture grasses and various agroforestry systems. Compared with non-flooded soils, soils on the floodplain (e.g. Gleissolos, Neossolos Fluvicos) show higher fertility and despite covering a smaller area in the Legal Amazon, they play an important role in annual crop cultivation, particularly rice (Alfaia and Falcão, 1993).

Most definitions of sustainability include the idea of increasing or maintaining the quality of the natural resource. In terms of soil fertility it suggests a management that avoids nutrient depletion by crop harvest, erosion, leaching and volatilization (Smyth, 1996).

Soil quality is defined as the capacity of soil to function within a specific kind of ecosystem in a manner that sustains plant and animal productivity, maintains or enhances water and air quality and supports human health and habitations (Karlen *et al.*, 1992). Thus, soil quality is needed in the development of more sustainable land management.

Smyth (1996) believes that knowing the nutrient dynamics in soils (e.g. phosphorus, nitrogen) is useful for indicating the sustainability of a production system. Smyth and Cassel (1995) associated different lime and fertilizer requirements for sustainable cultivation of Latossolos and Argissolos with different patterns of nutrient depletion of soils. However, sustainable use of different soil types in the Legal Amazon region may not be achieved only by knowing the physico-chemical environment of the soil and consequent interventions by farmers. Swift (1999) reported that integrated biological management is an additional component in the armoury of soil management practices. Hence, the benefits of N-fixing plants in rotation or combined with main crops, cover crop management to increase carbon sequestration and diverse living organisms must be included in strategies for sustainable production systems of different soils in the Legal Amazon.

Land use and land cover in the Amazon: history, technologies and tools

The spatial configuration of landscape elements can be attributed to a combination of environmental correlates and human forces that operate at different spatial and temporal scales (Forman and Godron, 1986; Dunn *et al.*, 1991), creating complex patterns of change (Di Castri and Hadley, 1988; Dunn *et al.*, 1991). The understanding of changing patterns and their consequences plays a key role for planning and managing natural resources, which involves integration and interpretation of various forms of data at spatial and temporal scales.

The assessment of land use and land cover (LULC) and the monitoring of its dynamics are essential requirements to better understand the patterns and processes of changes in vegetation and soil (Mendonça-Santos *et al.*, 1997; Mendonça-Santos, 1999; Mendonça-Santos and Claramunt, 2001), succession dynamics of natural vegetation and changes in biomass (Alves *et al.*, 1997), soil fertility and its correlation with forest regeneration (Moran *et al.*, 2000). Evaluation of changes in microbial populations with changes in land use is also needed in order to promote sustainable management of natural resources and environmental protection. In a global perspective land use changes are very relevant, influencing key aspects of the global terrestrial system such as biotic diversity (Sala *et al.*, 2000), climate (Houghton *et al.*, 1999), and soil degradation (Tolba and El Kholy, 1992).

The Committee on Global Change Research (1999) emphasizes the need to address the causes of land use changes. In most cases the causes are political and economic, with little concern for environmental issues. Lambin *et al.* (2001) reported that land use changes are due not only to population growth and poverty, but also to the population's response to economic opportunities mediated by institutional factors. Hence, local and national markets and policies bring opportunities and limitations to alternative land uses. However, global trends are the principal factors that control changes in land use, which are more or less reinforced by local circumstances.

Studies conducted by the Food and Agriculture Organization (FAO) show that most of the changes in land use in the tropics are from forest to agriculture or pasture systems (FAO, 1996). In Brazil, agriculture and pastures systems were responsible for 91% of the total deforestation, particularly in the 1980s, in which 51% of the deforested area was the result of conversion to annual and permanent crops and 40% to pasture systems (Amelung and Diehl, 1992). In the 1990s there was a decrease in the rate of deforestation as a consequence of the cancellation of subsidies to the expansion of the pasture area.

The large territory of Brazil and the high environmental diversity combined with different economic situations led to various land use patterns, which can be characteristic of each region as shown in Fig. 2.1.

The principal classes of vegetation that occur in the Amazon region are dense tropical rainforest, open tropical rainforest, seasonal semideciduous tropical forest, campinarana, savannah and early primary succession communities (IBGE, 1991).

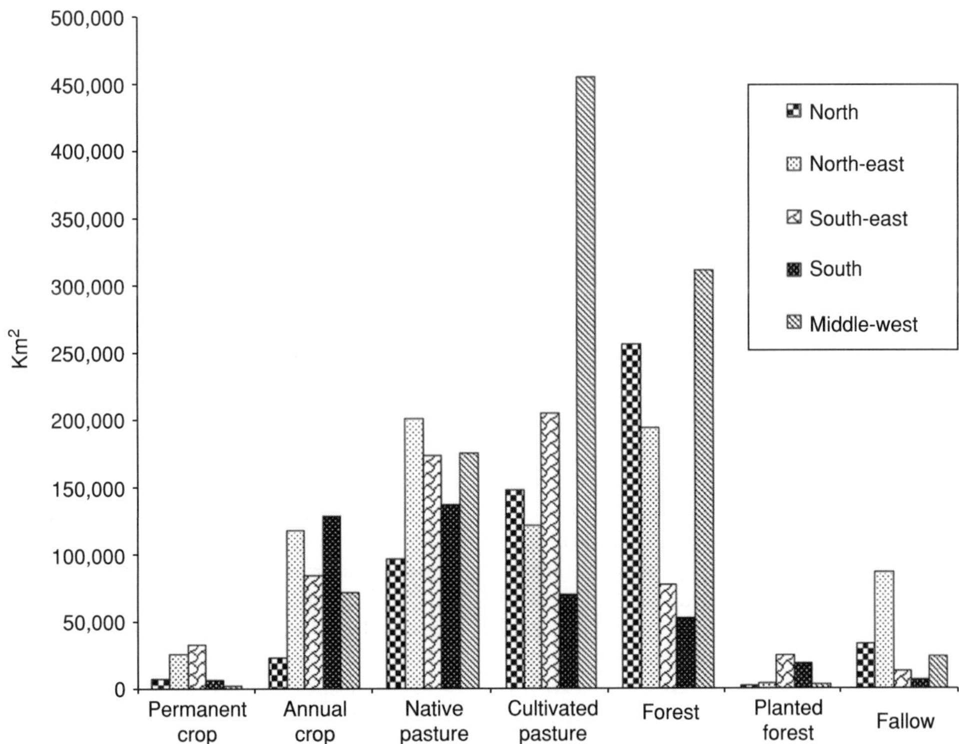

Fig. 2.1. Land use in Brazil (by regions). (Source: Manzatto *et al.*, 2002, after IBGE, 1997.)

The present-day situation of land use changes in the Brazilian Amazon is the result of different migration periods promoted by both federal and state governments (Mahar, 1979, 1988; Serrão *et al.*, 1996; Pedlowski *et al.*, 1997; Weinhold, 1999). Mahar (1979) reported that modern occupation of the Amazon region took place in five different periods starting in 1912 when the main activity was rubber exploitation. This was followed by the cultivation of special crops, such as pepper and manioc, promoted by the Superintendency for the Economical Development of the Amazon Region (SPVEA) in 1953. The main objective was to improve food self-sufficiency and additionally to expand the extraction of raw products for both international and domestic markets.

The third occupation period took place during the military dictatorship (1964–1985), which implemented a strong policy of economic development in the Amazon region, the so-called Operação Amazonia. During this period, development sectors were created, in which the federal government stimulated immigration and offered incentive schemes for private investments for infrastructure development. Also, scientific research on natural resources was supported, culminating in the creation of the Superintendency for the Development of the Amazon Region (SUDAM). The objective of SUDAM was to organize public investment in the Amazon region.

The fourth occupation period took place in 1970s with the creation of the National Integration Program (PIN), whose objective was to protect the Amazon region, promoting the migration of Brazilian citizens to the area. This would be accelerated with the construction of the BR-230 Highway (Transamazônica Highway) connecting the Atlantic coast to the Peruvian border. The Land Distribution Program (PROTERRA), whose objective was to facilitate land plot acquisition to improve rural work conditions and agroindustry in the Amazon region, complemented PIN. Fearnside (1986) reported the causes for the collapse of such programmes.

The fifth period was characterized by the Second National Development Plan, in which the federal government created the Program for Large Pasture-Based Cattle Raising, Logging and Mining (POLAMAZÔNIA). However, most of the credit provided by POLAMAZÔNIA was for promoting cattle raising.

These occupation efforts in the Amazon region were not successful in both economic and environmental aspects and one of the most serious consequences of these programmes is the yearly deforestation rates. Table 2.5 shows the extent and the average rate of deforestation in the Brazilian Amazon. The spatial distribution can be observed in Plate 9.

At present, the massive programme called 'Avança Brasil' (Forward Brazil) has been severely criticized, which consists of a package of 338 projects throughout Brazil, including the Amazon region (BRASIL, 2003). The projections of the impacts of 'Avança Brasil' and other recent projects in the Brazilian Amazon indicate tremendous problems concerning deforestation and consequent increases in carbon emissions (Fearnside, 2002).

It is important to understand changes that occur in the environment, particularly anthropogenic changes. LULC mapping combined with changes in time (succession) are well-known tools in scientific investigations. Thus the use of products generated by remote sensing has been widely adopted (Campbell, 1987; Mulders, 1987; Quattrochi and Pelletier, 1991).

Due to its multispectral and temporal aspects, which permit obtaining an overview of the landscape, remote sensing has become an unavoidable and relatively low-cost tool for environmental diagnosis, inventory, monitoring and planning, specially when combined with geographical information systems (GIS) and database technologies. This is particularly relevant to the Amazon region where access to remote locations is commonly restrained.

Acquiring of such an inventory is a laborious task, but remote sensing techniques enable the development of a consistent spatio-temporal database, which in turn enables combined analysis of data and the

Table 2.5. Deforested area and deforestation rate in the Amazon region.

Month /year	Jan/78	Apr/88	Aug/89	Aug/90	Aug/91	Aug/92	Aug/94	Aug/95	Aug/96	Aug/97	Aug/98	Aug/99	Aug/00
Deforested area (km^2)	152,200	377,500	401,400	415,200	426,400	440,186	469,978	497,055	517,069	532,086	551,782	569,269	587,727
Time period (year)	77/88	88/89	89/90	90/91	91/92	92/94	94/95	95/96	96/97	97/98	98/99	99/00	
Deforestation rate (km^2/year)	21,130	17,860	13,810	11,130	13,786	14,896	29,059	18,161	13,227	17,383	17,259	18,226	

Source: INPE (2002).

generation of new information that can be used in the process of decision making and problem resolution. The development of new sensors and the methodology for data analysis has boosted the potentialities of remote sensing and its use is very common in various scientific areas (Colwell, 1983).

Studies on LULC using remote sensing or airborne images are very common (Batistella, 2000). The most common imaging sensors are passive sensors, i.e. sensors able to detect solar radiation reflected or emitted by objects on the soil surface. A brief description of some sensor characteristics and tools is presented below.

Advanced Very High-Resolution Radiometers (AVHRRs) on the National Oceanic and Atmosphere Administration (NOAA) satellites (Table 2.6) provide estimates of cloud density and of the temperature on the sea surface. They are commonly used for environmental purposes including studies on land use at regional and global scales.

Thematic Mapper (TM) of Landsat-4 and -5 and Enhanced Thematic Mapper Plus (ETM+) of Landsat-7 are sensors of the Landsat series mostly used in studies on land use and cover, particularly TM sensor and presently ETM+ sensor of Landsat-7. Their characteristics are given in Tables 2.7 and 2.8.

High-Resolution Visible (HRV) of the Système Proboitoire de l'Observation de la Terre (SPOT) series is characterized by the ability to vary its angle of view, which is not limited to the perpendicular position of the satellite route (Table 2.9). This enables image overlapping and stereoscopy. A vegetation instrument was coupled to the SPOT-4 satellite, in which bands of 430–470 nm (blue) and 1580–1750 nm (mid-infrared) were added, with 1 km of spatial resolution.

Aerial photographs are obtained for different scales that were being used previous to the availability of images from remote sensors. Aerial photographs are a powerful tool for historical survey of land use and cover.

Besides passive imaging, studies on land use and cover in the Amazon region are being conducted with data provided from active imaging sensors and radars, particularly the Japanese Earth Resources Satellite (JERS) systems (L band), the space imaging radar C (SIR-C, C band), and the RADARSAT (C-band) (Batistella, 2000).

Presently, different sensors not listed here are available. The outputs can also be useful for studies on land use and cover such as the sensor in IKONOS II (Table 2.10) and the Moderate Resolution Imaging Spectroradiometer (MODIS) aboard *Terra* (EOS AM) and *Aqua* (EOS PM).

MODIS sampling frequency is 1–2 days for each satellite, acquiring 36 spectral bands at a radiometric resolution of 12 bits. Spatial resolution varies among bands: 250 m for bands 1 and 2; 500 m for bands 3–7; and 1000 m for bands 8–36. Some MODIS products like Land Cover/Land Cover Change may provide useful information about land use dynamics. The land cover parameter identifies 17 categories of land cover following the International Geosphere–Biosphere Programme (IGBP) global vegetation database, which defines nine classes of natural vegetation, three classes of developed lands, two classes of mosaic lands, and three classes of non-vegetated lands (snow/ice, bare soil/rocks, water). The land cover change parameter quantifies subtle and progressive land surface transformations as well as major rapid changes.

The detection of land use changes using digital images assumes that these changes lead to alterations in the reflectance from the Earth's surface. Digital techniques to detect spectral variations among several imaging data may be applied to the detection of land cover changes (use and vegetation). However, some factors related to the sensor system (differences among spectral bands, among spatial resolutions and variations in the radiometric response) or natural conditions (scattering variations and atmospheric absorption, presence of clouds and shadows, variations in the irradiance and solar angle, seasonal variations in the vegetation phenology and in soil moisture) may interfere in surveys of land cover change detection.

Table 2.6. Characteristics of AVHRR-NOAA.[a]

Band	Spectral range (nm)	Spectral region	Spatial resolution
1	580–680	Visible	1.1 or 4 km
2	725–1,100	Near infrared	1.1 or 4 km
3	3,550–3,930	Mid infrared	1.1 or 4 km
4	10,300–11,300	Normal infrared	1.1 or 4 km
5	11,500–12,500	Thermal infrared	1.1 or 4 km

[a]Temporal resolution: 12 h; radiometric resolution: 10 bits or 1024 grey levels.
Source: AVHRR-NOAA: http://edcdaac.usgs.gov/1KM/avhrr_sensor.html

Table 2.7. Characteristics of TM-Landsat-4 e 5.[a]

Band	Spectral range (nm)	Spectral region	Spatial resolution
1	450–520	Blue	30 × 30 m
2	520–600	Green	30 × 30 m
3	630–690	Red	30 × 30 m
4	760–900	Near-infrared	30 × 30 m
5	1,550–1,750	Mid-infrared	30 × 30 m
6	10,400–12,500	Thermal infrared	120 × 120 m
7	2,080–2,350	Mid-infrared	30 × 30 m

[a]Temporal resolution: 16 days; radiometric resolution: 8 bits or 256 grey levels.
Source: TM-Landsat-4 e 5: http://edc.usgs.gov/products/satellite/band.html

Table 2.8. Characteristics of ETM+Landsat-7.[a]

Band	Spectral range (nm)	Spectral region	Spatial resolution
1	450–520	Blue	30 × 30 m
2	530–610	Green	30 × 30 m
3	630–690	Red	30 × 30 m
4	780–900	Near-infrared	30 × 30 m
5	1,550–1,750	Mid-infrared	30 × 30 m
6	10,400–12,500	Thermal infrared	120 × 120 m
7	2,090–2,350	Mid-infrared	30 × 30 m
8	520–900	Visible and near-infrared	15 × 15 m

[a]Temporal resolution: 16 days; radiometric resolution: 8 bits or 256 grey levels.
Source: ETM+Landsat-7: http://edc.usgs.gov/products/satellite/band.html

Table 2.9. Characteristics of HRV-SPOT-1, -2 and -3.[a]

Band	Spectral range (nm)	Spectral region	Spatial resolution
XS1	500–590	Green	20 × 20 m
XS2	610–680	Red	20 × 20 m
XS3	790–890	Near-infrared	20 × 20 m
PAN	510–730	Visible and near-infrared	10 × 10 m

[a]Temporal resolution: 26 days; radiometric resolution: 8 bits or 256 grey levels.
Source: SPOT: http://www.spot.com/home/SYSTEM/IMEXPLO/imexplo.htm

Table 2.10. Characteristics of IKONOS II.[a]

Spectral range (nm)	Spectral region	Spatial resolution
450–520	Blue	4 × 4 m
520–600	Green	4 × 4 m
630–690	Red	4 × 4 m
760–900	Near-infrared	4 × 4 m
450–900	Panchromatic	1 × 1 m

[a]Temporal resolution: varies with latitude and bands; radiometric resolution: 11 bits or 2048 grey levels.
Source: IKONOS:http://www.spaceimaging.com/whitepapers_pdfs/IKONOS_Product_Guide.pdf; MODIS: http://modis.gsfc.nasa.gov

The influence of some of these factors may be partially minimized if images originating from the same sensor system are obtained in the same time of the year and without cloud effects. Normalization methods in the preprocessing of multitemporal images have been used to improve the results of the detection of changes (Singh, 1989; Almeida-Filho and Shimabukuro, 2002; Yuan and Elvidge, 2002).

Most of the methods for change detection may be grouped in two different approaches: comparison methods of postclassification and enhancement methods. The postclassification methods consider the identification of land cover classes in each image and the changes relate to alterations of land use observed in different periods of time. Enhancement methods are based on direct detection of spectral changes (Singh, 1989; Almeida-Filho and Shimabukuro, 2002; Yuan and Elvidge, 2002).

Enhancement methods include image transformation of different periods of time in new bands in which areas of change are highlighted. The processing using other analytical methods may be applied in the highlighted areas for the classification of the alterations. The accuracy of the results depends on the accuracy in the registration of the group of images involved in the analysis (Singh, 1989; Almeida-Filho and Shimabukuro, 2002; Yuan and Elvidge, 2002).

In the postclassification methods, change detection is done using a pair of images obtained at different dates. They are classified independently and the areas of changes are extracted directly through comparison of their results. The final accuracy depends on the accuracy of each individual classification as the product of the accuracy of each one (Singh, 1989). On the other hand, this approach depends on atmospheric conditions and the differences in the sensor's response, which may help to map classes of interest (Almeida-Filho and Shimabukuro, 2002). Mas (1999) observed that these methods are less sensitive to spectral variation due to differences in soil humidity and vegetation phenology, giving consistent results when the procedure involves images of different times of the year.

The resolution of the images used influences the results of the survey of land use and cover change detection. While analysing different spatial resolutions for the identification of classes of land cover in the Amazon region, Ponzoni *et al.* (2002) observed that for discrimination between forest and non-forest pixel size of less than 200 m has no interference. However, an effect was reported for the identification of secondary vegetation in its early or advanced stages of regeneration, whose occurrence in polygons demands resolutions higher than 100 m.

Image processing of remote sensing alone is not sufficient for the understanding of the land cover and use dynamics and its relation to environmental variables that need to be evaluated. Integrated analysis of different variables and their spatial relationships is necessary. In this case, Geographic Information Systems (GIS) have become an important tool for mapping LULC and for performing quantitative and qualitative analysis of changes (Mendonça-Santos, 1999) in addition to collecting, storing, retrieving, transforming and displaying of spatial data (Burrough and McDonnell, 1998).

Land Use Changes and Loss of Biodiversity

As discussed in the previous section, the Amazon occupation has generated many impacts, among them, large deforested areas in which the soils, with low natural fertility and high aluminium saturation, are quickly being degraded. In several occasions, the local inhabitants, inhibited by the low productivity of land and infrastructural problems in the Amazon, have been attracted to mining, which contributes to a tremendous loss of biodiversity.

Presently, a trend scenario can be designed by the Economical and Ecological zoning study – ZEE Brasil (BRASIL, 2002a) for the Legal Amazon. The result indicates that if the present infrastructural projects in the region continue to be implemented it will lead to the following major consequences:

1. Higher native and migrating population growth than the national average.
2. Increase in the number of municipalities and a high rate of urbanization.
3. Disparity between the enlargement of cities, particularly those included in governmental projects, and the rest of the cities in the Amazon region.
4. Difficulties in offering fixed positions to part-time workers normally active in major projects.
5. Expansion of pasture-based cattle raising and agricultural systems for export crop production.

Thus, as part of the National Integration and Development Axes of the Federal Government, it is envisaged to stimulate construction of transport corridors such as highways, railways, rivers and harbours when constructing thermo- and hydroelectric power plants. The reason is that such an infrastructure and logistic system is likely to affect the pace of soil use change with simultaneous creation of new conflicts and pressures on environments already fragile, if compensation measures are not implemented. It is also evident that as soon as infrastructure investments are consolidated they will condition the growth vectors of capital-intensive production systems (Table 2.11) as well as the structure of the regional urban network.

Therefore, this scenario of population growth and increasing demand for food production and the growing environmental awareness of society will lead, unavoidably, to significant changes in the concepts of regional development. In the recent paradigm of sustainable development, conciliation of quality and competitive targets with environmental conservation is a huge challenge in the Amazon region. Although there is sufficient knowledge of classification and mapping of soil, little information is available about sustainable use and management in comparison with ecosystems from other regions in Brazil.

In a natural environment, the first soil modifications start soon after deforestation, even if soils are not used for any purpose. These alterations are related mainly to changes in the quality and amount of organic matter deposited on the soil (and consequently, in the turnover rate of the organic matter of the soil) and in the moisture and thermic regimes of soils (larger exposure of soil to sun rays and rain and smaller evapotranspiration rates). These alterations are reflected to a larger or smaller degree in the soil biota and, variably, depending on climatic conditions, in vegetation cover, soil type and their relative position in the landscape. For example, in the dense tropical rainforests of Amazon's higher terraces (stable land), on clayey yellow Latossolos, generally the transformation of organic matter by soil biota occurs mainly in the thin layer of fallen leaves, branches and roots on the soil, in the first few centimetres of the soil surface. This may be observed in the occur-

Table 2.11. Public and private investments expected for the National Axes of Integration and Development Programme within Legal Amazon.

	National integration axes				$ (millions)	
Sector	North	Madeira Amazon	Araguaia Tocantins	West	Total Legal Amazon	% Total investments
Transport	317.0	1,585.7	5,742.1	2,168.9	9,813.7	23
Airports	42.6	191.8	581.0	38.4	853.8	2
Railways	–	–	3,307.8	1,174.5	4,482.3	10
Waterways	–	430.1	377.1	102.2	909.4	2
Ports	–	209.8	154.3	12.0	376.1	1
Highways	274.4	754.0	1,321.9	841.8	3,192.1	7
Energy	66.1	9,959.9	3,822.9	836.7	14,685.6	34
Gas pipeline	–	450.0	–	–	450.0	1
Hydroelectric dams	–	8,703.0	3,251.0	484.5	12,438.5	29
Thermo plants	57.0	685.0	–	215.0	957.0	2
Transmission lines	9.1	121.9	571.9	137.2	840.1	2
Communications	104.8	900.0	2,163.5	652.9	3,821.2	9
Total infrastructure	487.9	12,445.6	11,728.5	3,658.5	28,320.5	66
Social development	359.3	4,262.7	4,763.2	2,076.2	11,461.4	27
Knowledge	25.0	156.2	92.5	107.5	381.2	1
Environment	151.3	1,008.7	511.8	894.8	2,566.6	6
Total	1,023.5	17,873.2	17,096.0	6,737.0	42,729.7	100

Source: Estudo dos Eixos Nacionais de Integração e Desenvolvimento, Diaz *et al.* (2002).

rence of weakly or moderately developed surface horizons, and in the dark waters of some rivers, rich in soluble humic substances transported by the runoff of rainwaters. In alluvial and hydromorphic soils of the lowland areas, these alterations are slower, due to the lower rate of decomposition of organic matter in hydromorphic conditions and the additional deposition of organic matter from neighbouring upper areas.

Therefore, deforestation may result in soil degradation, either in cultivated areas or in natural vegetation, through water erosion. Guerra *et al.* (1999) consider erosion to be the result of fast and unplanned human occupation of newly cleared areas, fragile soils and heavy rainfall. Pereira (1977) reports that accelerated erosion begins with forest cutting and successive deforestation cycles, and increases with continuous land use with crops and pastures.

If the present situation continues and if the estimates of the trend scenario are confirmed, tropical forests of the Brazilian Amazon will continue to suffer tremendous anthropogenic modifications, as can be observed in Fig. 2.2, leading to continued loss of biodiversity and natural resources as observed in the past.

Although considering that there has been an increasing perception within society about environmental problems and their consequences in the area, soil degradation and its impacts on biodiversity have not received due attention.

Thus standard-setting studies and environmental planning must be carried out for the occupation of Amazônia, in order to establish relationships in the landscape among upper-terrace stable lands and lowlands under forests subjected to flooding and puddling. The wetlands of the Amazonian rainforest are fragile ecosystems depending on biogeochemical and hydrological processes. The general features of wetlands are the following:

1. The presence of water, either at the surface or within the root zone.
2. The anaerobic conditions leading to gleying or organic soil formation.
3. The presence of hydrophytic vegetation.
4. The absence of vegetation sensitive to seasonal flooding.

In the process of forest clearing and incorporation of new areas to agriculture, the impact on the soil may be even more severe if fire is used. In the Amazon region, as previously mentioned, the use of fire is very widespread in cattle raising and slash-and-burn agricultural systems. Under such circumstances, fire affects directly the physico-chemical characteristics of soils, such as loss of N and S by volatilization (Mackensen *et al.*, 1996; Hölscher, 1997), as well as soil biota, air quality, biodiversity and human health. The use of fire also leads to erosive processes by diminishing soil cover in the beginning of the rainy season.

Additionally, fires commonly escape control and cause extensive damage to wildlife, buildings and livestock. They also lead to changes in the atmosphere, increasing greenhouse gas emissions and causing global climate change (Diaz, 2002). Furthermore, as reported by Nepstad *et al.* (2001), if the historical relationship between road paving and forest alteration by humans continues, the Brazilian government's plan to pave, recuperate or construct 6245 km of roads in the Amazon may stimulate 120,000–270,000 km^2 of additional deforestation. Even without taking into account measurements of all losses, including those of biological origin and the loss of the productive potential of lands, estimates by Motta *et al.* (2001) illustrate the negative effects of forest burning. Tables 2.12 and 2.13 summarize the estimates of the physical and economic damage for the years 1996 and 1998.

After the conversion of land to agriculture, soil preparation and management become the main causes of land degradation in subtropical and tropical Brazilian environments. Their effects are observed mainly by the decrease in SOM content and its consequences, particularly by loss of the soil productive capacity. Figure 2.2 presents a holistic view of the effects of soil tillage on soil degradation, productivity

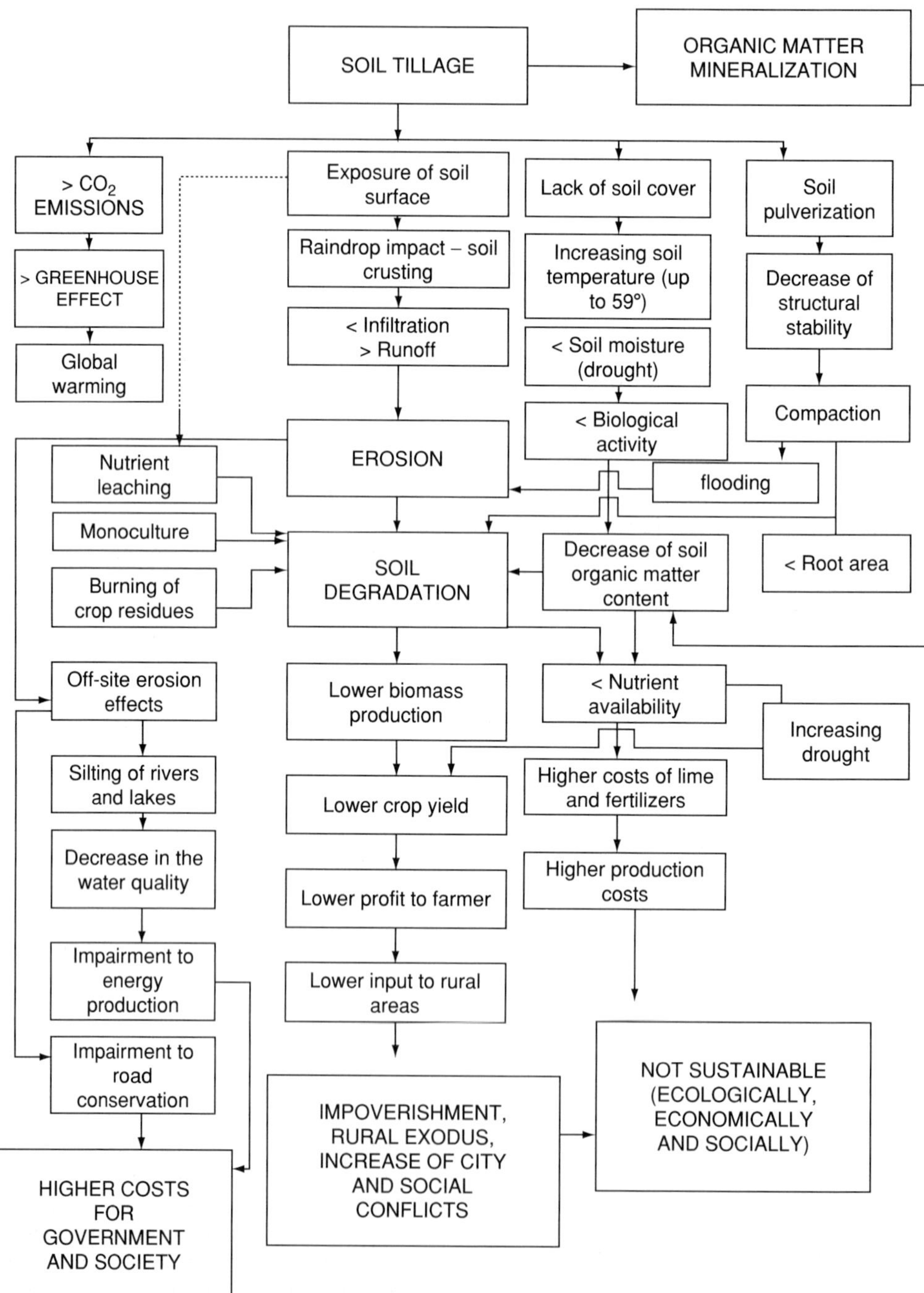

Fig. 2.2. Soil tillage influences on degradation, productivity loss as well as effects of conventional agricultural practices. (Source: Derpsch, 1998.)

Table 2.12. Physical damages caused by fire in the Amazon region.

Type of damage	Year 1996	Year 1998
Farm		
Pasture (km^2)	6,510	19,408
Primary forest (km^2)	7,250	21,614
Damage to buildings (km)	19,768	58,931
Carbon		
Primary forest (t/C)	88,162,999	265,510,230
Health		
Morbidity (number of internments)	4,319	12,875

Source: adapted from Motta *et al.* (2001).

loss as well as effects of conventional agricultural practices. Loss of SOM and organic horizon drastically reduce soil fertility and biological activity and enhance soil erosion, thus affecting the aquatic systems and wetlands.

SOM loss is undoubtedly the major form of soil degradation, affecting the aquatic systems and wetlands. It is a very complex process since this is a function of some factors linked to each other (Tommaselli *et al.*, 1999). D'Agostini (1999) emphasizes that the energy dynamics in the production of water erosion is associated with the dynamics of the hydrologic cycle, being the erosion expressed in energy that flows in the promotion of the cycle, partially converted into disrupting soil aggregates and soil particle transport.

Heavy rainfall may lead to runoff that can generate sheet erosion after aggregate disruption, followed by the reduction of the soil infiltration capacity, depending on soil type. Soil management can affect the form of aggregation of surface particles and the resistance to horizontal runoff transport in such a way that vertical movement may have implications for the soil infiltration rate. For example, if we compare two soil types under natural conditions and two soil management types, one with indigenous traditional technology and the other with modern technology, it is observed that even with existing soil physical differences, the indigenous soil management increased the natural soil infiltration capacity, whereas the modern technology decreased infiltration capacity, increasing susceptibility to erosion under an intensive rain regime (Table 2.14).

On the other hand, rainfall is one of the climatic factors of greater importance in soil erosion (Bertoni and Lombardi Neto, 1990). The heavy rainfall causes more

Table 2.13. Economical damage caused by fire in the Amazon region.

Type of damage	Monetary loss ($ (millions))	% of IGP[a] of the region
Farm		
1996	216	0.41
1998	594	1.04
Carbon[b]		
1996	309	0.59
1998	929	1.62
Health[c]		
1996	3	0.01
1998	10	0.02
Total 1996	528	1.01
Total 1998	1533	2.68

[a]Internal gross product.
[b]Net loss of carbon from native forest, estimated as carbon stock and costing at least $3 per tonne C present in simulating models of carbon trading of the Kyoto Protocol.
[c]Estimate between 1996 and 1998: from $3 million to 10 million based on the correlation between the burned area and the occurrence of respiratory diseases in the region and valued by annual costs of permanence in hospitals.
Source: adapted from Motta *et al.* (2001).

Table 2.14. Stablized infiltration rates (Ko (mm/h)) for two different soil classes under different land use types.

Soil classes	Soil use	Ko (mm/h)
Latossolo amarelo	Native vegetation	5.82
Latossolo amarelo	Pasture	0.52
Latossolo amarelo	Indigenous black earth	15.00
Plintossolo argilúvico	Savannah	1.55
Plintossolo argilúvico	Flooded rice (3 years cultivated with rice)	0.08–0.14

Source: EMBRAPA (2001).

erosion in soils, particularly those without vegetation than those with less intense rainfall, even if it lasts longer.

Finally, the knowledge of rain erosivity is of great value for recommendation of soil management practices that aim to reduce water erosion in regard to soil conservation planning (Alvarenga *et al.*, 1998). The integration of the knowledge about rain erosivity with soil types and properties, land use and vegetation cover is of utmost importance for the implementation of programmes and studies related to the protection of the Amazon biodiversity. However, it must be emphasized that the local inhabitants that migrated and the new farmers that are arriving are not fully familiar with the meaning of sustainable development commonly broadcast by the media and environmentalists. Nevertheless, they understand the need for development without destroying the environment, meaning that any programme or planning for the region must consider the expectations of the local communities.

Final Remarks

As previously shown in the sections on soil classes and properties, most soils in the Amazon region are not much different from those commonly seen in other regions in Brazil, but the actual knowledge of their behaviour under agricultural systems is still poorly understood. In traditional agricultural systems, such as forest clearing and introducing grasslands and annual or permanent crops (Jordan, 1985), plant nutritional disorders will soon appear. Native plant species develop mechanisms to recycle nutrients efficiently. The classic procedure undertaken worldwide in the past such as forest clearing and the subsequent introduction of agriculture is not sustainable in the Amazon region. Presently, there are major concerns that forest conversion to agricultural land releases stored carbon and reduces biodiversity. High levels of lime and fertilizers required for maintaining adequate crop yield are hardly economic in many remote areas of this region. Plans to pave highways may help the provision of fertilizers and machinery, but also greatly increase the accessibility of loggers, sawmills, primitive gold mining and hunting, which will cause massive deforestation and have a tremendous environmental impact (Carvalho *et al.*, 2001; Fearnside, 2002). Both local and federal authorities face difficulties in enforcing regulations and policies. Similar to Indonesia (Tomich *et al.*, 1998), the unique aspect of the Amazon region is that it contains large areas under forest where land use alternatives must be offered to pursue global environmental objectives with simultaneous consideration of agronomic sustainability, objectives of local farmers and policymakers at all levels and weaknesses in markets and other institutions that influence the adoption of land use alternatives by landowners. The knowledge of LULC and soil morphological, physical, chemical and mineralogical properties when combined with remote sensing, GIS and database technologies constitutes an important tool for the stratification of the environment and enables us to make correlations and interpretations of the spatial variability of soil biodiversity and soil quality changes through time. The major challenge of the studies on the Amazon ecosystems is the development of enough knowledge to define adequate procedures aiming at sustainable development of the region.

References

Alfaia, S.S. (1988) Correlação entre a capacidade de troca de cátions e outras propriedades de três solos da Amazônia Central. *Acta Amazônica* 18, 3–11.

Alfaia, S.S. and Falcão, N.P. (1993) Estudo da dinâmica de nutrientes em solos de várzea da Ilha do Careiro no Estado do Amazonas. *Amazoniana* 21, 1–9.

Almeida-Filho, R. and Shimabukuro, Y.E. (2002) Digital processing of a Landsat-TM series for mapping and monitoring degraded areas caused by independent gold miners, Roraima state, Brazilian Amazon. *Remote Sensing of Environment* 79, 42–50.

Alvarenga, R.C., Sans, L.M.A., Marques, J.J.G. de S. Melo and Curi, N. (1998) *Índices de erosividade da chuva, perdas de solo e fator erodibilidade para dois solos da Região de Sete Lagoas*. EMBRAPA-CNPMS, Sete Lagoas (Pesquisa em Andamento 24).

Alves, D.S., Soares, J.V., Amaral, S., Mello, E.M.K., Almeida, S.A.S., Silva, O.F. and Silveira, A.M. (1997) Biomass of primary and secondary vegetation in Rondônia, western Brazilian Amazon. *Global Change Biology* 3, 451–561.

Amelung, T. and Diehl, M. (1992) *Deforestation of Tropical Rainforest – Economic Causes and Impact on Development*. Tubingen, Germany (Kieler Studien 241).

Baligar, V.C. and Fageria, N.K. (1997) Nutrient use efficiency in acid soils: nutrient management and plant use efficiency. In: Moniz, A.C., Furlani, A.M.C., Schaeffert, R.E., Fageria, N.K., Rosolem, C.A. and Cantarella, H. (eds) *Plant–Soil Interactions at Low pH*. Brazilian Soil Science Society, Campinas, SP, Viçosa, MG, Brazil, pp. 75–96.

Batistella, M. (2000) Extracting Earth surface feature information for land-use/land-cover classifications in Amazônia: the role of remote sensors and processing techniques. In: *GIS Brasil 2000, VI Show de Geotecnologias, Salvador, Brazil. Anais*. Fatorgis, Curitiba. CD-ROM.

Bertoni, J. and Lombardi Neto, F. (1990) *Conservação do Solo*. Ícone, São Paulo, Brazil.

Braakhekke, W.G., Stuurman, H.A., Reuler, H. and Van Janssen, B.H. (1993) Relations between nitrogen and phosphorus immobilization during decomposition of forest litter. In: Fragoso, M.A.C. and Beusichem, M.L. van (eds) *Optimization of Plant Nutrition*. Kluwer, Dordrecht, The Netherlands, pp. 117–123.

BRASIL (1975) Ministério das Minas e Energia. Departamento Nacional da Produção Mineral. *Projeto RADAMBRASIL*. Folha SB 21 Tapajós: Geologia, Geomorfologia, Solos, Vegetação e Uso Potencial da Terra. Rio de Janeiro (Levantamento de Recursos Naturais 7).

BRASIL (1977a) Ministério das Minas e Energia. Departamento Nacional da Produção Mineral. *Projeto RADAMBRASIL*. Folha SA. 19. Içá: Geologia, Geomorfologia, Solos, Vegetação e Uso Potencial da Terra. Rio de Janeiro (Levantamento de Recursos Naturais 14).

BRASIL (1977b) Ministério das Minas e Energia. Departamento Nacional da Produção Mineral. *Projeto RADAMBRASIL*. Folha SB. 19. Juruá: Geologia, Geomorfologia, Solos, Vegetação e Uso Potencial da Terra. Rio de Janeiro (Levantamento de Recursos Naturais 15).

BRASIL (1978a) Ministério das Minas e Energia. Departamento Nacional da Produção Mineral. *Projeto RADAMBRASIL*. Folha SA. 20. Manaus: Geologia, Geomorfologia, Solos, Vegetação e Uso Potencial da Terra. Rio de Janeiro (Levantamento de Recursos Naturais 18).

BRASIL (1978b) Ministério das Minas e Energia. Departamento Nacional da Produção Mineral. *Projeto RADAMBRASIL*. Folha SB. 20. Purus: Geologia, Geomorfologia, Pedologia, Vegetação e Uso Potencial da Terra. Rio de Janeiro (Levantamento de Recursos Naturais 17).

BRASIL (1978c) Ministério das Minas e Energia. Departamento Nacional da Produção Mineral. *Projeto RADAMBRASIL*. Folha SC. 20. Porto Velho: Geologia, Geomorfologia, Solos, Vegetação e Uso Potencial da Terra. Rio de Janeiro (Levantamento de Recursos Naturais 16).

BRASIL (2002a) *Cenários para a Amazônia Legal. Bases para discussão*. Ministério do Meio Ambiente – SDS, Brasília. CD-ROM.

BRASIL (2002b) *Cenários para a Amazônia Legal. Sistematização de dados*. Ministério do Meio Ambiente – SDS, Brasília. CD-ROM.
BRASIL (2003) *Avança Brasil*. Ministério do Planejamento, Brasília, DF. Available at: www.abrasil.gov.br
Burrough, P.A. and McDonnell, R.A. (1998) *Principles of Geographical Information Systems: Spatial Information and Geostatistics*. Oxford University Press, Oxford, UK.
Campbell, J.B. (1987) *Introduction to Remote Sensing*. The Guilford Press, New York.
Carvalho, G., Barros, A.C., Moutinho, P. and Nepstad, D. (2001) Sensitive development could protect Amazonia instead of destroying it. *Nature* 409, 131.
Cochrane, T.T. and Sanchez, P. (1982) Land resources, soils, and their management in the Amazon region. In: Hecht, S.B. (ed.) *Amazonia: Agriculture and Land-Use Research*. CIAT, Cali, Columbia, pp. 137–209.
Coelho, M.R., Santos, H.G. dos, Silva, E.F. and Áglio, M.L.D. (2002) O Recurso Natural Solo. In: Manzatto, C.V., Freitas Junior, E. and Peres, J.R.R. (eds) *Uso agrícola dos solos brasileiros*. Embrapa Solos, Rio de Janeiro, pp. 1–11.
Colwell, R.N. (1983) *Manual of Remote Sensing*, 2nd edn. American Society for Photogrammetry and Remote Sensing, Falls Church, Virginia.
Committee on Global Change Research (1999) *Global Environmental Change: Research Pathways for the Next Decade*. National Academy, Washington, DC.
Correa, J.C. and Reichardt, K. (1995) Efeito do tempo de uso das pastagens sobre as propriedades de um latossolo amarelo da Amazônia Central. *Pesquisa Agropecuária Brasileira* 30, 107–114.
Craswell, E.T. and Lefroy, R.D.B. (2001) The role and function of organic matter in tropical soils. In: Martius, C., Tiessen, H. and Vlek, P.L.G. (eds) *Managing Organic Matter in Tropical Soils: Scope and Limitations*. Kluwer, Dordrecht, The Netherlands, pp. 7–18.
Cravo, M.S. and Smyth, T.J. (1997) Manejo sustentado da fertilidade de um latossolo da Amazônia Central sob cultivos sucessivos. *Revista Brasileira de Ciência do Solo* 21, 607–616.
Cuevas, E. and Medina, E. (1986) Nutrient dynamics within Amazonian forest ecosystems I. Nutrient flux in fine litter fall and efficiency of nutrient utilization. *Oecologia* 68, 466–472.
Cuevas, E. and Medina, E. (1988) Nutrient dynamics within Amazonian forests II. Fine root growth, nutrient availability, nutrient availability and leaf litter decomposition. *Oecologia* 76, 222–235.
D'Agostini, L.R. (1999) *Erosão: o problema mais que o processo*. UFSC, Florianópolis.
Demattê, J.L.I. (1988) Manejo de solos ácidos dos trópicos úmidos: região Amazônica. Fundação Cargill, Campinas.
Derpsch, A. (1998) Agricultura sustentável. In: Saturnino, H.M. and Landers, J.N. (eds) *O meio ambiente e o plantio direto*. EMBRAPA-SPI, Brasília, pp. 29–48.
Di Castri, F. and Hadley, M. (1988) Enhancing the credibility of ecology: interacting along and across hierarchical scales. *GeoJournal* 17, 5–35.
Diaz, M.C.V. (2002) Visões e Perspectivas Futuras para o Meio Amazônico. IPAM, Manaus, Brazil.
Dunn, C.P., Sharpe, D.M., Guntenspergen, G.R., Stearns, F. and Yang, Z. (1991) Methods for analyzing temporal changes in landscape pattern. In: Turner, M.G. (ed.) *Quantitative Methods in Landscape Ecology: The Analysis and Interpretation of Landscape Heterogeneity*. Spring-Verlag, New York, pp. 173–198 (Ecological studies 82).
Egler, P.C.G. (2001) *Avaliação Ambiental Estratégica – Considerações sobre métodos para sua realização*. Centro de Desenvolvimento Sustentável, Brasília.
EMBRAPA. Serviço Nacional de Levantamento e Conservação de Solos (1976) *Levantamento de reconhecimento de solos de três áreas prioritárias na Rodovia Transamazônica*. EMBRAPA-SNLCS, Recife (Boletim Técnico 48).
EMBRAPA. Serviço Nacional de Levantamento e Conservação de Solos (1978) *Estudo expedito de solos na área da pré-Amazônia Maranhense e na parte oeste do Piauí*. EMBRAPA-SNLCS, Recife. Internal filed document.
EMBRAPA. Serviço Nacional de Levantamento e Conservação de Solos (1980a) *Estudo expedito de solos no Estado do Maranhão para fins de classificação, correlação e legenda preliminar*. EMBRAPA-SNLCS, Rio de Janeiro (Boletim Técnico 61). SUDENE, Recife (Série Recursos de Solos 13).
EMBRAPA. Serviço Nacional de Levantamento e Conservação de Solos (1980b) *Levantamento exploratório-reconhecimento de alta intensidade e aptidão agrícola dos solos da área compreendida entre os km 18 e 152 da Rodovia Santarém – Cuiabá e do rio Curuá – Una*. EMBRAPA-SNLCS, Rio de Janeiro (Boletim Técnico 70).
EMBRAPA. Serviço Nacional de Levantamento e Conservação de Solos (1981a) *Levantamento de reconhecimento de média intensidade e aptidão agrícola dos solos da área do Pólo Altamira, PA*. EMBRAPA-SNLCS, Rio de Janeiro (Boletim Técnico 77).

EMBRAPA. Serviço Nacional de Levantamento e Conservação de Solos (1981b) *Mapa de Solos do Brasil.* Escala 1:5.000.000. EMBRAPA-SNLCS, Rio de Janeiro.

EMBRAPA. Serviço Nacional de Levantamento e Conservação de Solos (1982a) *Levantamento de reconhecimento de alta intensidade dos solos e avaliação da aptidão agrícola das terras de área ao longo da BR-174, na região do rio Anauá, no município de Caracaraí, Território Federal de Roraima.* EMBRAPA-SNLCS, Rio de Janeiro (Boletim Técnico 79).

EMBRAPA. Serviço Nacional de Levantamento e Conservação de Solos (1982b) *Levantamento de reconhecimento de média intensidade dos solos e de uma área sob influência dos rios Araguari, Falsino e Tartarugal Grande, Território Federal do Amapá.* EMBRAPA-SNLCS, Rio de Janeiro (Boletim de Pesquisa 7).

EMBRAPA. Serviço Nacional de Levantamento e Conservação de Solos (1983a) *Levantamento de reconhecimento de média intensidade dos solos e avaliação da aptidão agrícola das terras da área do Pólo Tapajós.* EMBRAPA-SNLCS, Rio de Janeiro (Boletim de Pesquisa 20).

EMBRAPA. Serviço Nacional de Levantamento e Conservação de Solos (1983b) *Levantamento exploratório dos solos que ocorrem ao longo da rodovia Manaus-Porto Velho.* EMBRAPA-SNLCS, Rio de Janeiro (Boletim de Pesquisa 21).

EMBRAPA. Serviço Nacional de Levantamento e Conservação de Solos (1986) *Levantamento exploratório-reconhecimento de solos do Estado do Maranhão.* EMBRAPA-SNLCS, Rio de Janeiro (Boletim de Pesquisa 35), SUDENE, Recife (Série Recursos de Solos 17).

EMBRAPA. Centro Nacional de Pesquisa em Solos (1992) *Delineamento macroagroecológico do Brasil,* 1:5.000.000. Rio de Janeiro (1 map).

EMBRAPA. Centro Nacional de Pesquisa em Solos (1999) *Sistema Brasileiro de Classificação de Solos.* Embrapa Produção de Informação, Brasília, Embrapa Solos, Rio de Janeiro.

EMBRAPA. Centro Nacional de Pesquisa em Solos (2001) *Estudos pedológicos e suas relações ambientais.* Embrapa Solos, Rio de Janeiro (Relatório Técnico. Contrato IPAAM/Embrapa Solos).

Ewel, J.J. (1986) Designing agricultural ecosystems for the humid tropics. *Annual Review of Ecological Systems* 17, 245–271.

Falesi, I.C. (1972) O estado atual dos conhecimentos sobre os solos da Amazônia brasileira. In: *Zoneamento agrícola da Amazônia.* IPEAN, Belém, pp. 17–67 (Boletim Técnico do Instituto de Pesquisa Agropecuária do Norte 54).

FAO (1996) *Forest Resources Assessment 1990: Survey of Tropical Forest Cover and Study of Change Processes.* FAO, Rome (FAO Forestry Paper, 130).

FAO (1998) World reference base for soil resources. FAO/ISSS/ISRIC, Rome (World Soil Resources Reports 84).

Fearnside, P.M. (1986) *Human Carrying Capacity of the Brazilian Rainforest.* Columbia University Press, New York.

Fearnside, P.M. (2002) Avança Brasil: environmental and social consequences of Brazil's planned infrastructure in Amazonia. *Environmental Management* 30, 735–747.

Forman, R.T.T. and Godron, M. (1986) *Landscape Ecology.* John Wiley & Sons, New York.

Guerra, A.J.T., Silva, A.S. da and Botelho, R.G.M. (1999) *Erosão e Conservação dos Solos: conceitos, temas e aplicações.* Bertrand Brasil, Rio de Janeiro.

Herrera, R. and Jordan, C.F. (1981) Nitrogen cycle in a tropical Amazonian rain forest: the caatinga of low mineral nutrient status. *Ecology Bulletin* 3, 493–505.

Hölscher, D., Möller, R.F., Denich, M. and Fölster, H. (1997) Nutrient input–output budget of shifting agriculture in eastern Amazonia. *Nutrient Cycling in Agroecosystems* 47, 49–57.

Houghton, R.A., Hackler, J.L. and Lawrence, K.T. (1999) The U.S. carbon budget: contribution from land-use change. *Science* 285, 574–578.

IBGE (1991) *Manual técnico da vegetação brasileira.* IBGE, Rio de Janeiro (Manuais Técnicos de Geociências 1).

IBGE (1997) *Censo Agropecuário do Brasil 1995–1996.* IBGE, Rio de Janeiro, v.1.

INPE (2002) Monitoramento da floresta amazônica por satélite 2000–2001. Available at: http://sputnik.dpi.inpe.br:1910/col/dpi.inpe.br/lise/2002/06.12.13.16/doc/capa.htm

Jordan, C.F. (1985) Ciclagem de nutrientes e silvicultura de plantações na Bacia Amazônica. In: Cabala-Rosand, P. (ed.) *Simpósio sobre reciclagem de nutrientes e agricultura de baixos insumos nos Trópicos,* CEPLAC-SBCS, Ilhéus, pp. 187–202.

Karlen, D.L., Eash, N.S. and Unger, P.W. (1992) Soil and crop management effects on soil quality indicators. *American Journal of Alternative Agriculture* 7, 48–55.

Kato, M.S.A., Kato, O.R., Denich, M. and Vlek, P.L.G. (1999) Fire-free alternatives to slash-and-burn for shifting cultivation in the eastern Amazon region: the role of fertilizers. *Field Crops Research* 62, 225–237.

Kern, D.C. (1996) Geoquímica e pedogeoquímica de sítios arqueológicos com Terra Preta na Floresta Nacional de Caxiuanã (Portel-Pará). Tese de Doutorado em Geoquímica. Universidade Federal do Pará, Belém, Brasil.
Kern, D.C., D'aquino, G., Rodrigues, T.E., Frazão, F.J.L., Sombroek, W. and Neves, E.G. (2003) Distribution of Amazonian dark earths. In: Lehmann, J., Kern, D., Glaser, B. and Woods, W. (eds) *Amazonian Dark Earths – Origin, Properties and Management.* Kluwer, Dordrecht, The Netherlands.
Knox, E.G. (1965) Soil individuals and soil classification. *Soil Science Society of America Proceedings* 29, 79–84.
Lambin, E.F., Turner, B.L., Geist, H.J., Agbola, S.B., Angelsen, A., Bruce, J.W., Coomes, O.T., Dirzo, R., Fischer, G., Folke, C., George, P.S., Homewood, K., Imbernon, J., Leemans, R., Li, X., Moran, E.F., Mortimore, M., Ramakrishnan, P.S., Richards, J.F., Skanes, H., Steffen, W., Stone, G.D., Svedin, U., Veldkamp, T.A., Vogel, C. and Xu, J. (2001) The causes of land-use and land-cover change moving beyond the myths. *Global Environmental Change* 11, 261–269.
Lathwell, D.J. and Grove, T.L. (1986) Soil–plant relationship in the tropics. *Annual Review of Ecological Systems* 17, 1–16.
Luna-Orea, P. and Wagger, M.G. (1996) Management of tropical legume cover crops in the Bolivian Amazon to sustain crops yields and soil productivity. *Agronomy Journal* 88, 765–776.
Mackensen, J., Hölscher, D., Klinge, D. and Fölster, H. (1996) Nutrient transfer to the atmosphere by burning of debris in eastern Amazonia. *Forest Ecology and Management* 86, 121–128.
Madari, B., Benites, V.M. and Cunha, T.J.F. (2003) The effect of management on the fertility of Amazonian anthropogenic dark earth soils. In: Lehman, J., Kern, D., Glaser, B. and Woods, W. (eds) *Amazonian Dark Earths – Origin, Properties and Management.* Kluwer, Dordrecht, The Netherlands.
Mahar, D. (1988) Government policies and deforestation in Brazil's Amazon region. The World Bank, Washington, DC (Environment Department Working Paper 7).
Mahar, D.J. (1979) *Frontier Development Policy in Brazil: A Study of Amazonia.* Praeger Publishers, New York.
Manzatto, C.V., Ramalho Filho, A., Costa, T.C.C., Mendonça-Santos, M.L., Coelho, M.R., Silva, E.F. and Oliveira, R.P. (2002) Potencial de uso e uso atual das terras. In: Manzatto, C.V., Freitas Júnioe, E. and Peres, J.R.R. (eds) *Uso agrícola dos solos brasileiros.* Embrapa Solos, Rio de Janeiro, pp. 13–21.
Martins, P.F.S., Cerri, C.C., Volkoff, B., Andreux, F. and Chauvel, A. (1991) Consequences of clearing and tillage on the soil of a natural Amazonian ecosystem. *Forest Ecology and Management* 38, 273–302.
Mas, J.F. (1999) Monitoring land-cover changes: a comparison of change detection techniques. *International Journal of Remote Sensing* 20, 139–152.
McGrath, D.A., Duryea, M.L. and Cropper, W.P. (2001) Soil phosphorus availability and fine root proliferation in Amazonian agroforests 6 years following forest conversion. *Agriculture, Ecosystems & Environment* 83, 271–284.
Mendonça-Santos, M.L. (1999) GIS and spatio-temporal modelling for the study of alluvial soil and vegetation evolution. PhD thesis, École Polytechnique Fédérale de Lausanne, Switzerland.
Mendonça-Santos, M.L. and Claramunt, C. (2001) An integrated landscape and local analysis of land cover evolution in an alluvial zone. *Computers, Environment and Urban Systems* 25, 557–577.
Mendonça-Santos, M.L., Guenat, C., Thevoz, C., Bureau, F. and Vedy, J.C. (1997) Impacts of embanking on the soil–vegetation relationships in a floodplain ecosystem of a pre-alpine river. *Global Ecology and Biogeography Letters* 6, 339–348.
Moran, E.F. and Brondízio, E.S. (1998) Land-use change after deforestation in Amazônia. In: Liverman, D., Moran, E.F., Rindfuss, R.R. and Stern, P.C. (eds) *People and Pixels.* National Academy Press, Washington, DC, pp. 94–120.
Moran, E.F., Brondízio, E.S., Tucker, J., Silva-Forsberg, M.C. and Falesi, I.C. (2000) Effects of soil fertility and land use on forest succession in Amazônia. *Forest Ecology and Management* 139, 93–108.
Motta, R.S., Mendonça, M.J.C., Nespstad, D., Diaz, M.C.V., Alencar, A., Gomes, J.C. and Ortiz, R.A. (2001) *O custo do uso do fogo na Amazônia.* IPEA/IPAM, Rio de Janeiro (Texto para Discussão 912).
Mulders, M.A. (1987) *Remote sensing in soil science.* Elsevier Science, Amsterdam.
Myers, R.J.K., Palm, C.A., Cuevas, E., Gunatilleke, I.U.N. and Brossard, M. (1994) The synchronisation of nutrient mineralisation and plant nutrient demand. In: Woomer, P.L. and Swift, M.J. (eds) *The Biological Management of Tropical Soil Fertility.* John Wiley & Sons, Chichester, UK, pp. 81–116.
Nepstad, D., Carvalho, G., Barros, A.C., Alencar, A., Capobianco, J.P., Bishop, J., Moutinho, P., Lefebvre, P., Silva, U.L. Jr and Prins, E. (2001) Road paving, fire regime feedbacks, and the future of Amazon forests. *Forest Ecology and Management* 154, 395–407.
Novais, R.F. and Smyth, T.J. (1999) Fósforo em solo e planta em condições tropicais. Universidade Federal de Viçosa, Viçosa, Brazil.

Oliveira, J.B., Jacomine, P.K.T. and Camargo, M.N. (1992) *Classes gerais de solos do Brasil*, 2nd edn. FUNEP, Jaboticabal, São Paulo, Brazil.
Palmieri, F., Santos, H.G. dos, Gomes, I.A., Lumbreras, J.F. and Aglio, M.L.D. (2003) The Brazilian soil classification system. In: Eswaran, H., Rice, T., Ahrens, R. and Stewart, B.A. (eds) *Soil Classification: A Global Desk Reference*. CRC Press, Boca Raton, Florida, pp. 127–146.
Pedlowski, M.A., Dale, V.H., Matricardi, E.A.T. and Silva Filho, E.P. (1997) Patterns and impacts of deforestation in Rondônia, Brazil. *Landscape and Urban Planning* 38, 149–157.
Pereira, W. (1977) Avaliação da erosividade das chuvas em diferentes locais do Estado de Minas Gerais. Tese de Mestrado, Universidade Federal de Viçosa, Viçosa, Minas Gerais, Brazil.
Pereira, W.L.M., Veloso, C.A.C. and Gama, J.R.N.F. (2000) Propriedades químicas de um Latossolo Amarelo cultivado com pastagens na Amazônia Oriental. *Scientia Agricola* 57, 531–537.
Ponzoni, F.J., Galvão, L.S. and Epiphanio, J.C.N. (2002) Spatial resolution influence on the identification of land cover classes in the Amazon environment. *Anais da Academia Brasileira de Ciências* 74, 717–725.
Quattrochi, D.A. and Pelletier, R.E. (1991) Remote sensing for analysis of landscape: an introduction. In: Turner, G.M. and Gardner, R.H. (eds) *Quantitative Methods in Landscape Ecology: The Analysis and interpretation of landscape heterogeneity*. Springer-Verlag, New York, pp. 51–76.
Raij, B. van (1969) Capacidade de troca de frações orgânicas e minerais dos solos. *Bragantia* 28, 85–112.
Resende, M., Curi, N. and Santana, D.P. (1988) *Pedologia e fertilidade do solo: interações e aplicações*. ESAL, Lavras, POTAFOS, Piracicaba.
Rodrigues, T.E. (1996) Solos da Amazônia. In: Alvarez, V.H., Fontes, L.E.F and Fontes, M.P.F. (eds) *O solo nos grandes domínios morfoclimáticos do Brasil e o desenvolvimento sustentado*. SBCS, UFV, DPS, Viçosa, Minas Gerais, Brazil, pp. 251–260.
Sala, O.E., Chapin, F.S., Armesto, J.J., Berlow, E., Bloomfield, J., Dirzo, R., Huber-Sanwald, E., Huenneke, L.F., Jackson, R.B., Kinzig, A., Leemans, R., Lodge, D.M., Mooney, H.A., Oesterheld, M., Poff, N.L., Sykes, M.T., Walker, B.H., Walker, M. and Wall, D.H. (2000) Biodiversity: global biodiversity scenarios for the year 2100. *Science* 287, 1770–1774.
Sanchez, P.A. (1976) *Properties and Management of Soil in Tropics*. John Wiley & Sons, New York.
Sanchez, P.A. and Uehara, G. (1980) Management consideration for acid soils with high phosphorus fixation capacity. In: Khaswana, F.E., Sample, E.C. and Kamprath, E.J. (eds) *The Role of Phosphorus in Agriculture*. American Society of Agronomy, Madison, Wisconsin, pp. 471–514.
Sanchez, P.A., Villachica, J.H. and Bandy, D.E. (1983) Soil fertility dynamics after clearing of a tropical rainforest in Peru. *Soil Science Society of America Journal* 47, 1171–1178.
Sanchez, P.A., Palm, C.A., Szott, L.T., Cuevas, E. and Lal, R. (1989) Organic input management in tropical agroecosystems. In: Coleman, D.C., Oades, J.M. and Uehara, G. (eds) *Dynamics of Soil Organic Matter in Tropical Ecosystems*. University of Hawaii Press, Honolulu, Hawaii, pp. 125–152.
Serrão, E.A.S., Nepstad, D. and Walker, R. (1996) Upland agricultural and forestry development in the Amazon: sustainability, criticality and resilience. *Ecological Economics* 18, 3–13.
Seubert, C.E., Sanchez, P.A. and Valverde, C. (1977) Effects of land clearing methods on soil properties of an ultisol and crop performance in the Amazon jungle of Peru. *Tropical Agriculture* 54, 307–321.
Singh, A. (1989) Digital change detection techniques using remotely-sensed data. *International Journal of Remote Sensing* 10, 989–1003.
Smyth, T.J. (1996) Manejo da fertilidade do solo para a produção sustentada de cultivos na Amazônia. In: Alvarez, V.H., Fontes, L.E.F. and Fontes, M.P.F. (eds) *O solo nos grandes domínios morfoclimáticos do Brasil e o desenvolvimento sustentado*. SBCS, UFV, DPS, Viçosa, Minas Gerais, Brazil, pp. 71–93.
Smyth, T.J. and Cassel, D.K. (1995) Synthesis of long-term soil management research on ultisols and oxisols in Amazon. In: Lal, R. and Stewart, B.A. (eds) *Soil Management: Experimental Basis for Sustainability and Environmental Quality*. Lewis Publishers, Boca Raton, Florida, pp. 13–59.
Soil Survey Staff. Department of Agriculture. Soil Survey Division. Soil Conservation Service (1999) *Soil Taxonomy: A Basic System of Soil Classification for Making and Interpreting Soil Surveys,* 2nd edn. USDA, Washington, DC (Agriculture Handbook 436).
Stark, N.M. and Jordan, C.F. (1978) Nutrient retention by the root mat of an Amazonian rain forest. *Ecology* 59, 434–437.
Swift, M.J. (1999) Towards the second paradigm: integrated biological management of soil. In: Siqueira, J.O., Moreira, F.M.S., Lopes, A.S., Guilherme, L.R.G., Faquin, V., Furtini Neto, A.E. and Carvalho, J.G. (eds) *Inter-relação fertilidade, biologia do solo e nutrição de plantas*. SBCS, Viçosa; UFLA/DCS, Lavras, Brazil, pp. 11–24.
Szott, L.T. and Kass, D.C.L. (1993) Fertilizers in agroforestry systems. *Agroforestry Systems* 23, 157–176.

Togman, A.A., Demattê, J.L.I. and Demattê, J.A.M. (1998) Tear e distribuição da matéria orgânica em Latossolos das regiões da floresta amazônica e dos cerrados do Brasil Central. *Scientia Agricola* 55, 343–354.

Tolba, M.K. and El-Kholy, O.A. (1992) *The World Environment 1972–1992: Two Decades of Challenge*. Chapman & Hall, London.

Tomich, T.P., Van Noordwijk, M., Budidarsono, S., Gillison, A., Kusumanto, T., Murdiyarso, D., Stolle, F. and Fagi, A.M. (1998) *Alternatives to Slash-and-Burn in Indonesia – Summary Report and Synthesis of Phase II*. ICRAF, Nairobi (Report 8).

Tommaselli, J.T.G., Freire, O. and Carvalho, W.A. (1999) Erosividade da chuva da Região Oeste do Estado de São Paulo. *Revista Brasileira de Agrometeorologia* 7, 269–276.

Vieira, L.S. and Santos, P.C.T. dos (1987) *Amazônia: seus solos e outros recursos naturais*. Editora Agronômica Ceres, São Paulo.

Vitorello, V.A., Cerri, C.C., Andreux, F., Feller, C. and Victória, R.L. (1989) Organic matter and natural carbon-13 distribution in forested and cultivated oxisols. *Soil Science Society of America Journal* 53, 773–778.

Vitousek, P.M. and Matson, P.A. (1988) Nitrogen transformations in a range of tropical forest soils. *Soil Biology and Biochemistry* 20, 361–367.

Watson, R.T., Noble, I.R., Bolin, B., Ravindranath, N.H., Verardo, D.J. and Dokken, D.J. (2000) *Land Use, Land-Use Change and Forestry: A Special Report of the IPCC*. Cambridge University Press, Cambridge, UK.

Weinhold, D. (1999) Estimating the loss of agricultural productivity in the Amazon. *Ecological Economics* 31, 63–76.

Woods, W.I. (2003) Development of anthrosol research. In: Lehmann, J., Kern, D., Glaser, B. and Woods, W. (eds) *Amazonian Dark Earths – Origin, Properties and Management*. Kluwer, Dordrecht, The Netherlands.

Woomer, P.L., Martin, A., Albrecht, A., Resck, D.V.S. and Scharpenseel, H.W. (1994) The importance and management of soil organic matter in the tropics. In: Woomer, P.L. and Swift, M.J. (eds) *The Biological Management of Tropical Soil Fertility*. John Wiley & Sons, Chichester, UK, pp. 47–80.

Yuan, D. and Elvidge, C. (2002) NALC land cover change detection pilot study: Washington D.C. area experiments. *Remote Sensing of Environment* 66, 166–178.

3 Soil Macrofauna Communities in Brazilian Amazonia

E. Barros,[1] J. Mathieu,[2] S. Tapia-Coral,[1]
A.R.L. Nascimento[1] and P. Lavelle[2]
[1]*Instituto Nacional de Pesquisas da Amazônia – Agronomia, Av. André Araújo, 2936, Manaus-AM, 69083-000, Brazil;* [2]*Institut de Recherche pour le Développement – UMR 137 BIOSOL, 32 Avenue Henri Varagnat, 93143, Bondy Cedex, France, e-mail: patrick.lavelle@bondy.ird.fr*

Introduction

In tropical areas, soil macroinvertebrates play an important role in the provision of many ecosystem services through their action on soil processes (Fragoso and Lavelle, 1995; Lavelle *et al.*, 1995, 1997). They participate in the regulation of decomposition and nutrient cycling processes (Lavelle *et al.*, 1992), and in the maintenance of soil physical properties suitable for plant growth (Lee and Foster, 1992; Oades, 1993; Blanchart *et al.*, 1997). They can modulate the mineralization rate of soil organic matter by selectively activating several functional groups of microflora in the soil, at distinct temporal and spatial scales (Beare *et al.*, 1994; Lavelle *et al.*, 1995; Wardle and Lavelle, 1997). However, these processes greatly depend on the composition of soil macrofauna, and understanding the effects of human activities on these communities is of utmost importance.

In Amazonia, forest is currently cleared at the rate of 2 million ha per year (Laurance *et al.*, 2001), and 53 million ha had already been deforested in 1997. The great majority of the deforested area has been transformed into pastures for extensive cattle ranching (Fearnside and Barbosa, 1998; INPE Brazil, 1998). Forest clearing deeply modifies the amount of soil nutrients and soil organic matter and the physical properties (Grimaldi *et al.*, 1993; Moraes *et al.*, 1996; Fearnside and Barbosa, 1998; Barros *et al.*, 2001; McGrath *et al.*, 2001; Desjardins *et al.*, 2004). The soil macrofauna communities are also strongly modified by forest clearing. In some cases, such modification can lead to a complete change of the soil functioning. For instance, in central Amazonia, near Manaus, forest transformation to pasture led to the formation of a permanent soil crust on the surface. The crust was so hard that water could not enter the soil any more, and most plants died from water deficit. This crust was due to the massive invasion by an opportunist earthworm, *Pontoscolex corethrurus*, which compacts the soil very strongly (Chauvel *et al.*, 1999; Barros *et al.*, 2004).

This chapter summarizes the general patterns of the soil macrofauna communities in the most common land use types of the Brazilian Amazon. Modifications in the abundance and species diversity of communities according to land use changes and scales are particularly addressed. In the section 'Regional Patterns', community patterns are detailed in a number of subregional situations of eastern, central and western Amazonia.

Soil macroinvertebrate communities have been assessed in 118 sites of the Amazonian region with the same standardized sampling recommended by the Tropical Soil Biology and Fertility Programme (Anderson and Ingram, 1993). In each plot, communities were sampled in ten (sometimes five) soil monoliths 30 cm deep, and 25 × 25 cm large, at every 5 m along a transect. Soil macrofauna was extracted separately from four different soil layers – litter, 0–10 cm, 10–20 cm and 20–30 cm – and stored in 75% alcohol, except for earthworms that were fixed in 4% formalin before being stored in alcohol. In the laboratory, 17 main taxonomic groups of organisms were separated. In a limited number of sites, identifications up to the morphospecies level (and real species for a few orders) have been performed.

Community Structure

Species richness and endemism

Soil macrofauna communities generally comprise 15–18 orders with highly contrasting ecologies: Gastropoda, Oligochaeta, Isopoda, Arachnida, Diplopoda, Chilopoda, Blattaria, Orthoptera, Dermaptera, Hemiptera, Lepidoptera larvae, Diptera larvae, Coleoptera larvae, Coleoptera adults, Formicidae, Isoptera and others.

In primary forests the overall local richness of macroinvertebrates was estimated at 156 and 270 species, respectively, in central (Barros *et al.*, 2004) and eastern Amazonia (Mathieu *et al.*, 2004) in the soil litter system. These data, however, are only indications since their evaluation is highly subject to the collection effort and also the accuracy of separation of morphospecies used as a surrogate for species in the absence of sufficient taxonomic expertise. This richness, however, was very unevenly distributed among groups. Some, like Coleoptera or Araneidae, may comprise locally 60–80 species whereas termites would only have 20–30 and earthworms less than 15 species at the most.

At the very small scale of 25 × 25 × 30 cm soil monoliths, the pattern was different. Approximately 15 different species have been found at the Benfica primary forest site on average. Ants were the richest group with 3.8 species on average. Insect larvae, Coleoptera, spiders and earthworms were the other more diverse groups, with at least 1.8 species per sample, on average; Chilopoda, termites and Thysanoptera had at least one species on average in each sample (Fig. 3.1).

Another important feature in these communities was the large proportion of very rare species. Among the 270 species collected at Benfica (Para, Brazil), 99 had been collected only once, 61 twice and 200 less than five times on a total of 17 m^2 sampled (i.e. 270 samples each of 1/16th m^2) (Fig. 3.2).

Finally, these species had highly variable distribution ranges. Earthworms are known to be highly endemic, with the notable exception of a dozen peregrine species that behave as invasive species in deforested areas. Once invasives have established in disturbed areas, native species have little chance to recover, even when the original forest is restored (Lapied and Lavelle, 2003; Lavelle and Lapied, 2003). Most earthworm species have such small distributional ranges that the ratio of the number of species found at one single site to the number of species found in the whole Amazonian region has been estimated at less than 1%. Termites and ants also have relatively high rates of endemism with ratios of 23% and 28%, respectively (Lavelle and Lapied, 2003).

These high rates of endemism and the large numbers of rare species make soil macrofauna communities highly susceptible to species losses when subjected to disturbances.

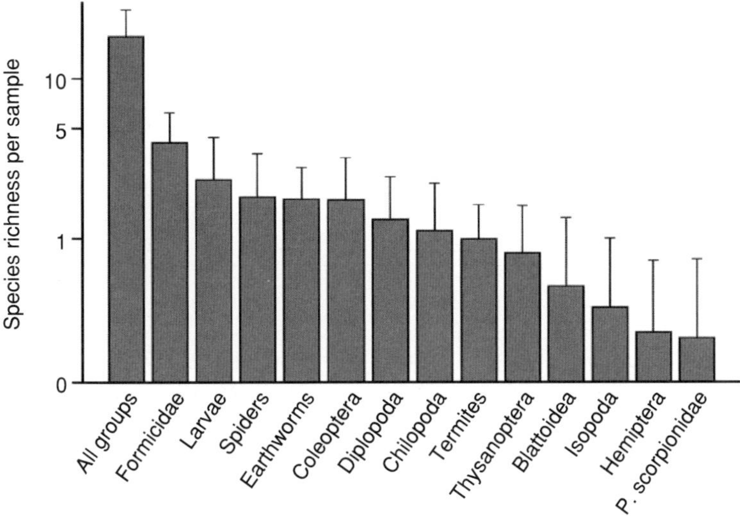

Fig. 3.1. Average species richness of the different groups of soil macrofauna in 25 × 25 cm sampling units from a primary forest at Benfica (Para, Brazil). (Source: Mathieu *et al.*, 2005.)

Response of Soil Macroinvertebrate Communities to Land Use Practices

Diversity and species richness

Species richness generally decreases severely after deforestation. In the Manaus region (central Amazonia), 156 macroinvertebrate morphospecies were recorded in forests. After conversion to pastures, only 29–48 morphospecies were found, of which 15–30% had not been found previously in the forest soils (Barros *et al.*, 2004).

In sites investigated in western Amazonia, diversity evaluated by the Shannon index calculated on the number of large orders was higher (2.22) in the forest than in any other system. Diversity decreased gradually with increasing intensification of land use from fallow (2.14), to agroforestry (1.92), pasture (1.73) and annual crop systems (1.63) (Table 3.1). In this area, termites, being the most abundant group, were identified separately at the level of genera and when possible at the species level. The same trend observed in total faunal diversity across the land use intensification gradient was also seen for termites. Ten genera of Isoptera were identified in forests, nine in agroforestry systems, seven in fallows, two in pastures and four in annual crops (Barros *et al.*, 2002).

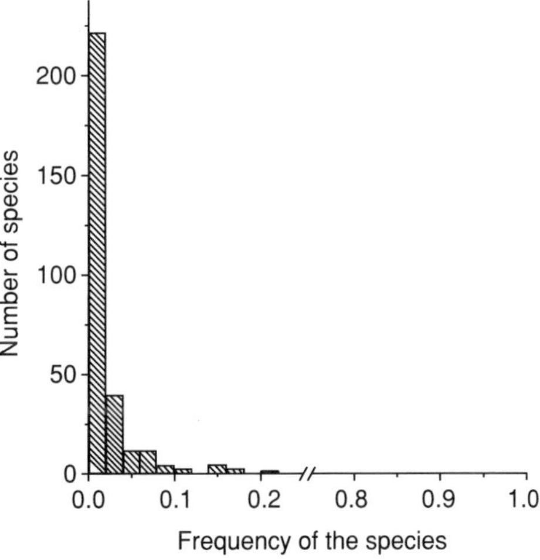

Fig. 3.2. Frequency of macrofauna species in the sampling at Benfica (Para, Brazil). Note the huge dominance of species representing less than 2% of the density. (Source: Mathieu, 2004.)

In eastern Amazonia, clearing of the primary forest had a very strong effect on

Table 3.1. Shannon index for soil macrofauna diversity in different land use systems in western Amazonia.

	Richness	Shannon index	Evenness
Disturbed forest	16	2.22	0.62
Fallow	17	2.14	0.55
Agroforestry	13	1.92	0.53
Pasture	10	1.73	0.52
Annual crop	16	1.63	0.43

Source: Barros *et al.* (2002).

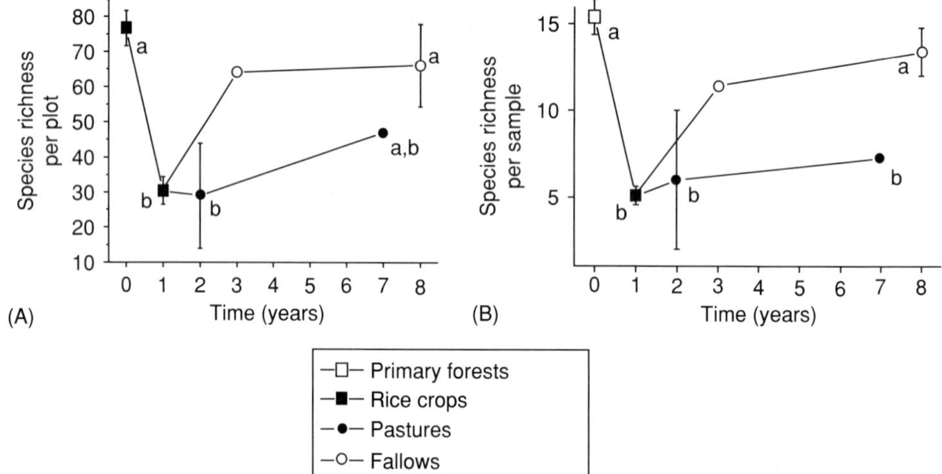

Fig. 3.3. Variation of soil macroinvertebrate species richness in different plots forming a chronosequence of land use types in eastern Amazonia at Benfica (Para, Brazil). (A) Species richness per plot, (B) species richness per sample. Data that do not have common letters are different (Scheffé test, $P < 0.05$). Bars indicate SE when several plots had been sampled. (Source: Mathieu *et al.*, 2005.)

soil macrofauna species richness, both at the sampling point and at the plot level (Fig. 3.3). Species richness per plot was halved due to forest clearing (76 species per plot in forests and 30 in rice fields, 10 soil samples per plot). In old pastures, planted after rice cultivation, and maintained for 5 years, species increased again to 47 species per plot (Mathieu, 2004). In a young fallow following 1 year of rice cultivation, the initial species richness was almost restored after 2 years (64 species per plot), a value very close to that in the primary forest. In old fallows, species richness per plot was also high (66 species per plot).

Species richness per sample showed the same pattern at the scale of 1/16 m^2 sampling units. Highest values were recorded in the primary forests with 15 species per sample on average whereas rice fields installed right after deforestation and burning only had 5 species and pastures, 7.2 in 6-year-old plots. Species richness was rapidly restored in the fallows, reaching 11.4 species per sample in 2-year-old fallows and 13.4 species per sample in 7-year-old fallows, respectively. Such a positive effect of fallows on the restoration of soil invertebrate communities is, however, not found everywhere. In places where deforestation occurred 10–15 years before and where continuity with native forest is lost, restoration may not occur. In the region of Manaus, Barros (1999) found only 29 species of macroinvertebrate in a 20-year-old fallow, in a region where native forest had 156 and pastures of different ages between 28 and 49 species.

Community structure in different land use systems

As many as 118 sites have been investigated across the whole Amazonian region, including sites in Peru and Colombia, using the same standardized methodology (Fig. 3.4). Communities exhibited rather large variations between sites that are not fully explained up to now. Although land use systems often explain local variations (Lavelle and Pashanasi, 1989; Decaëns

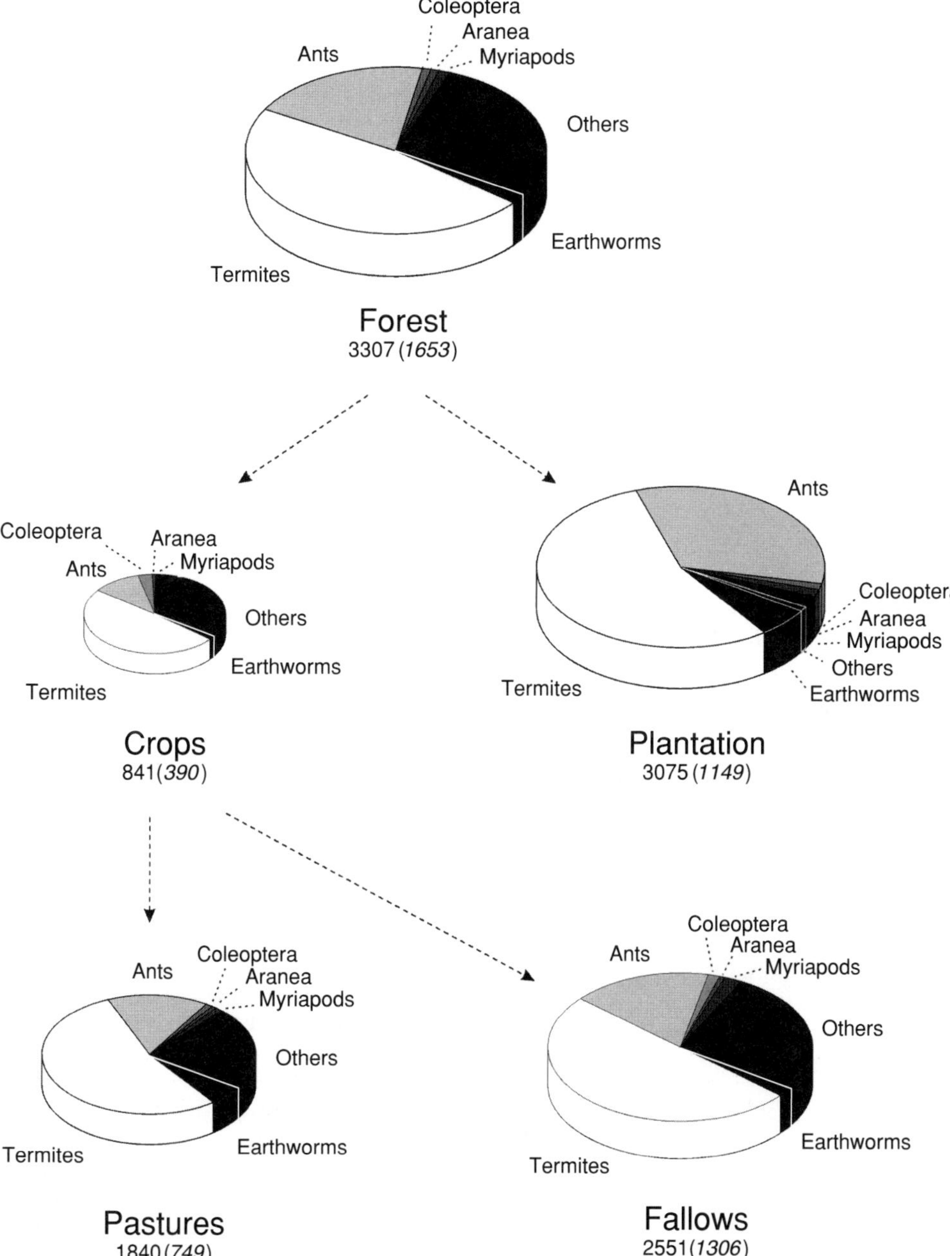

Fig. 3.4. Soil macrofauna density (individuals/m^2) in different land use systems in Amazonia.

et al., 1994; Mathieu, 2004), at a larger scale, determinants may be different as climate and soil parameters and landscape features add their effects (Barros *et al.*, 2002). A general multivariate analysis did not show a significant effect of land use systems across Amazonia or any other clear pattern in their determination. Despite this lack of statistical significance, some general trends can be observed.

Across all sites, soil macrofauna density was higher in forest plots than in other land use systems, with 3300 individuals per square metre (ind/m^2) on average, with a large dominance of social insects. Plantations of perennial plants and fallows presented densities close to the forest (3075 and 2551 ind/m^2, respectively). Rice crops had much lower density (841 ind/m^2). Termites and ants accounted for at least half of the individuals in all land use systems. Earthworms and Coleoptera were the other two important groups, although their density was more variable between land use types. Crops had a higher proportion of Coleoptera than the other land use systems. Communities in plantations were strongly dominated by termites and ants and had rather abundant earthworm populations. In pastures, termites largely dominated whereas the proportion of ants was lower than in most other land use systems and that of earthworms generally greater. In fallows, species densities were very similar to the forest (Fig. 3.4).

Regional Patterns

Brazilian Amazonia is large and interregional variations of geological substrate, climate, biogeographical patterns and land use strategies may have large impacts on macroinvertebrate communities. The following case studies illustrate the differences that may be observed across Brazilian Amazonia at large (Fig. 3.5).

Eastern Amazon

In the region of Marabá (state of Para) a survey of soil macrofauna has been conducted in a forest, three pastures aged 2, 11 and 16 years and a fallow (Desjardins *et al.*, 2004). The total macroinvertebrate density was high in the forest (17.246 ind/m^2) due to large densities of termites and much lower in pastures (1294–4803 ind/m^2). Biomass was 36.2 g fresh weight per square metre (g fw/m^2) in the forest, decreased in pastures (6.0–7.1 g fw/m^2) and was even lower in fallow (3.5 g). Species richness maximum in the forest (63 morphospecies recognized) was almost halved in the pastures (30–46) and in the fallow system (30). Termites were dominant in all systems, especially in the forest (76%). The ants were the second most important group (20% in the forest; 36% in the pasture). The density of earthworms ranged from 100 to 265 ind/m^2. In pastures, total macrofauna biomass was lower than in the forest. The termites (8–45%) and principally the earthworms (36–79%) were responsible for these values. There was no significant difference between the pastures and the other systems. Isopoda, Coleoptera, Arachnida, Diplopoda, Chilopoda and Heteroptera were present at all sites.

At the Benfica site, some 100 km away from Marabá, in a recently deforested area, the overall density was much lower, with values ranging from 134 to 1707 ind/m^2, depending on the type of land use and site (Mathieu, 2004; Mathieu *et al.*, 2004). Rice fields grown after deforestation and burning of the primary forest had especially depressed communities. Pastures, fallows and secondary forests represented successive steps towards the original forest community. Termites and ants had much lower densities than at the Marabá site.

Central Amazon

In central Amazonia, 80 km north of Manaus, faunal density and biomass did not differ significantly among systems.

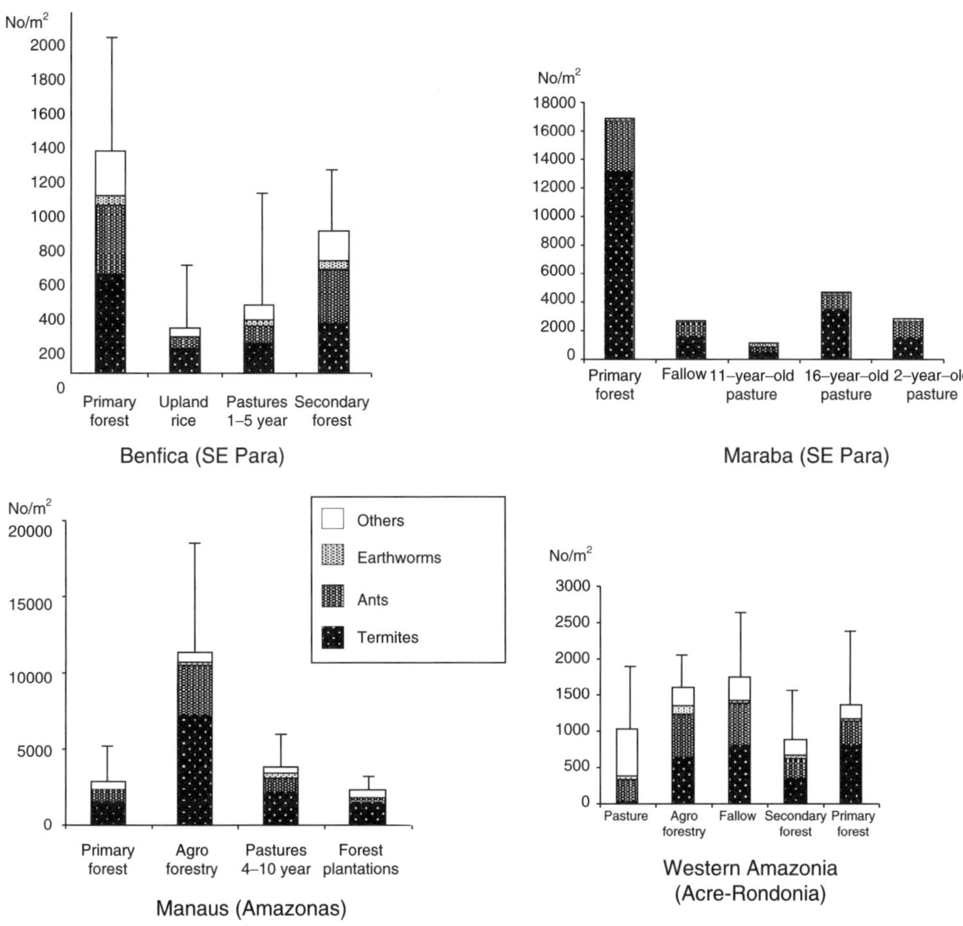

Fig. 3.5. Soil macrofauna communities in four different regions of Brazilian Amazonia. (Source: Bandeira and Harada (1998), Barros *et al.* (1999, 2002), Nascimento and Barros (2002), Desjardins *et al.* (2004), Mathieu (2004), Mathieu *et al.* (2004, 2005).)

However, silvopastoral systems had a larger diversity (14–15 different orders) than agrosilvicultural systems (AS) (10–11). The absence of a continuous litter layer in the AS was also reflected in the vertical distribution of soil invertebrates. The two silvopastoral systems had similar faunal densities in the litter layer than the fallow, with 338 ind/m^2 (12% of the total fauna), 205 ind/m^2 (9%) and 352 ind/m^2 (13%), respectively. In contrast, no litter fauna was found in the AS during the first 3 years after instalment of the system. In the upper 5 cm of soil, there were 807 (29%) and 693 ind/m^2 (34%), respectively, in the silvopastoral systems with high and low input.

In Manaus, total density was 6670 ind/m^2 in the forest. The density was much lower in pastures with highest values in the old pastures (2950 ind/m^2) and lowest values in the young pastures (1060 ind/m^2). Termites were dominant in all sites (41–85% of total individual density). Ant density decreased after pasture installation, and increased when the pasture was degraded or abandoned. The total biomass was 53.3 g fw/m^2 in the forest plot. Similar values were recorded in pastures, except for the 15-year-old plot (not degraded), that presented values ten times lower. The earthworms were responsible for these high values like in the Marabá sites (67–97%).

Western Amazon

In western Amazon, lowest densities were recorded in secondary forest (884 ind/m^2) and pasture (840 ind/m^2) plots. These values were significantly different from those found in fallow, agroforestry systems and annual crops, where 1737, 1745 and 1761 ind/m^2, respectively, had been collected. Fallow and agroforestry systems contained very high densities of termites and ants. Cropping systems also had very high termite densities, probably because sampling had been performed straight after harvest when termite activity on crop residues is high. In pastures, a large number of Coleoptera were observed (395 ind/m^2). Most of these Coleoptera were rhizophagous, as already observed in Mexican pastures (Villalobos and Lavelle, 1990). Their high densities are probably related to the high root biomass in pastures.

Total biomass was highest in pasture (56 g/m^2) and less than 11 g/m^2 in all other land use systems. The Oligochaeta group was dominant in all systems except fallow, where the most prevalent group was Diplopoda, with 5 g/m^2. The Oligochaeta biomass in the pasture system was 53 g/m^2, i.e. nine times larger than in the fallow.

Local Distribution: Single Tree and Grass Tuft Effects

Macroinvertebrate communities exhibited large local variability following the distribution of plants and the quality of the litter and environment provided by different plant species. In the silvopastoral system, and the AS studied in central Amazon, invertebrate communities were sampled separately in the tree rows and in the *Desmodium* fodder crop between the rows. There was an insignificant tendency for

Table 3.2. Mean abundance (ind/m^2) and biomass (g/m^2) of macroinvertebrates in the silvopastoral system (high input) (ASPh) and silvopastoral system (low input) (ASPl) (±SE).

	ASPh		ASPl	
	Tree	*Desmodium*	Tree	*Desmodium*
Total density (ind/m^2)	7,493 (4,225)	5,714 (3,878)	17,480 (15,422)	3,284 (2,952)
Total biomass (ind/m^2)	35.6 (28.2)	23.6 (15.4)	45.5 (25.3)	7.9 (4.1)
Number of taxa	9	11	10	12
Diplopoda (ind/m^2)	91 (15)	169 (31)	80 (45)	82 (26)
Diplopoda (g/m^2)	2.4 (0.9)	3.4 (1.2)	1.8 (1.0)	1.4 (0.4)
Isopoda (ind/m^2)	437 (232)	510 (114)	264 (85)	142 (68)
Isopoda (g/m^2)	4.8 (3.2)	3.8 (1.9)	2.1 (0.6)	1.2 (0.8)
Oligochaeta (ind/m^2)	152 (56)	53 (32)	67 (47)	41 (28)
Oligochaeta (g/m^2)	12.5 (8.5)	3.9 (1.8)	4.1 (3.6)	1.4 (0.9)

Source: Barros et al. (2003).

the soil under the trees to have a higher faunal density and biomass than under the *Desmodium* whereas soil under the *Desmodium* had a larger number of faunal groups (Table 3.2).

Among the litter feeders, diplopods and isopods responded significantly to the input level in the silvopastoral systems (high and low input). The earthworms would also be favoured by fertilizer inputs, although the effect was mainly observed under tree rows. We speculate that the effect of the input level on these faunal groups was mainly indirect and was principally mediated by the faster growth of the plants in the silvopastoral system (high input) plots with increased litter production and improved conditions of temperature and moisture in the litter layer and the topsoil.

In the AS associating trees with a legume cover, the faunal biomass was much greater under Brazil nut and mahogany than under cupuassu and passion fruit ($P = 0.09$) (Table 3.3). This trend was mainly caused by a higher earthworm biomass under the former two species. Earthworms belonged to the endogeic category (i.e. living in the soil), with meso-humic[1] species dominating under Brazil nut and cupuassu and polyhumic[2] species under mahogany and passion fruit. Fragoso *et al.* (1997) had actually observed a similar increase in the earthworm biomass,

[1] Geophages eating soil as it is, with no particle selection.
[2] Geophages eating soil rich in organic matter.

Table 3.3. Total number of taxa and mean biomass (g/m^2) of macroinvertebrates in the agrosilvicultural system (palm-based) (AS1) and agrosilvicultural system (high-diversity tree crop) (AS2) (±SE).

	AS1		AS2			
	Cupuassu	Peach palm	Cupuassu	Brazil nut	Mahogany	Passion fruit
Number of taxa	10	8	5	7	7	6
Gastropoda	0.17 (0.08)	0.15 (0.07)	0.00 (0)	0.00 (0)	0.01 (0)	0.00 (0)
Oligochaeta	17.95 (12.09)	21.81 (15.56)	2.44 (1.10)	3.22 (2.03)	6.76 (4.42)	0.65 (0.08)
Isopoda	5.54 (3.98)	0.22 (0.10)	0.09 (0.06)	0.00 (0)	0.00 (0)	0.03 (0.01)
Diplopoda	0.00 (0)	0.00 (0)	0.09 (0.02)	0.02 (0)	0.00 (0)	0.00 (0)
Chilopoda	0.14 (0.06)	0.09 (0.03)	0.00 (0)	0.03 (0)	0.05 (0.01)	0.13 (0.07)
Hemiptera	0.04 (0.01)	0.00 (0)	0.00 (0)	0.06 (0.02)	0.00 (0)	0.00 (0)
Coleoptera	0.01 (0.01)	0.00 (0)	0.73 (0.20)	0.31 (0.12)	0.00 (0)	0.13 (0.02)
Hymenoptera	0.07 (0.02)	0.77 (0.56)	0.11 (0.10)	0.32 (0.15)	0.02 (0.01)	0.05 (0.01)
Isoptera	0.10 (0.06)	2.57 (1.22)	6.64 (4.20)	3.05 (2.16)	2.50 (1.34)	0.09 (0.05)
Thysanoptera	0.00 (0)	0.00 (0)	0.00 (0)	0.00 (0)	0.00 (0)	0.00 (0)
Orthoptera	0.00 (0)	0.06 (0.02)	0.00 (0)	0.00 (0)	0.00 (0)	0.00 (0)
Total	24.01 (20.54)	25.67 (21.06)	10.10 (8.02)	7.01 (4.12)	9.35 (7.10)	1.09 (0.92)

Source: Barros *et al.* (2003).

and also in the number of ecological categories, i.e. endogeic, epigeic and anecic, under peach palm in comparison with other vegetation types. This confirms the observation that trees and palms with relatively fast growth favour the development of the soil macrofauna, presumably through their effects on litter and microclimate.

In a chronosequence from degraded pastures to fallows and agroforestry systems in central Amazonia the fastest regeneration of soil fauna was observed in the agroforestry systems (Barros, 1999). There were no significant differences between the tree species with respect to faunal density and species richness, although some groups (Diploda, Isopoda, Araneidae and Chilopoda) would be absent from a number of systems, especially when litter cover was not continuous. Differences were also observed in the depth distribution of invertebrates, probably reflecting differences in conditions of the habitat, especially the quality and abundance of litter deposited at the soil surface.

Vegetation cover also influences soil macrofauna at very small scales. In eastern Amazonia, Mathieu *et al.* (2004) showed that species richness was twice as high under herb tufts (nine to ten species per sample) than in nearby bare ground (four to five species per sample). The overall density was almost three times higher in covered ground (768 ind/m^2) than in bare ground (272 ind/m^2). Moreover, this effect was significant within all soil macrofauna groups.

Discussion

Soil macrofauna communities present several peculiarities that distinguish them from many other organisms of the aboveground communities. They have a relatively high species richness and broad functional diversity. The three major functional groups, litter transformers, ecosystem engineers and predators, are represented by a wide range of taxonomic groups and ecologies. The structure of communities is globally characterized by great differences in species richness among orders, a large proportion of rare species and rather high rates of endemism. The sum of these specificities makes forest soil communities very vulnerable to deforestation and land use intensification. Many forest species do not adapt to conditions of open land and invasive species (especially of earthworms) may replace native species almost irreversibly. A rather low proportion of species actually adapts to conditions of cropped land although agroforestry systems that better mimic the original environment are less detrimental. At the landscape level, conservation of species seems to depend highly on the conservation of sizeable patches of forest where native forest can survive and recolonize neighbouring deforested systems, when ecological conditions are suitable.

At present, the total number of soil macroinvertebrate species in the Amazon region is not known. Data are rare, even at short scales. This situation probably reflects the difficulty of soil macrofauna species identification and the lack of taxonomists. Some simulations based on accumulation curves suggest that total soil macrofauna species richness could reach 2200 species on a surface of 11 km^2, in eastern Amazonia (Mathieu, 2004). Authors warn that this estimation is probably higher than in reality. Nevertheless, the magnitude of the estimation is sufficient to understand the very great diversity of soil macrofauna, even at regional scales. A few studies have provided data on termite communities. Constantino (1992) identified 35 genera and 64 species in the Marãa region in western Amazonia. Bandeira and Torres (1985) found 63 termite species in the primary forest of eastern Amazonia. In this study a gradient was observed, with termite diversity decreasing with land use intensification. The type of land use seems to be highly relevant as regards the conservation of soil macroinvertebrate communities, although some regional factors may sometimes override their effects.

The fallow and agroforestry systems had a great abundance of ants and termites in common, without a clear dominance of either, probably because of the more diversified organic inputs that allowed colonization by different organisms. In the fallow systems, biomass was mainly represented by Diplopoda. Stork and Brindell (1993) had already mentioned the importance of this group (in terms of biomass and density) in a study in the natural forest of Seram in Indonesia. Barros (1999) also observed a significant increase in density and biomass of Diplopoda, in a chronosequence in central Amazonia, as weeds substituted for grasses with the ageing of pastures. This phenomenon may be due to the preference of Diplopoda for litter in forest areas. Tapia-Coral *et al.* (1999) showed that Diplopoda was the second most important group, after Isopoda, in a study of litter macroinvertebrate communities in agroforestry systems in central Amazonia. In semiarid forests of Guadeloupe (Caribbean Islands), Loranger (1999) also recorded a high abundance of diplopods.

Vertical distribution of macrofauna significantly varied with the type of land use, depending on the abundance and quality of the surface litter layer. The original forest had the largest proportion of invertebrates in litter (19%). In the fallow and agroforestry system 10% and 13% of total fauna were extracted from the litter, respectively. Barros *et al.* (2003) found values of 12% and 9%, respectively, in litter of plots with high and low inputs in central Amazonian agroforestry systems. This emphasizes the fact that studies of macrofauna limited to the litter layer can only give a limited idea of the system's pattern since this community only represents a small and variable percentage of the total fauna (Vohland and Schroth, 1999). In the best of cases, as in natural or modified forest areas, these values can reach approximately 20%.

The results show that soil macroinvertebrates are sensitive indicators of the nature of land use and management. The agroforestry systems had higher abundance and diversity than any other land use type. This is an indication that these systems may sustain sufficiently abundant and diverse communities to optimize the effects of these beneficial organisms (Brown *et al.*, 1999; Chauvel *et al.*, 1999). Agroforestry systems, however, cover a wide array of highly diverse practices. The quality of organic matter produced and the effects of vegetation on soil water and temperature regimes may vary considerably, with significant effects on the soil macrofauna community (Tian *et al.*, 1995, 1997). However, more detailed studies are needed to identify the best possible combinations of plant species and spatial arrays to allow optimal production and sustainability.

References

Anderson, J.M. and Ingram, J.S.I. (1993) *Tropical Soil Biology and Fertility: A Handbook of Methods.* CAB International, Wallingford, UK.

Baindaira, A.G. and Harada, A.Y. (1998) Densidade e Distribuição vertical de macroinvertebrades em solas argilosos e arenosos na Amazonia Central. *Acta Amazonica* 28(2), 191–204.

Bandeira, A.G. and Torres, M.F.P. (1985) Abundância e distribuição de invertebrados do solo em ecossistemas da Amazônia Oriental. O papel ecológico dos cupins. *Boletim do Museu Paraense Emílio Goeldi, Zoologia* 2, 13–38.

Barros, E. (1999) Effet de la macrofaune sur la structure et les processus physiques du sol des pâturages dégradés d'Amazonie. Thesis Université Paris VI.

Barros, E., Curmi, P., Hallaire, V., Chauvel, A. and Lavelle, P. (2001) Role of macrofauna in the transformation and reversibility of soil structure of an oxisol during forest to pasture conversion. *Geoderma* 100, 193–213.

Barros, E., Pashanasi, B., Constantino, R. and Lavelle, P. (2002) The soil macrofauna community in land use systems in Amazonia after slash and burn. *Biology and Fertility of Soils* 35, 338–347.

Barros, E., Neves, A., Fernandes, E.C.M., Wandelli, E., Blanchart, E. and Lavelle P. (2003) Soil macrofauna community of Amazonian agroforestry systems. *Pedobiologia* 47(3), 273–280.

Barros, E., Grimaldi, M., Sarrazin, M., Chauvel, A., Mitja, D., Desjardins, T. and Lavelle, P. (2004) Soil physical degradation and changes in macrofaunal communities in central Amazon. *Applied Soil Ecology* 26, 157–168.

Beare, M.H., Coleman, D.C., Crossley, D.A., Hendrix, P.F. and Odum, E.P. (1994) A hierarchical approach to evaluating the significance of soil biodiversity to biogeochemical cycling. *Plant and Soil* 31, 1–18.

Blanchart, E., Lavelle, P., Braudeau, E., Le Bissonais, Y. and Valentin, C. (1997) Regulation of soil structure by geophagous earthworm activities in humid savannas of Ivory Coast. *Soil Biology and Biochemistry* 29, 431–439.

Brown, G., Pashanasi, B., Gilot-Villenave, C., Patron, J.C., Senapati, B.K., Giri, S., Barois, I., Lavelle, P., Blanchart, E., Blakemore, R.J., Spain, A.V. and Boyer, J. (1999) Effects of earthworms on plant growth in the tropics. In: Lavelle, P., Brussaard, L. and Hendrix, P. (eds) *The Management of Earthworms in Tropical Agroecosystems*. CAB International, Wallingford, UK, pp. 87–148.

Chauvel, A., Grimaldi, M., Barros, E., Blanchart, E., Desjardins, T., Sarrazin, M. and Lavelle, P. (1999) Pasture damage by an Amazonian earthworm. *Nature* 398, 32–33.

Constantino, R. (1992) Abundance and diversity of termites (Insecta: Isoptera) in two sites of primary rain forest in Brazilian Amazonia. *Biotropica* 24, 420–430.

Decaëns, T., Lavelle, P., Jimenez Jaen, J.J., Escobar, G. and Rippstein, G. (1994) Impact of land management on soil macrofauna in the Oriental Llanos of Colombia. *European Journal of Soil Biology* 30(4), 157–168.

Desjardins, T., Barros, E., Sarrazin, M., Girardin, C. and Mariotti, A. (2004) Effects of forest conversion to pasture on soil carbon content and dynamics in Brazilian Amazonia. *Agriculture, Ecosystems and Environment* 103, 365–373.

Fearnside, P.M. and Barbosa, R.I. (1998) Soil carbon changes from conversion of forest to pasture in Brazilian Amazonia. *Forest Ecology and Management* 108, 147–166.

Fragoso, C. and Lavelle, P. (1995) Are earthworms important in the decomposition of tropical litter. In: Reddy, M.V. (ed.) *Soil Organisms and Litter Decomposition in the Tropics*. Oxford & IBH, Delhi, pp. 103–112.

Fragoso, C., Brown, G.G., Patrón, J.C., Blanchart, E., Lavelle, P., Pashanasi, B., Senapati, B. and Kumar, T. (1997) Agricultural intensification, soil biodiversity and agroecosystem function in the tropics: the role of earthworms. *Applied Soil Ecology* 6, 17–35.

Grimaldi, M., Sarrazin, M., Chauvel, A., Luizao, F.J., Nunes, N., Rodrigues, M.R.L., Amblard, P. and Tessier, D. (1993) Effet de la deforestation et des cultures sur la structure des sols argileux d'Amazonie brésilienne. *Cahiers Agricultures* 2, 36–47.

INPE Brazil (1998) Amazonia: deforestation 1995–1997. Instituto Nacional de Pesquisas Espaciais (INPE). Available at http://www.obt.inpe.br/prodes/index.html

Lapied, E. and Lavelle, P. (2003) The peregrine earthworm *Pontoscolex corethrurus* in the east coast of Costa Rica. *Pedobiologia* 47, 471–474.

Laurance, F.W.A., Albernaz, K.M. and Costa, C. (2001) Is deforestation accelerating in the Brazilian Amazon? *Environmental Conservation* 28, 305–311.

Lavelle, P. and Lapied, E. (2003) Endangered earthworms of Amazonia: an homage to Gilberto Righi. *Pedobiologia* 47, 419–427.

Lavelle, P. and Pashanasi, B. (1989) Soil macrofauna and land management in Peruvian Amazonia (Yurimaguas, Loreto). *Pedobiologia* 33, 283–291.

Lavelle, P., Blanchart, E., Martin, A., Spain, A.V. and Martin, S. (1992) The impact of soil fauna on the properties of soils in the humid tropics. In: Sanchez, P.A. and Lal, R. (eds) *Myths and Science of Soils of the Tropics*. SSSA Special Publication, Madison, Wisconsin, pp. 157–185.

Lavelle, P., Lattaud, C., Trigo, D. and Barois, I. (1995) Mutualism and biodiversity in soils. *Plant and Soil* 170, 23–33.

Lavelle, P., Bignell, D., Lepage, M., Wolters, V., Roger, P., Ineson, P., Heal, O.W. and Dhillion, S. (1997) Soil function in a changing world: the role of invertebrate ecosystem engineers. *European Journal of Soil Biology* 33, 159–193.

Lee, K.E. and Foster, R.C. (1992) Soil fauna and soil structure. *Australian Journal of Soil Research* 29, 745–746.

Loranger, G. (1999) Déterminants de la décomposition de la litière dans une forêt semi décidue de la Guadeloupe. Doctorat en Sciences de la terre et Pédologie de l' Université Paris VI.

Mathieu, J. (2004) Etude de la macrofaune du sol dans une zone de déforestation en Amazonie du sud est, dans le contexte de l'agriculture familiale. Thesis Université Paris VI, 238 p.

Mathieu, J., Rossi, J.P., Grimaldi, M., Mora, P., Lavelle, P. and Rouland, C. (2004) A multi-scale study of soil macrofauna biodiversity in Amazonian pastures. *Biology and Fertility of Soils* 40, 300–305.

Mathieu, J., Rossi, J.P., Mora, P., Lavelle, P., Martins, P.S., Rouland, C. and Grimaldi, M. (2005) Recovery of soil macrofauna communities after forest clearance in eastern Amazonia, Brazil. *Conservation Biology* 19(5), 1598–1605.
McGrath, D.A., Smith, C.K., Gholz, H.L. and Assis Oliveira, F. (2001) Effects of land-use change on soil nutrient dynamics in Amazônia. *Ecosystems* 4, 625–645.
Moraes, J.F.L., Volkoff, B., Cerri, C.C. and Bernoux, M. (1996) Soil properties under Amazon forest and changes due to pasture installation in Rondônia, Brazil. *Geoderma* 70, 63–81.
Nascimento, A.R.L. and Barros, E. (2002) Macrofauna do solo em sistemas agroflorestais do projeto RECA (RO). In: *IV Agroforestry System Brazilian Symposium*. Ilhevs (BA), Brazil. CD-ROM.
Oades, J.M. (1993) The role of biology in the formation, stabilization and degradation of soil structure. *Geoderma* 56, 377–400.
Stork, N.E. and Brindell, M.J.D. (1993) Arthropod abundance in lowland rainforest of Seram. In: Edwards, I.D., Macdonald, A.A. and Proctor, J. (eds) *Natural History of Seram*. Intercept, Andover, UK, pp. 115–130.
Tapia-Coral, S., Luizão, F. and Wandelli, E.V. (1999) Macrofauna da liteira em sistemas agroflorestais sobre pastagens abandonadas na Amazônia Central. *Acta Amazônica* 29, 477–495.
Tian, G., Brusaard, L. and Kang, B.T. (1995) Breakdown of plant residues with contrasting chemical compositions; effects of earthworms and millipedes. *Soil Biology and Biochemistry* 27, 277–280.
Tian, G., Kang, B.T. and Brussaard, L. (1997) Effect of mulch quality on earthworm activity and nutrient supply in the humid tropics. *Soil Biology and Biochemistry* 29, 369–373.
Villalobos, F.J. and Lavelle, P. (1990) The soil coleoptera community of a tropical grassland from Laguna Verde, Veracruz (Mexico). *Revue D'Ecologie et de Biologie du Sol* 27(1), 73–93
Vohland, K. and Schroth, G. (1999) Distribution patterns of the litter macrofauna in agroforestry and monoculture plantations in central Amazonia as affected by plant species and management. *Applied Soil Ecology* 13, 57–68.
Wardle, D.A. and Lavelle, P. (1997) Linkages between soil biota, plant litter quality and decomposition. In: Cadisch, G. and Giller, K.E. (eds) *Driven by Nature: Plant Litter Quality and Decomposition*. CAB International, Wallingford, UK, pp. 107–124.

4 Earthworm Ecology and Diversity in Brazil

S.W. James[1] and G.G. Brown[2]

[1]*Kansas University Natural History Museum and Biodiversity Research Center, Lawrence, Kansas, 66045, USA, e-mail: sjames@ku.edu;* [2]*Embrapa Soja, Rod. Carlos João Strass acesso Orlando Amaral, C.P. 231, Londrina, PR, 86001-970, Brazil, e-mail: browng@cnpso.embrapa.br*

Introduction

The diversity of life in Brazil is renowned among biologists, environmentalists and the educated public worldwide because the Amazon Basin, Cerrados and the Atlantic rainforest (*Mata Atlantica*) are famous as areas of great biological wealth endangered by human activity. Of the large nations encompassing a significant fraction of a continent, Brazil is clearly the most biodiverse, followed by Australia, Mexico and China in uncertain order. When we look at particular biotic elements, this ranking may change, but for earthworms, it is most likely to hold true. We can say this in spite of the inadequate knowledge of earthworms in all the large nations. Canada and Russia consist largely of territory devoid of native earthworms and thus have low species diversity. The USA has an unknown number of species probably in excess of 200, but unlikely to be significantly larger than that (Fender, 1995; James, 1995). Australia has more than 560 known native species (Blakemore, 2000) and considerable underexplored areas in which more may be found. The count of 128 species known for Mexico is probably less than half of the total (Fragoso, 2001), given the diversity of habitats and the topographic complexity of the country. In China, some 300 nominal species have been found (Qiu and Wu, unpublished compilation), but the dissected tropical and subtropical terrain of the southern half is poorly known.

Given that among these large nations, Brazil is the only one whose entire territory, with the exception of a semiarid sector in the north-east, is habitable by earthworms and has no history of glaciation to remove earthworms from habitable areas, it is very probable that Brazil's earthworm fauna is the most diverse in the world. Against this background we present a preliminary account of the current state of knowledge of Brazilian earthworm taxonomy, ecology, economic impact and exploitation by humans.

Taxonomy of Brazilian Worms: a Bit of History

No one knows when the first biologist (or the earlier equivalent, the naturalist) encountered an earthworm in Brazil, but it was probably early in colonial history. The *Mata Atlantica* region along the coast of Brazil was the first colonized and still harbours many native species, some of impressive dimensions. Europeans familiar with the modest size of their homeland earthworms

 Soil Biodiversity in Amazonian and Other Brazilian Ecosystems (eds F.M.S. Moreira *et al.*)

could not have failed to notice such animals, and this is reflected in the scientific reporting on earthworms in the 19th and early 20th century.

In fact, the first Brazilian earthworm to be described in the preserved state was over a metre long. In 1835, Leuckart established the genus *Glossoscolex* and then described the large earthworm *Glossoscolex giganteus* from Rio de Janeiro in 1836, probably collected in the forest on the slopes of the Corcovado mountain.

A few decades later, the naturalist Fritz Müller stumbled upon abundant populations of the extremely common and widespread earthworm *Pontoscolex corethrurus* (the bristle-tailed worm), describing the species in 1857, from specimens of Itajaí (Santa Catarina).[1] He also commented on the distribution of this species, writing, 'the brush-tail, the commonest of earthworms of this country (Brazil), ... may be found in almost every clod of arable land ...' In 1877 and 1878 Darwin and Müller exchanged correspondence regarding earthworms, while Darwin was gathering data for his famous earthworm book (Darwin, 1881). Part of Müller's response to Darwin's request regarding the abundance of earthworms and their castings in Brazilian forests is published in Darwin's book (pp. 67–68, in a 1976 edition). Unfortunately, we could not secure a copy of Müller's letter to obtain all the details. Nevertheless, we know from Müller and Darwin (1881) that several species of earthworms were common in Santa Catarina and that 'in most parts of the forests and pasture lands, the whole soil, to a depth of a quarter of a metre, looks as if it had passed repeatedly through the intestines of earthworms, even where hardly any castings are to be seen on the surface'. These castings are probably the work of *P. corethrurus*, abundant in the area, as Müller himself had stated earlier. Furthermore, a very large and rare, still undescribed species (at the time) produced very large burrows (2 cm diameter), penetrating the soil to a very great depth.

In the late 19th century, most new records and new species came from the state of Rio de Janeiro south to Rio Grande do Sul. These were the regions favoured by many colonists from western and central Europe, whose homelands had active biological research programmes. Consequently, most of the collections were handed over to specialists in Germany (Michaelsen, Ude, Horst, Kinberg), Italy (Cognetti, Rosa), England (Benham) and France (Perrier). Many of these species were rather large. This taxonomic work (no ecological studies had been conducted so far) continued into the early 20th century but dropped off sharply by the 1920s. But it was principally the work of W. Michaelsen, both at the end of the 19th and at the beginning of the 20th century, that was fundamental in expanding the understanding of the biodiversity of Brazilian earthworms. He described 34 species of Brazilian earthworms up to his death in 1937. The other European taxonomists mentioned above also contributed to the knowledge of Brazilian earthworms, but altogether they described fewer than 12 species.

The first paper on earthworm biodiversity in Brazil was that of Perrier (1877), who listed five species in four genera: *Perichaeta dicystis* and *Perichaeta tricystis*,[2] *Urochaeta corethrura* (*P. corethrurus*), a Eudrilidae (probably *Eudrilus eugeniae*) and *Titanus brasiliensis* (*G. giganteus*). A few years later Moreira (1903) listed 22 species, although two were later considered synonyms (*Pheretima barbadensis* and *Pheretima hawayana*, both = *Amynthas gracilis*; *Rhinodrilus papillifer* = *Urobenus brasiliensis*), so that this list actually had only 20 valid species. Of these, 9 were exotic and 11 were native species. In 1927, Michaelsen published the last synthesis (before the

[1]This was the first species that Müller described in Brazil. It was found coexisting with *Geobia subterranean*, which used the galleries to find and feed on the earthworms.

[2]Both *Perichaeta* spp. (possibly *Amynthas* spp.) are nomen dubium/incertum according to Michaelsen (1900a). Therefore, we do not know what species they are until someone looks at the specimens in the collection at the Musée d'Histoire Naturelle in Paris.

present one) of earthworm biodiversity in Brazil in his paper *Die Oligochätenfaua Brasiliens*. The list contains 51 valid species, of which 15 were exotic, widespread species (29%).

Cernosvitov (1934a,b, 1935, 1938, 1939) also published on Brazilian earthworms, describing 12 native earthworm species (nine glossoscolecids and three acanthodrilids) collected on various expeditions (of other scientists) to the country. After that, little was done in terms of megadrile taxonomy in Brazil, until Cordero published on systematics of the Glossoscolecidae from his base in Uruguay in the 1940s. After a 3-month visit to Ceará and with the help of colleagues in Brazil, he described three new glossoscolecid species from mainland north-eastern Brazil (Cordero, 1943, 1944), a region from which no earthworms were known at the time.

Finally, in the early 1960s Gilberto Righi began to work on Brazilian earthworms, among other invertebrates, and eventually worked exclusively on earthworms until his death in 1999. We owe most of the described species in Brazil to Righi and his students (they described a total of 145 species/subspecies in 41 genera), and the collection he left is the most extensive of all holdings of Brazilian earthworms. This collection is now at the Museu de Zoologia of the Universidade de São Paulo (MZUSP), and still contains undescribed material. The collection at the MZUSP contains approximately 1300 lots and dozens of boxes with histological preparations (Moreno and Mischis, 2003).

Other permanent collections,[3] mostly much smaller, are at the National Institute for Amazonian Research (INPA) in Manaus, AM (about 200 lots, mostly native species); Universidade do Rio dos Sinos (UNISINOS) in São Leopoldo, RS (about 5200 test tubes, mostly exotic species); Museu Paraense Emílio Goeldi in Belém (MPEG), PA (~15 lots); Museu Nacional in Rio de Janeiro, RJ (~7 lots).[4] Other 'unofficial' collections of which we are aware, some of them temporary deposits, are located at Embrapa Soybean in Londrina, PR (~100 lots), Minhobox (~40 lots) and the Universidade Federal de Juiz de Fora, Juiz de Fora, MG (~20 lots). All these other collections also contain unidentified material.

Recently, Zicsi and Csuzdi (1987, 1999) and Zicsi *et al.* (2001) identified earthworms from various parts of Brazil, mostly in Amazonia, and dedicated a new genus *Righiodrilus* (Zicsi, 1995), with 20 species in Brazil (Table 4.1) and a new species (*Cirodrilus righii*), in recognition of Righi's immense efforts towards the better understanding of Neotropical megadrile taxonomy and biodiversity.

With Righi's death, no active taxonomist remains in Brazil, and very few active taxonomists remain in Latin America to take on this great challenge. Fragoso *et al.* (2003) estimated that taking the rate of description of new species by Righi as 6.4 species per year, we would need 46 taxonomists working full time for 10 years or 10 taxonomists working full time for 46 years to describe the remainder of the world's estimated earthworm biodiversity (about 3000 species). A more realistic estimate, based on full-time effort producing 50 species descriptions per year would require six taxonomists for 10 years, assuming they do not do anything else, plus several teams of collectors. However, this human resource is not available, and the current trend of taxonomic training and the priorities of governments of various Latin American countries give little hope of abating this problem. A proposed Brazilian government programme to stimulate the training and capacity building of Brazilian scientists and students in taxonomy was recently halted before it had even begun. Therefore, given the relatively small number of earthworm taxonomists and ecologists in Latin America, and the relatively large number of species that still need to be described both in this continent

[3]Most of them registered with the Conselho de Gestão do Patrimônio Genético (CGEN) as faithful depositories of the Brazilian genetic resources (http://www.mma.gov.br/port/cgen/index.cfm).

[4]The former collection was much larger but was destroyed by vandalous acts to the museum on numerous occasions.

Table 4.1. List of earthworm genera and species found in Brazil, together with their distribution and origin. Large-bodied earthworms (minhocuçu) are identified with an asterisk.[1]

Number	Family (Genus species)	Sites/states found[2]	Origin	References[3]
	Glossoscolecidae			
1	*Alexidrilus littoralis*[4] Ljungström, 1972	Tenente Portela, RS	Native	Ljungström (1972a), Knäpper (1977)
2	*Alexidrilus lourdesae* Righi, 1971	Estrela, RS	Native	Righi (1971a)
3	*Andiodrilus icomi* Righi *et al.*, 1976	Near Manaus, AM; Serra do Navio, AP	Native	Righi (1971a), Righi *et al.* (1976)
4	*Andiodrilus* n. sp. 1	Itupiranga, PA	Native	J&B
5	*Andiorrhinus*[5] *amaparis* Righi, 1971	Serra do Navio, AP	Native	Righi (1971a)
6	*Andiorrhinus amazonius* Michaelsen, 1918	Various sites near Manaus, AM; Porto Velho, RO	Native	Righi *et al.* (1976), Righi (1988a), Adis and Righi (1989)
7	*Andiorrhinus bucki* Righi, 1986	Bataguassu, MS	Native	Righi (1986a)
8	*Andiorrhinus caudatus* Righi *et al.*, 1976	In and near Manaus, Sucunduri, AM; PN Amazônia, PA; 5 sites in N RO	Native	Righi (1982a, 1988a), Adis and Righi (1989)
9	*Andiorrhinus evelineae* Righi, 1986	Near Porto Velho, around Itapuã do Oeste, RO	Native	Righi (1986a)
10	*Andiorrhinus holmgreni* Michaelsen, 1918	Cacoal, Presidente Médici, RO	Native	Righi (1986a)
11	*Andiorrhinus paraguayensis* Rosa, 1895	Bataguaçu, MS	Native	MZUSP
12	*Andiorrhinus pauate* Righi, 1986	Pimenta Bueno, Cacoal, Espigão d'Oeste, N of Vilhena, RO	Native	Righi (1986a)
13	*Andiorrhinus pictus* Michaelsen, 1925	Manacapurú, AM	Native	Michaelsen (1925)
14	*Andiorrhinus planaria*[6] Michaelsen, 1934	Upper river Jaú, river Negro, AM	Native	Michaelsen (1934), Adis and Righi (1989)
15	*Andiorrhinus proboscideus* Cernosvitov, 1939	Óbidos, PA	Native	Cernosvitov (1939)
16	*Andiorrhinus rondoniensis* Righi, 1986	Near Porto Velho, RO	Native	Righi (1986b)
17	*Andiorrhinus rubescens* Michaelsen, 1925	Manacapurú, AM	Native	Michaelsen (1925)
18	*Andiorrhinus samuelensis** Righi, 1986	Along river Jamari, Samuel, RO	Native	Righi (1986a)
19	*Andiorrhinus tarumanis*[7] Righi *et al.*, 1976	Various near Manaus, AM; Ilha de Maracá, RR	Native	Righi *et al.* (1976), Righi (1986a), Adis and Righi (1989, 1997), Zicsi *et al.* (2001)
20	*Andiorrhinus torquemadai* Righi, 1984	Cáceres, Vila Bela da Santíssima Trindade, Pontes e Lacerda, MT	Native	Righi (1984d, 1986a)

Continued

Table 4.1. List of earthworm genera and species found in Brazil, together with their distribution and origin. Large-bodied earthworms (minhocuçu) are identified with an asterisk.[1] – cont'd

Number	Family (Genus species)	Sites/states found[2]	Origin	References[3]
21	*Andiorrhinus* n. sp. 1	Buri, SP	Native	J&B
22	*Andiorrhinus* n. sp. 2	Lerroville, São Jerônimo da Serra, PR	Native	J&B
23	*Andiorrhinus* n. sp. 3*[8]	Ponta Grossa, PR	Native	J&B
24	*Andiorrhinus* n. sp. 4*	Itararé, SP	Native	J&B
25	*Andiorrhinus* spp.*[9]	Mauá, Faxinal, Curitiba, Ortigueira, Irati, PR	Native	J&B
26	*Anteoides pigy* Righi, 1982	PN Amazônia, PA	Native	Righi (1982a)
27	*Atatina gatesi* Righi *et al.*, 1978	Near Reserva Ducke, AM	Native	Righi *et al.* (1978)
28	*Atatina puba* Righi, 1971	Belém, PA	Native	Righi (1971a)
29	*Cirodrilus aidae* Righi, 1994	João Pessoa, PB	Native	Righi (1994)
30	*Cirodrilus angeloi* Righi, 1975	Serra do Navio, lower river Matapi, AP	Native	Righi (1975)
31	*Cirodrilus righii* Zicsi *et al.*, 2001	Manaus, AM	Native	Zicsi *et al.* (2001)
32	*Chibui*[10] *bari** Righi and Guerra, 1985	Rio Branco, AC	Native	Righi and Guerra (1985), Guerra (1988a,b, 1994a)
33	*Diachaeta aceoca* Righi, 1982	PN Amazônia, PA	Native	Righi (1982a)
34	*Diachaeta adisi* Righi, 1989	Near Manaus, AM	Native	Righi (1989a)
35	*Diachaeta adnae* Righi, 1986	9 sites in RO	Native	Righi (1986b)
36	*Diachaeta arawak* Righi, 1989	Near Manaus, AM	Native	Righi (1989a)
37	*Diachaeta atroaris* Righi *et al.*, 1978	Reserva Ducke, AM	Native	Righi *et al.* (1978)
38	*Diachaeta carsevenica*[11] Cernosvitov, 1934	Upper river Calçoene, AP	Native	Cernosvitov (1934a, 1935)
39	*Diachaeta juli* Righi *et al.*, 1978	Near (N) of Manaus, AM	Native	Righi *et al.* (1978)
40	*Diachaeta kannerae* Righi, 1984	Poconé, MT	Native	Righi (1984b)
41	*Diachaeta mura* Righi, 1989	Near Manaus, AM	Native	Adis and Righi (1989), Righi (1989b)
42	*Diachaeta nia* Righi *et al.*, 1976	Sucunduri, AM	Native	Righi *et al.* (1976)
43	*Diachaeta xecatu* Righi *et al.*, 1978	Sucunduri, AM	Native	Righi *et al.* (1978)
44	*Diaguita vivianeae* Righi, 1984	6 sites in MT	Native	Righi (1984d), Righi and Guerra (1985)
45	*Enantiodrilus borelli* Cognetti, 1902	Ilha do Marajó, PA	Native?[12]	Michaelsen (1927)
46	*Fimoscolex angai* Righi, 1971	Salesópolis, SP	Native	Righi (1971a)
47	*Fimoscolex inurus* Cognetti, 1913	Cotia, Itatiba, Mogi das Cruzes, Ribeirão Pires, Salesópolis, São Paulo, SP; Joinville, SC	Native	Luederwaldt (1927), Righi (1974), Righi (1986a), J&B, MZUSP

48	*Fimoscolex ohausi*[13] Michaelsen, 1900	Macaé, Petrópolis, RJ	Native	Michaelsen (1900b, 1925), Luederwaldt (1927)
49	*Fimoscolex sacii** Righi, 1971	Ibiúna, Itapecerica, Jacupiranga, Juquiá, Juquitiba, Miracatu, Registro, São Bernardo, Vargem Grande, SP, Rio de Janeiro, RJ	Native	Righi (1971a), Righi and Ayres (1975), Zicsi and Csuzdi (1999)
50	*Fimoscolex sporadochaetus*[14] Michaelsen, 1918	Near Belo Horizonte, Conselheiro Lafaiete, MG	Native	Michaelsen (1918), Righi (1971b)
51	*Fimoscolex tairim* Righi, 1974	Itatiaia, RJ	Native	Righi (1974)
52	*Fimoscolex thayeri* Cernosvitov, 1934	Mendes, RJ	Native	Cernosvitov (1934a, 1935)
53	*Fimoscolex* n. sp. 1*	Salesópolis, SP	Native	J&B
54	*Fimoscolex* n. sp. 2	Jaguapitã, PR	Native	J&N
55	*Fimoscolex* n. sp. 3	Ponta Grossa, PR	Native	J&B
	Fimoscolex sp.	Ponta Grossa, PR	Native	J&B
56	*Glossodrilus antunesi*[15] Righi, 1971	Serra do Navio, AP; Caxias, MA; Ilha do Maracá, Bonfim, RR	Native	Righi (1975), Hamoui and Donatelli (1984), Righi (1998a, 1990b)
57	*Glossodrilus bresslaui*[16] Michaelsen, 1918	Barreira, Rio de Janeiro, RJ; Cananéia, SP	Native	Righi (1975, 1999)
58	*Glossodrilus geayi*[17] Cernosvitov, 1934	Upper river Calçoene, AP	Native	Cernosvitov (1934a, 1935)
59	*Glossodrilus motu* Righi, 1990	Ilha de Maracá, RR	Native	Righi (1990b, 1998a)
60	*Glossodrilus parecis* Righi and Ayres, 1975	Seropédica, RJ; Parecis, RS	Native	Righi and Ayres (1975), Righi (1980a)
61	*Glossodrilus* n. sp. 1	Itupiranga, PA	Native	J&B
62	*Glossoscolex amomee** Righi, 1971	Cotia, Cubatão, Jarinu, Peruíbe, Santo André, São Paulo, São Vicente, SP; Rio de Janeiro, RJ	Native	Righi (1971a), Righi and Lobo (1979), Zicsi and Csuzdi (1999)
63	*Glossoscolex bergi** Rosa, 1900	Foz do Iguaçú, PR	Native	Zicsi and Csuzdi (1987)
64	*Glossoscolex bondari* Michaelsen, 1925	Piracicaba, Rio Claro, SP; Sidrolândia, MS	Native	Michaelsen (1925), Righi and Lobo (1979), Righi (1984b)
65	*Glossoscolex catharinensis** Michaelsen, 1918	Near Joinville, river Itapocu, SC; São Sebastião do Caí, RS; Ribeirão Pires, Paranapiacaba, SP	Native	Michaelsen (1918), Righi (1974), MZUSP
66	*Glossoscolex colonorum* Michaelsen, 1918	Near Joinville, river Itapocu, SC	Native	Michaelsen (1918)

Continued

Table 4.1. List of earthworm genera and species found in Brazil, together with their distribution and origin. Large-bodied earthworms (minhocuçu) are identified with an asterisk.[1] – cont'd

Number	Family (Genus species)	Sites/states found[2]	Origin	References[3]
67	*Glossoscolex fachinii** Righi, 1971	Araras, SP	Native	Righi (1971a)
68	*Glossoscolex fasold** Michaelsen, 1918	Paranapiacaba, Piracicaba, SP	Native	Michaelsen (1918), MZUSP
69a	*Glossoscolex giganteus giganteus** Leuckart, 1836	Paranapiacaba, Campos do Jordão, SP; Agulhas Negras, Itatiaia, Rio de Janeiro, Seropédica, Teresópolis, RJ	Native	Michaelsen (1918, 1925), Luederwaldt (1927), Righi (1980a)
69b	*Glossoscolex giganteus australis** Righi and Lobo, 1979	Near Apiaí, SP	Native	Righi and Lobo (1979)
70	*Glossoscolex gordurensis* Michaelsen, 1918	Gorduras, near Belo Horizonte, MG; Campos do Jordão, Itanhaém, Ribeirão Pires, Santo André, SP	Native	Michaelsen (1918, 1925), Righi (1999)
71a	*Glossoscolex grandis** Michaelsen, 1892	Passo Fundo, RS	Native	Michaelsen (1892, 1918)
71b	*Glossoscolex grandis ibirai* Righi, 1971	Ibirá, SP	Native	Righi (1971a), Caballero (1973)
72	*Glossoscolex grecoi* Righi and Lobo, 1979	Pirassununga, Vassununga SP	Native	Righi and Lobo (1979)
73	*Glossoscolex jimi* Righi, 1972	Cedros, SC (near Blumenau?)	Native	Righi (1972a)
74	*Glossoscolex klossae** Righi, 1972	Ilha da Gipóia, RJ	Native	Righi (1972a)
75	*Glossoscolex matogrossensis* Righi, 1984	Foz do Iguaçú, PR; Sidrolândia, Maracajú, MS	Native	Righi (1984b), Zicsi and Csuzdi (1987)
76	*Glossoscolex montagneri** Righi, 1972	Itaguaí, RJ; São Sebastião, SP	Native	Righi (1972a), Zicsi and Csuzdi (1999)
77	*Glossoscolex mrazi* Cernosvitov, 1934	São Paulo (?),[18] SP	Native	Cernosvitov (1934b)
78	*Glossoscolex paulistus** Michaelsen, 1925	Piracicaba, Rio Claro, Araras, Sumaré, SP	Native	Michaelsen (1925), Righi (1971a, 1997), MZUSP
79	*Glossoscolex robustus** Cernosvitov, 1938	Teresópolis, RJ	Native	Cernosvitov (1938)
80	*Glossoscolex sazimai** Righi and Lobo, 1979	Caraguatatuba, SP	Native	Righi and Lobo (1979)
81	*Glossoscolex taunayi* Michaelsen, 1925	Serra da Bocaina, SP–RJ border	Native	Michaelsen (1925)
82	*Glossoscolex tocape* Righi, 1980	Ribeirão Preto, SP	Native	Righi (1980a)
83	*Glossoscolex truncatus* Rosa, 1895	Uruguayana, RS; Itajaí, SC	Native	Michaelsen (1925), Luederwaldt (1927)
84	*Glossoscolex tupii* Righi, 1971	Engenheiro Marsilac, SP (near São Paulo)	Native	Righi (1971a)
85	*Glossoscolex umijiae** Righi and Lobo, 1979	Cotia, SP	Native	Righi and Lobo (1979)
86a	*Glossoscolex uruguayensis uruguayensis** Cordero, 1943	São Leopoldo, RS	Native	Righi (1974)

86b	*Glossoscolex (uruguayensis) corderoi*[19]* Righi, 1968	São Manuel, Botucatú, Buri, SP	Native	Righi (1968a, 1974), Ljungström (1972b), J&B
87	*Glossoscolex vizottoi*[20]* Righi, 1971	27 counties in NW and W SP; Dourados, MS	Native	Righi (1971a, 1980b), Caballero (1973), J&B
88	*Glossoscolex wiengreeni** Michaelsen, 1897	Eldorado, Piquete, Santo André, São Paulo, Ilha Bela, SP; Itatiaia, Serra da Bocaina, RJ; Porto Alegre, Santa Maria, Guaíba, RS; Near Joinville, river Itapocu, SC	Native	Michaelsen (1897, 1918), Moreira (1903), Luederwaldt (1927), Righi (1971a), Knäpper and Porto (1979)
89	*Glossoscolex* n. sp. 1	Taciba, SP	Native	J&B
90	*Glossoscolex* n. sp. 2	Salesópolis, SP	Native	J&B
91	*Glossoscolex* n. sp. 3	Lupionópolis, PR	Native	J&B
92	*Glossoscolex* n. sp. 4	Morretes, PR	Native	J&B
93	*Glossoscolex* n. sp. 5	São Jerônimo, PR	Native	J&B
94	*Glossoscolex* n. sp. 6	Antonina, PR	Native	J&B
95	*Glossoscolex* n. sp. 7*	São Jerônimo, PR	Native	J&B
96	*Glossoscolex* n. sp. 8	Bandeirantes, PR	Native	J&B
97	*Glossoscolex* n. sp. 9*	Salesópolis, SP	Native	J&B
98	*Glossoscolex* n. sp. 10	Jaguapitã, PR	Native	J&B
99	*Glossoscolex* n. sp. 11	Buri, SP	Native	J&B
100	*Glossoscolex* n. sp. 12	Londrina, Sertanópolis, PR	Native	J&B
101	*Glossoscolex* n. sp. 13*	Ilha Bela, SP	Native	J&B
102	*Glossoscolex* n. sp. 14	Mauá, Faxinal, PR	Native	J&B
103	*Glossoscolex* n. sp. 15	Buri, SP	Native	J&B
104	*Glossoscolex* n. sp. 16	Ortigueira, PR	Native	J&B
105	*Glossoscolex* n. sp. 17	Ponta Grossa, PR	Native	J&B
106	*Glossoscolex* n. sp. 18*	Campina Grande do Sul, PR	Native	J&B
107	*Glossoscolex* n. sp. 19	Lages, Campo Belo do Sul, SC	Native	J&B
108	*Glossoscolex* n. sp. 20*	Rio de Janeiro, RJ	Native	J&B
109	*Glossoscolex* n. sp. 21[21]	Lupionópolis, Centenário do Sul, Londrina, PR	Native	J&B
110	*Glossoscolex* n. sp. 22*	Itaguajé, PR	Native	J&B
111	*Glossoscolex* n. sp. 23	Primeiro de Maio, PR	Native	J&B
	Glossoscolex spp.	Jaguapitã, Cafeara, PR; Assistência, SP; Camaquã, RS	Native	J&N, R&R

Continued

Table 4.1. List of earthworm genera and species found in Brazil, together with their distribution and origin. Large-bodied earthworms (minhocuçu) are identified with an asterisk.[1] – cont'd

Number	Family (Genus species)	Sites/states found[2]	Origin	References[3]
112	*Goiascolex cabrelli* Righi, 1971	Porangatu, Paraíso do Norte (near Brasília), GO; Near Oriente Novo, RO	Native	Righi (1971a, 1988a)
113	*Goiascolex edgardi* Righi, 1986	Near Pimenta Bueno, RO	Native	Righi (1986b)
114	*Goiascolex pepus* Righi, 1972	7 sites in MT, Various sites in RO	Native	Righi (1972b, 1984b,d, 1986b), Righi and Guerra (1985)
115	*Goiascolex vanzolinii* Righi, 1984	Vila Bela de Santíssima Trindade, Mato Grosso, MT	Native	Righi (1984b, 1990a)
116	*Goiascolex* n. sp.1[22]	Sorocaba, SP	Native	J&B
117	*Holoscolex caramuru* Righi, 1975	Porto Velho, Ouro Preto d'Oeste, RO; lower river Matapi, AP; PN Amazônia, PA; Lago Calado, AM	Native	Righi (1975, 1982a, 1988a), Righi *et al.* (1978)
118	*Holoscolex nemorosus tacoa* Righi *et al.*, 1978	Near Manaus, AM	Native	Righi *et al.* (1978)
119	*Maipure*[23] *matapi* Righi, 1969	Lower river Matapi, AP; near Manaus, AM; near Rio Branco, AC	Native	Righi (1969, 1971a, 1996), Adis and Righi (1989)
120	*Martiodrilus duodenarius* Michaelsen, 1918	Serra do Navio, AP	Native	Righi (1971a)
121	*Onychochaeta serieia* Righi, 1971	Porangatu, Paraíso do Norto, GO	Native	Righi (1971a, 1972c)
122	*Opisthodrilus adneae* Righi, 1984	Cáceres, MT	Native	Righi (1984d)
123a	*Opisthodrilus*[24] *borelli borelli* Rosa, 1895	Near Cuiabá, Cáceres, Sonho Azul, MT; Sidrolândia, Miranda, Bela Vista, MS	Native	Righi (1972b, 1984b,d)
123b	*Opisthodrilus borelli tuberculiferus* Righi, 1984	Poconé, MT	Native	Righi (1984b)
124	*Opisthodrilus rhopalopera* Cognetti, 1906	Rio Preto, MG	Native	Cognetti de Martiis (1906)
125	*Pontoscolex corethrurus* Müller, 1857	Most widespread sp. in Brazil; found in AC, AM, AP, BA, DF, ES, GO, MA, MG, MS, MT, PA, PB, PE, PR, RJ, RO, RR, RS, SC, SE, SP	Native	Benham (1890), Cognetti de Martiis (1900), Moreira (1903), Michaelsen (1918), Luederwaldt (1927), Cernosvitov (1934a, 1935), Vanucci (1953), Lenko (1972), Caballero (1973), Righi (1971a, 1980b,1982a, 1984b–e, 1988a,b,1990b, 1997, 1998a),

				Righi *et al.* (1976), Knäpper (1972, 1979), Knäpper and Porto (1979), Righi and Guerra (1985), Zicsi and Csuzdi (1987), Guerra (1988b, 1982, 1994a), Guerra and Silva (1994), AG, JR, Peneireiro (1999), Zicsi *et al.* (2001), J&B, J&N
126	*Pontoscolex cuasi* Righi, 1984	Serra do Navio, AP; Belém, PA; Boa Vista, Bonfim and Ilha do Maracá, RR	Native	Righi (1984c, 1988b, 1990b, 1998a)
127	*Pontoscolex eudoxiae* Righi *et al.*, 1978	Reserva Ducke, AM	Native	Righi *et al.* (1978), Righi (1984c)
128	*Pontoscolex franzi* Zicsi and Czusdi, 1999	Capitão Poço, PA	Native	Zicsi and Csuzdi (1999)
129	*Pontoscolex maracaensis* Righi, 1984	Area near Ilha do Maracá, RR	Native	Righi (1984c, 1990b)
130	*Pontoscolex marcusi* Righi and Ayres, 1976	Rio Preto da Eva, 2 sites N of Manaus, AM	Native	Righi and Ayres (1976), Righi (1984c)
131	*Pontoscolex nogueirai* Righi, 1984	In and around Ilha do Maracá, near Bonfim, Boa Vista, RR; Capitão Poço, PA	Native	Righi (1984c, 1998a, 1990b), Zicsi and Csuzdi (1999)
132	*Pontoscolex pydanieli**[25] Righi, 1988	4 sites in N RO	Native	Righi (1988c, 1990a)
133	*Pontoscolex roraimensis*[26] Righi, 1984	Around Ilha do Maracá, near Boa Vista, Bonfim, RR	Native	Righi (1984c, 1998a), Guerra (1994)
134	*Pontoscolex vandersleeni* Michaelsen, 1933	Manaus, AM	Native	Zicsi *et al.* (2001)
135	*Rhinodrilus adelae* Cordero, 1943	Acarapé, CE	Native	Cordero (1943)
136	*Rhinodrilus alatus*[27]* Righi, 1971	Paraopeba, Sete Lagoas, MG	Native	Righi (1971a)
137	*Rhinodrilus annulatus*[28] Cernosvitov, 1934	Upper river Calçoene, AP	Native	Cernosvitov (1934a, 1935)
138	*Rhinodrilus bursiferus*[29] Righi, 1971	Serra do Navio, AP; RJ	Native	Righi (1971a), Zicsi and Csuzdi (1999)
139	*Rhinodrilus contortus* Cernosvitov, 1938	Manaus, AM	Native	Cernosvitov (1938), Zicsi *et al.* (2001)
140	*Rhinodrilus curiosus* Righi *et al.*, 1976	Near Manaus, AM; in and near Rio Branco, AC	Native	Righi *et al.* (1976), Righi and Guerra (1985), Guerra (1988b, 1994a)
141	*Rhinodrilus duseni** Michaelsen, 1918	Pilar do Sul, Itapeva, SP; Curitiba, PR; Fátima do Sul, MT	Native	Michaelsen (1918), Righi (1971a, 1974, 1984b)
142	*Rhinodrilus elisianae* Righi *et al.*, 1976	Sucunduri, AM; many sites in RO; Capitão Poço, Belém, PA	Native	Righi *et al.* (1976), Righi (1986b, 1988a, 1990a), Zicsi and Csuzdi (1999)

Continued

Table 4.1. List of earthworm genera and species found in Brazil, together with their distribution and origin. Large-bodied earthworms (minhocuçu) are identified with an asterisk.[1] – cont'd

Number	Family (Genus species)	Sites/states found[2]	Origin	References[3]
143	*Rhinodrilus evandroi** Righi, 1971	Brasília, DF	Native	Righi (1971a)
144	*Rhinodrilus fafner*[30]* Michaelsen, 1918	Near Belo Horizonte, MG	Native	Michaelsen (1918)
145	*Rhinodrilus fransisci* Cordero 1944	Sabiucá, PE	Native	Cordero (1944)
146	*Rhinodrilus garbei*[31]* Michaelsen, 1925	Pirapora, PE Ibitipoca, MG; Botucatú, SP	Native	Michaelsen (1925), Luederwaldt (1927), Castro and d'Agosto (1999)
147	*Rhinodrilus hoeflingae* Righi, 1980	Cachoeira dos Macacos, Caetanópolis, PE Ibitipoca, MG	Native	Righi (1980b), Castro and d'Agosto (1999)
148	*Rhinodrilus horsti*[32]* Beddard, 1892	MG? (site unknown)	Native	Michaelsen (1918), Beddard (1892)
149	*Rhinodrilus jucundus* Righi, 1985	Paraíso do Norte (near Brasília), GO; lower river Tocantins, PA	Native	Righi (1985, 1989b)
150	*Rhinodrilus lakei** Michaelsen, 1934	Catrimani, RR, around Manaus, AM	Native	Michaelsen (1934)
151	*Rhinodrilus longus** Cernosvitov, 1934	Lower river Calçoene, AP	Native	Cernosvitov (1934a, 1935)
152	*Rhinodrilus lourdesae* Righi, 1986	Near Ouro Preto d'Oeste, RO	Native	Righi (1986b)
153	*Rhinodrilus lucilleae* Righi *et al.*, 1976	Sucunduri, AM	Native	Righi *et al.* (1976)
154	*Rhinodrilus mamita* Cordero, 1943	Maranguape, CE	Native	Cordero (1943)
155	*Rhinodrilus marcusae* Righi, 1985	Porto de Mandioca (near Cruz das Almas), BA	Native	Righi (1985)
156	*Rhinodrilus mortis* Righi, 1972	São José da Serra, MT	Native	Righi (1972b)
157	*Rhinodrilus motucu*[33]* Righi, 1971	Cuiabá, Poconé, MT; Porangatu, Uruaçu, GO; Una, Itajubá, Itagibá, Jequié, BA; Umbaúba, SE	Native	Righi (1971a,b, 1984b, 1985)
158	*Rhinodrilus panxin*[34] Righi, 1971	Marabá, Geladinho, São João do Araguaia, PA; Porangatu, GO	Native	Righi (1971a, 1974)
159	*Rhinodrilus pitun** Righi, 1989	Buritizal da Corrente (near Recife), PE	Native	Righi and Moraes (1990)
160	*Rhinodrilus priollii** Righi, 1967	Various sites in and near Manaus, AM	Native	Righi (1967b), Zicsi *et al.* (2001)
161	*Rhinodrilus romani* Michaelsen, 1928	Upper river Negro, AM	Native	Michaelsen (1928)
162	*Rhinodrilus senckenbergi* Michaelsen, 1931	Region of the river Doce, ES; PE Ibitipoca, MG? (record uncertain)	Native	Michaelsen (1931), Castro and d'Agosto (1999)
163	*Rhinodrilus xeabaibus** Righi, 1969	Itatiaia, Mauá, RJ	Native	Righi (1985)
164	*Righiodrilus aioca* Righi, 1975	Serra do Navio, AP	Native	Righi (1975)

165	*Righiodrilus amazonius* Zicsi and Csuzdi, 1999	Capitão Poço, PA	Native	Zicsi and Csuzdi (1999)
166	*Righiodrilus arapaco* Righi, 1982	Ilha do Maracá, RR	Native	Righi (1982b)
167	*Righiodrilus cigges* Righi, 1970	Serra do Navio, AP	Native	Righi (1975)
168	*Righiodrilus dithecae* Righi, 1988	River Matapi, AP	Native	Righi (1988b)
169	*Righiodrilus fontebonensis* Righi, 1988	Fonte Boa, AM	Native	Righi (1988b)
170	*Righiodrilus freitasi* Righi, 1971	Serra do Navio, AP	Native	Righi (1971a, 1975)
171	*Righiodrilus itajo* Righi, 1971	Araras, SP	Native	Righi (1971a, 1975)
172	*Righiodrilus mairaro* Righi, 1982	Bonfim, RR	Native	Righi (1982b)
173	*Righiodrilus marcusae* Righi, 1969	Ilha do Marajó, PA	Native	Righi (1968b)
174	*Righiodrilus mucupois* Righi, 1970	Serra do Navio, AP	Native	Righi (1970)
175	*Righiodrilus oliveirae* Righi, 1982	Boa Vista, Ilha de Maracá, RR	Native	Righi (1982b, 1998a)
176	*Righiodrilus ortonae* Righi, 1988	Belém, PA	Native	Righi (1988b)
177	*Righiodrilus schubarti* Righi *et al.*, 1978	Novo Airão, AM	Native	Righi *et al.* (1978)
178	*Righiodrilus sucunduris* Righi *et al.*, 1976	Sucunduri, Coari, AM; Serra do Navio, AP; PN Amazônia, PA	Native	Righi (1982a, 1988b), Righi *et al.* (1976)
179	*Righiodrilus tico* Righi, 1982	Tefé, Tabatinga, AM; Ilha de Maracá, RR	Native	Righi (1982b, 1988b, 1998a)
180	*Righiodrilus tinga* Righi, 1971	Serra do Navio, AP	Native	Righi (1971a, 1975)
181a	*Righiodrilus tocantinensis tocantinensis* Righi, 1972	Cametá, Geladinho, Itupiranga, Marabá, Mocajuba, Tucuruí, Pato, PA; Pontes e Lacerda, MT; João Pessoa, PB	Native	Righi (1975), Righi and Guerra (1985), Guerra and Silva (1994)
181b	*Righiodrilus tocantinensis pola* Righi, 1984	Vila Bela de Santíssima Trindade, Pontes e Lacerda, MT	Native	Righi (1984d)
182	*Righiodrilus uete* Righi, 1988	Mirante da Serra, RO	Native	Righi (1988a)
183	*Righiodrilus venancioi* Righi, 1982	PN Amazônia, PA	Native	Righi (1982a)
184	*Thamnodrilus ohausi* Michaelsen, 1918	Manaus, AM	Native	Michaelsen (1918, 1934)
185	*Thamnodrilus salathei** Michaelsen, 1934	Catrimani, RR; around Manaus, AM	Native	Michaelsen (1934), Righi (1971a)
186	*Tuiba dianae*[35] Righi *et al.*, 1976	Various sites near Manaus, AM	Native	Righi *et al.* (1976), Adis and Bogen (1982), Adis and Righi (1989), Righi (1989a, 1997), Zicsi *et al.* (2001)
187	*Tupinaki*[36] *bokermanni* Righi, 1971	Paranapiacaba, SP	Native	Righi (1971a)
188	*Tupinaki parini* Righi, 1969	Tripuí (near Ouro Preto), MG	Native	Righi (1968b)

Continued

Table 4.1. List of earthworm genera and species found in Brazil, together with their distribution and origin. Large-bodied earthworms (minhocuçu) are identified with an asterisk.[1] – cont'd

Number	Family (Genus species)	Sites/states found[2]	Origin	References[3]
189a	*Urobenus brasiliensis*[37] Benham, 1887	Antonina, Campina Grande do Sul, Faxinal, Foz do Iguaçu, Londrina, Mauá, Sertanópolis, PR; Nova Friburgo, Petrópolis, Teresópolis, Itatiaia, Mendes, RJ; Botucatú, Cubatão, Jundiaí, Santo André, São Bernardo do Campo, SP; Itaquí, Taquara, Porto Alegre, Turuçu, RS; São Luís, MA; Gorduras (near Belo Horizonte), Conceição do Mato Dentro (Serra do Cipó), Chapéu de Sol, MG; São José da Serra, MT; Itupiranga, PA; Manaus, AM; Near Joinville, Rio dos Cedros, river Itapocu, SC	Native	Benham (1887), Üde (1893), Luederwaldt (1927), Cernosvitov (1934a, 1935), Righi (1971a,b, 1972b, 1974, 1980a, 1985), Zicsi *et al.* (2001), J&B, MZUSP
189b	*Urobenus* sp.[38]	Londrina, Sertanópolis, PR	Native	J&B
190	*Urobenus buritis Righi et al.*, 1976	Manaus area, AM	Native	Righi *et al.* (1976, 1985)
191	*Urobenus gitus* Righi, 1971	Belém, PA	Native	Righi (1971a)
192	*Urobenus igpigpuera* Righi, 1982	PN Amazônia, PA	Native	Righi (1982a)
193	*Urobenus petrerei* Righi, 1985	São Luís, MA; Bagagem, PA	Native	Righi (1985, 1989b)
Almidae				
194	*Criodrilus lacuum* Hoffmeister, 1845	Triunfo, RS	Exotic	Knäpper and Porto (1979)
195	*Criodrilus* (?)[39] n. sp. 1	Camaquã, RS	Native?	R&R
196	*Criodrilus* (?) n. sp. 2	Camaquã, RS	Native?	R&R
197	*Drilocrius dreheri* Michaelsen, 1925	Franca, Tanabi, Neves Paulista, São José do Rio Preto, Mirassol, SP	Native	Michaelsen (1925), Caballero (1973)
198	*Drilocrius iheringi*[40] Michaelsen, 1895	Piracicaba, SP	Native	Michaelsen (1925), Luederwaldt (1927)
199	*Drilocrius* n. sp. 1	Jaguapitã, PR	Native	J&B
200	*Drilocrius* n. sp. 2	Bandeirantes, PR	Native	J&B
201	*Glyphidrilocrius ehrhardti*[41] Michaelsen, 1925	Manaus, Manacapuru, AM	Native	Michaelsen (1925), Adis and Righi (1989)
Ocnerodrilidae				
202	*Bauba santosi* Righi, 1980	Umbaúba, SE	Native	Righi (1980a)
203	*Belladrilus arua* Righi, 1984	Pontes e Lacerda, MT	Native	Righi (1984d)

204	*Belladrilus otarion* Righi, 1995	Iporanga, SP	Native	Righi (1995b)
205	*Belladrilus pocaju* Righi, 1984	Maracajú, Terenos, MS; Poconé, MT	Native	Righi (1984a)
206	*Belladrilus* n. sp. 1	Jaguapitã, PR	Native	J&N
207	*Brunodrilus angeloi* Righi, 1971	Serra do Cipó, MG	Native	Righi (1971c)
208	*Dariodrilus ferrarius* Righi *et al.*, 1978	Sucunduri, AM	Native	Righi *et al.* (1978)
209	*Eukerria asilis*[42] Righi, 1967	Ilha de Marajó, PA; Cabo, PE	Native?	Righi (1967a, 1971b)
210	*Eukerria cuca* Righi, 1984	Cuibá, MT	Native	Righi (1984a)
211	*Eukerria eiseniana*[43] Rosa, 1895	Near Pontes e Lacerda, Cuiabá, Cáceres, MT; Ilha de Maracá, near Bonfim, RR; Presidente Médici, Pimenta Bueno, RO; Ledário, Bela Vista, Terenos, MS; Botucatú, SP; Camaquã, RS; Jaguapitã-PR	Native?	Righi (1972b, 1984a,d, 1988a), Righi and Guerra (1985), MZUSP, R&R, J&B
212	*Eukerria emete* Righi and Guerra, 1985	4 sites in MT; Paiquerê, PR	Native	Righi and Guerra (1985), J&B
213	*Eukerria garmani argentinae* Jamieson, 1970	Estrela, Camaquã, RS	Native?	Righi and Ayres (1975), R&R
214	*Eukerria guamais* Righi, 1971	Manaus, AM; Belém and PN Amazônia, PA; Pimenta Bueno, RO	Native	Righi (1971b, 1983, 1988a)
215	*Eukerria kukenthali* Michaelsen, 1908	Codajás, AM	Exotic	Righi (1988b)
216	*Eukerria mucu* Righi, 1988	Pimenta Bueno, Mirante da Serra, RO	Native	Righi (1988a)
217	*Eukerria saltensis*[44] Beddard, 1895	São Paulo, SP; Blumenau, SC; MG; Camaquã, RS; SC; Jaguapitã, PR	Exotic	Michaelsen (1927), Gates (1972), Righi (1968b, 1971b, 1999), Ljungström *et al.* (1975), R&R, J&N
218	*Eukerria stagnalis* Kinberg, 1867	Estrela, Porto Alegre, Camaquã, RS; Ilha Bela, SP	Native? probably peregrine	Michaelsen (1927), Righi and Ayres (1975), MZUSP, R&R
219	*Eukerria subandina*[45] Rosa, 1895	Corumbá, MS; Mirante da Serra, RO	Native? wide-spread	Righi (1984a, 1988a), Cognetti de Martiis (1900)
220	*Eukerria taisa* Righi, 1983	PN Amazônia	Native	Righi (1983)
221	*Eukerria urna* Righi, 1967	Codajás, AM; Itajubá, BA; Ilha do Marajó (near Cachoeira do Arari), PA, Pimenta Bueno, RO; Bonfim, RR	Native	Righi (1967a, 1971b, 1988a,b), Righi and Guerra (1985)
222	*Exsidrilus rarus* Righi *et al.*, 1978	Sucunduri, AM	Native	Righi *et al.* (1978)
223	*Gordiodrilus habessinus* Michaelsen, 1913	Near Vilhena, near Ariquemes, Pimenta Bueno, RO; Near Pontes e Lacerda, Cáceres, Nova Alvorada, MT; Ladário, MS	Exotic	Righi (1984a,d, 1988a), Righi and Guerra (1985)

Continued

Table 4.1. List of earthworm genera and species found in Brazil, together with their distribution and origin. Large-bodied earthworms (minhocuçu) are identified with an asterisk.[1] – cont'd

Number	Family (Genus species)	Sites/states found[2]	Origin	References[3]
224	*Gordiodrilus marcusi* Righi, 1968	São Paulo, Birigui, Rio Claro, SP; Britânia, GO	Native?	Righi (1968a), MZUSP
225	*Gordiodrilus paski* Stephenson, 1928	São Paulo, SP	Exotic	Righi (1968b)
226	*Haplodrilus amazonicus* Righi, 1983	PN Amazônia	Native	Righi (1983)
227	*Haplodrilus iheringi* Michaelsen, 1925	Piracicaba, SP	Native	Michaelsen (1925)
228	*Haplodrilus michaelseni*[46] Cognetti, 1900	Corumbá, MS; Londrina, PR	Native	Michaelsen (1927), J&N
229	*Haplodrilus tagua* Righi *et al.*, 1978	Sucunduri, AM	Native	Righi *et al.* (1978)
230	*Haplodrilus* n. sp. 1	Londrina, PR	Native	J&B
231	*Kerriona garbei** Michaelsen, 1924	Porto Cachoeira, ES	Native	Michaelsen (1924), Luederwaldt (1927)
232	*Kerriona limae*[47] Righi, 1980	Salesópolis, SP	Native	Righi (1980b)
233	*Kerriona luederwaldti*[48] Michaelsen, 1924	Itatiaia, RJ	Native	Luederwaldt (1927)
234	*Kerriona* sp. 1	Antonina, PR	Native	J&B
235	*Kerriona* sp. 2[49]	Matinhos, PR	Native	J&B
236	*Kerriona* sp. 3[50]	Morretes, PR	Native	J&B
237	*Liodrilus ipu* Righi, 1975	Belém, PA	Native	Righi (1975)
238	*Liodrilus mendesi* Righi, 1994	João Pessoa, PB	Native	Righi (1994)
239	*Lourdesia paraibaensis* Righi, 1994	João Pessoa, PB	Native	Righi (1994)
240	*Nematogenia lacuum* Beddard, 1893	Cacoal, Pimenta Bueno, Espigão d'Oeste, Ouro Preto do Oeste, RO; In and N of Pontes e Lacerda, Tabuleta, Cáceres, Vila Bela da Santíssima Trindade, MT	Exotic?	Righi (1984d, 1988a), Righi and Guerra (1985)
241	*Nematogenia panamaensis* Eisen, 1900	Botucatu, SP; Salvador, BA	Exotic	MZUSP
242	*Ocnerodrilus ibemi* Righi, 1968	São Sebastião, SP	Native	Righi (1968b)
243	*Ocnerodrilus occidentalis*[51] Eisen, 1878	Ilha de Marajó, PA; Codajás, AM; Poconé, MT; Bela Vista, MS; Lauro Müller, SC, São Paulo, SP; Jaguapitã, PR	Exotic	Righi (1968b, 1984a, 1988b), MZUSP, J&N
244	*Ocnerodrilus potyuara* Righi, 1994	Mari, PB	Native	Righi (1994)
245	*Ocnerodrilidae* sp. 1	Centenário do Sul, PR	Native?	J&B
246	*Ocnerodrilidae* sp. 2	Jaguapitã, PR	Native	J&N
247	*Paulistus taunayi*[52]* Michaelsen, 1925	Itabuna, BA	Native	Michaelsen (1925)
248	*Pygmaeodrilus amapaensis* Righi, 1988	Serra do Navio, AP	Native	Righi (1988b)

Eudrilidae				
249	*Eudrilus eugeniae*[53] Kinberg, 1867	Itajubá, Jequié, Ilha de Itaparica, BA; Petrópolis, Rio de Janeiro, Nova Friburgo, RJ; Ponta de Pedras, Recife, PE; Maiautá, PA; São Sebastião, Boituva, Campinas, Vinhedo, São Paulo, SP; Primeiro de Maio, Londrina, Ibiaci, PR; Areia, PB; Juiz de Fora, MG (vermiculture); Aracajú, SE; São Luís, MA	Exotic	Beddard (1891), Moreira (1903), Luederwaldt (1927), Gates (1954), Righi (1967e, 1968b, 1971b, 1972b), Guerra and Silva, J&B, MZUSP
250	*Hyperiodrilus africanus* Beddard, 1891	Ponta de Pedras, PE	Exotic	Righi (1972b)
Lumbricidae				
251	*Aporrectodea caliginosa*[54] Savigny, 1826	Canela, Estrela, Guaíba, Nova Petrópolis, Porto Alegre, Rolante, São Leopoldo, Sapucaia do Sul, Santa Cruz do Sul, São Francisco de Paula, Sobradinho, Viamão, RS	Exotic	Righi (1967c), Knäpper and Hauser (1969), Knäpper and Porto (1979), MZUSP
252	*Aporrectodea rosea*[55] Savigny, 1826	Porto Alegre, RS	Exotic	Michaelsen (1892)
253	*Aporrectodea trapezoides* Dugès, 1828	Porto Alegre, RS	Exotic	Michaelsen (1892)
254	*Bimastos parvus* Eisen, 1874	Buri, Anhembi, SP; Nova Teutônia, Camaquã, RS	Exotic	Cernosvitov (1942), Righi (1968a), J&B, R&R
255	*Dendrobaena veneta* Rosa, 1886	Porto Alegre, RS	Exotic	Knäpper and Porto (1979)
256	*Dendrodrilus rubidus rubidus* Savigny, 1826	Itatiaia, Petrópolis, Rio de Janeiro, RJ	Exotic	Michaelsen (1927), Gates (1972), Righi (1980b)
257	*Eisenia andrei*[56] Bouché, 1972	Various sites in SP, PR, RJ and MG (vermiculture)	Exotic	GB, AG
258	*Eisenia fetida*[57] Savigny, 1826	Guaíba, Ivotí, Lageado, Porto Alegre, São Leopoldo, Sapucaia do Sul, Tramandaí, RS; perhaps various sites in SP, PR, RJ, MG and SC (vermiculture)	Exotic	Knäpper (1872a), Michaelsen (1892), Righi (1967c), Knäpper and Porto (1979)
259	*Eisenia lucens* Waga, 1857	Santo Ângelo, Fontoura Xavier, Porto Alegre, São Francisco de Paula, RS	Exotic	Knäpper and Porto (1979)
260	*Eiseniella tetraedra* Savigny, 1826	Several counties in the regions of Itá Machadinho and Campos Novos, SC and RS	Exotic	Pacheco *et al.* (1992)
261	*Octolasion cyaneum* Savigny, 1826	Pelotas, Porto Alegre, São Leopoldo, RS	Exotic	Righi (1967c), MZUSP

Continued

Table 4.1. List of earthworm genera and species found in Brazil, together with their distribution and origin. Large-bodied earthworms (minhocuçu) are identified with an asterisk.[1] – cont'd

Number	Family (Genus species)	Sites/states found[2]	Origin	References[3]
Sparganophilidae				
262	*Areco reco*[58] Righi *et al.*, 1978	Reserva Ducke, AM	Native	Righi *et al.* (1978)
Megascolecidae				
263	*Amynthas aeruginosus* Kinberg, 1867	Prudentópolis, PR	Exotic	J&B, MZUSP
264	*Amynthas aspergillum*[59] Perrier, 1872	São Paulo, SP	Exotic	Righi (1967d)
265	*Amynthas corticis*[60] Kinberg, 1867	Serra do Cipó, Juiz de Fora, MG; 16 counties in PR; PN Itatiaia, Seropédica, Nova Friburgo, RJ; 9 counties in RS; 6 counties in SP	Exotic	Gates (1954), Knäpper (1977), Knäpper and Porto (1979), Righi (1980b), Voss (1986), Krabbe *et al.* (1993), GB, Zicsi and Csuzdi (1999) Ressetti (2004), J&B
266	*Amynthas gracilis*[61] Kinberg, 1867	Manaus, AM; Ituberá, BA; 5 counties in MG; Belém, PA; Areia, PB; 15 counties in PR; 6 counties in RJ; 23 counties in RS; Blumenau, Schroeder, SC; 39 counties in SP	Exotic	Beddard (1891), Michaelsen (1892, 1900, 1903), Rosa (1894), Moreira (1903), Luederwaldt (1927), Cernosvitov (1934a, 1935), Vanucci (1953), Gates (1954), Righi and Knäpper (1965), Righi (1967d, 1980b,1997), Knäpper (1972a,b), Lenko (1972), Caballero (1973), Knäpper and Porto (1979), Voss (1986), Krabbe *et al.* (1993), Guerra and Silva (1994), Chang (1997), Peneireiro (1999), Zicsi and Csuzdi (1999), Ressetti (2004), R&R, AG, GB, MZUSP, J&B
267	*Amynthas morrisi* Beddard, 1892	Salvador, BA; 15 counties in RS; Curitiba, Castro, PR	Exotic	Righi (1971b), Knäpper (1972a,b), Knäpper and Porto (1979), Krabbe *et al.* (1993), Chang (1997), Ressetti (2004)

268	*Metaphire californica* Kinberg, 1867	Piracicaba, São Paulo, SP; Caetanópolis, MG; Castro, Curitiba, PR; Rio de Janeiro, RJ; 20 counties in RS; Salvador, BA; Lauro Müller, SC	Exotic	Moreira (1903), Luederwaldt (1927), Righi (1971b, 1980b), Knäpper (1972a,b), Knäpper and Porto (1979), Krabbe *et al.* (1993), Chang (1997), Ressetti (2004), MZUSP
269	*Metaphire schmardae*[62] Horst, 1883	Porto Alegre, Estância Velha, Canoas, São Leopoldo, RS; Pomerode, Blumenau, SC; Curibita, PR; Colina, Cotica, São Paulo, SP; Teresópolis, PN Itatiaia, RJ	Exotic	Michaelsen (1927), Knäpper (1972a,b), Hauser *et al.* (1975), Knäpper and Porto (1979), Righi (1967d, 1980b), MZUSP
270	*Pheretima darnleiensis*[63] Fletcher, 1886	Campos do Jordão, São Sebastião, São Paulo, São José do Rio Preto, Engenheiro Marsilac, Salesópolis, SP; Curitiba, PR; 15 counties in RS; Conceição de Mato Dentro, Tripuí (near Ouro Preto), MG	Exotic	Righi and Knäpper (1965, 1966), Righi (1967d, 1980b), Knäpper (1972a,b), Caballero (1973), Chang (1997), MZUSP
271	*Polypheretima elongata* Perrier, 1872	Caetanópolis, Curvelo, Cachoeira dos Macacos, MG; Recife, PE; Itajubá, BA; Anhembi, SP	Exotic	Righi (1971b, 1980b), J&B
272	*Polypheretima taprobanae* Beddard, 1892	Rio de Janeiro, RJ; São Paulo, Piracicaba, Paranapiacaba, SP; Santa Cruz do Sul, São Leopoldo, RS	Exotic	Moreira (1903), Luederwaldt (1927), Righi (1967d)
273	*Pontodrilus litoralis*[64] Grubbe, 1855	Several sites along S coast in SP, SC, RJ, RS; Ilha de Itamaracá, PE	Exotic	Michaelsen (1900, 1910), Moreira (1903), Luederwaldt (1927), Righi (1968b), MZUSP
Acanthodrilidae				
274	*Chilota* sp.[65]	Barueri, SP	Exotic	MZUSP, Lenko (1972)
275	*Dichogaster affinis* Michaelsen, 1890	Areia, PB; Itaguaí, RJ; Jaguapitã, Arapotí, Londrina, PR; around Manaus, AM; Calçoene and Lower river Calçoene, AP; Poconé, Pontes e Lacerda, Chapada dos Guimarães, MT; Inhaúma, Curvelo, MG; Jequié, BA	Exotic	Cernosvitov (1934a, 1935), Righi (1968b, 1971b, 1980b, 1984a,d,e, 1990a), Righi *et al.* (1978), Guerra and Silva (1994), G.G. Brown, personal observation, R&N, J&N

Continued

Table 4.1. List of earthworm genera and species found in Brazil, together with their distribution and origin. Large-bodied earthworms (minhocuçu) are identified with an asterisk.[1] – cont'd

Number	Family (Genus species)	Sites/states found[2]	Origin	References[3]
276	*Dichogaster andina*[66] Cognetti, 1904	Rio Parú do Oeste, Jacundá, Canoal, PA; Rio Preto da Eva, several sites near Manaus, river Negro (border AM-RR), AM	Exotic?	Righi *et al.* (1978), Righi (1988b), Adis and Righi (1989), Zicsi and Csuzdi (1999)
277	*Dichogaster annae*[67] Horst, 1893	Blumenau, Florianópolis, SC; Uruçucá, BA; São Paulo, Osasco, SP; Chapada dos Guimarães, MT; RS	Exotic	Luederwaldt (1927), Righi (1968b, 1984a,e, 1999), Righi and Ayres (1975)
278	*Dichogaster badajos* Righi *et al.*, 1978	Lake Badajós region, AM	Native	Righi *et al.* (1978)
279	*Dichogaster bolaui*[68] Michaelsen, 1891	Rio Branco, AC; In and near Manaus, Huitanaã (on river Purús), AM; Corumbá, Urucúm, Carandazinho, MS; Lower river Calçoene, AP; Itabuna, Itajubá, Jequié, BA; Caxias, MA; Cachoeira dos Macacos, Jabuticatubas, Paraopeba, Tripuí, MG; 9 sites in MT; Belém, Mocajuba, Cocal (no river Tocantins), PN Amazônia, PA; Castro, Jaguapitã, Arapotí, PR; Ariquemes, Mirante da Serra, Jí-Paraná, RO; Ilha de Maracá, Bonfim, RR; Anhembi, Botucatú, Campos do Jordão, Mirassol, Paraibuna, Guarujá, São Paulo, Taciba, SP; Florianópolis, Lauro Müller, SC; Ilha de Itamaracá, PE	Exotic	Cognetti de Martiis (1900), Cernosvitov (1934a, 1935), Righi (1968b, 1971b, 1972b, 1980b, 1984a,d,e, 1988b, 1990a, 1997), Lenko (1972), Caballero (1973), Righi *et al.* (1978), Righi and Guerra (1985), Zicsi and Csuzdi (1999), Ressetti (2004), R&N, MZUSP, J&B
280	*Dichogaster gracilis* Michaelsen, 1892	Jí-Paraná, Riozinho, Vilhena, Pimenta Bueno, RO; João Pessoa, PB; Manaus, AM; Pontes e Lacerda, Serra de Campina, Vila Bela de Santíssima Trindade, MT, Cafeara, PR	Exotic	Michaelsen (1928), Righi and Guerra (1985), Righi (1988a,b, 1984d), Guerra and Silva (1994), J&B
281	*Dichogaster ibaia* Righi *et al.*, 1978	60 km N of Manaus, AM	Native	Righi *et al.* (1978)
282	*Dichogaster modiglianii* Rosa, 1896	In and N of Pontes e Lacerda, Serra da Campina, MT; Manaus, Chicago (on river Japurá), AM; Arapotí, PR; Ilha de Maracá, RR	Exotic	Righi *et al.* (1978), Righi (1984d, 1990a, 1998a), Righi and Guerra (1985)

283	*Dichogaster saliens* Beddard, 1892	Rio Branco, AC; In and near Manaus, Tefé, AM; Itajubá, Jequié, BA; Caxias, MA; Cachoeira dos Macacos, Prado, Paraopeba, MG; Bataguaçu, Terenos, MS; Chapada dos Guimarães, Cuiabá, Poconé, Pontes e Lacerda, Vila Bela da Santíssima Trindade MT; Belém, Mocajuba, PA; Jaguapitã, Cafeara, PR; Itaguaí, RJ; Pimenta Bueno, Cacoal, RO; Ibirubá, Fontoura Xavier, RS; Mirassol, Botucatú, São Paulo, Colina, SP	Exotic	Righi (1968b, 1971b, 1972b, 1980b, 1984a,d,e, 1988b, 1990a), Caballero (1973), Righi *et al.* (1978), Knäpper and Porto (1979), Righi and Guerra, (1985), J&B, R&N
284	*Microscolex dubius* Fletcher, 1887	São Paulo, SP; Taquara, Camaquã, RS	Exotic	Üde (1893), Luederwaldt (1927), Ljungström *et al.* (1975), R&R
285	*Microscolex phosphoreus* Dugès, 1837	RS (site not specified)	Exotic	Moreira (1903), Cognetti de Martüs; (1905), Michaelsen (1927)
286	*Neogaster aidae* Righi, 1975	Lower river Matapi, AP	Native	Righi (1975)
287	*Neogaster americana*[69] Cernosvitov, 1934	Lower river Calçoene, AP	Native	Cernosvitov (1934a, 1935)
288	*Neogaster angeloi* Righi, 1988	Serra do Navio, AP	Native	Righi (1988b)
289	*Neogaster gavrilovi* Righi and Caballero, 1970	Serra do Navio, AP	Native	Righi and Caballero (1970)
290	*Pickfordia*[70] *divergens itapecu* Righi *et al.*, 1978	Sucunduri, AM; Jacundá, PA; Chapada dos Guimarães, MT	Native	Righi *et al.* (1978), Righi (1984e, 1989b)
291	*Pickfordia tocaya* Righi *et al.*, 1978	Reserva Ducke, AM	Native	Righi *et al.* (1978)
292	*Wegeneriona belenensis* Righi, 1988	Belém, PA	Native	Righi (1988b)
293	*Wegeneriona brasiliana* Cernosvitov, 1939	Óbidos, PA	Native	Cernosvitov (1939)
294	*Wegeneriona cernosvitovi* Righi and Caballero, 1970	Serra do Navio, AP	Native	Righi and Caballero (1970)
295	*Wegeneriona michaelseni* Cernosvitov, 1934	Lower river Calçoene, AP	Native	Cernosvitov (1934a, 1935)

[1] Large-bodied earthworm (minhocuçú), i.e. greater than 30-cm length and around 1 cm diameter or more.

[2] States are abbreviated according to official abbreviations adopted in Brazil; PN = National Park; PE = State Park.

[3] MZUSP = Museu de Zoologia, Universidade de São Paulo, collection or Righi's collection, now deposited at MZUSP; personal observations as follows: J&B = S.W. James and G.G. Brown; GB = G. Brown; J&N = S.W. James and D.H. Nunes; R&N = C. Rodriguez and D.H. Nunes; AG = A. Guimarães; R&R = A.C. Rodrigues and C. Rodriguez; JR = Jöerg Römbke.

[4] The full description of this earthworm was not found, and the summary published by Ljungström (1972a) must be expanded to provide futher information on this species and its differentiation with the only other known *Alexidrilus* (*lourdesae*).

[5] Righi (1993) separated the genus *Andiorrhinus* into four subgenera: *Amazonidrilus* (containing the Brazilian species *A. amazonius, A. planaria, A. tarumanis, A. rondoniensis, A. paraguayensis, A. pauate, A. bucki, A. holmgreni, A. evelineae* and *A. torquemadai*), *Turedrilus* (containing the Brazilian species *A. samuelensis, A. caudatus* and *A. amaparis*), *Andiorrhinus* (containing the Brazilian species *A. pictus, A. proboscideus* and *A. rubescens*) and *Meridrilus* (with no Brazilian representatives).

Continued

Table 4.1. List of earthworm genera and species found in Brazil, together with their distribution and origin. Large-bodied earthworms (minhocuçu) are identified with an asterisk.[1] – cont'd

[6] Found in bromeliad by Michaelsen (1934). May be a permanent epiphyte inhabitant (Adis and Righi, 1989).
[7] This earthworm species has been found to migrate vertically from litter and topsoil into trees with the flooding of the forest in the lower river Negro region at the confluence with the river Solimões (Amazonas) (Adis and Righi, 1989).
[8] Species very similar to *Andiorrhinus* n. sp. 2.
[9] Large greenish worms still unidentified. Similar to *Andiorrhinus* n. sp. 2 and 3.
[10] The name *Chibui* is a local indigenous name for minhocuçú (large earthworm).
[11] Stated as 'Haute Carsevenne', Venezuela. The river is now called river Calçoene, and is in AP.
[12] This species was originally described from the province of Jujuy, Argentina. Its status as being a native or exotic is uncertain.
[13] Found in bromeliad in Serra de Macaé, RJ.
[14] This species was considered to be in danger of extinction by Righi (1998b), even though Righi (1971b) collected it in 1969 near Conselheiro Lafaiete, MG. The species was mistakenly considered as extinct in the last meeting (2002) on endangered species of Brazil (MMA, 2003). It should still be considered endangered and not extinct.
[15] Synonym: *G. baiuca* Hamoui and Donatelli, 1983.
[16] Described by Michaelsen (1918) as *Glossoscolex bresslaui*, then transferred to *Andioscolex* and finally to *Glossodrilus*.
[17] See note number 11.
[18] Exact species location in SP state is uncertain.
[19] Originally separated out by Righi (1968a) as *G. corderoi*. On examination of various *G. uruguayensis* (*uruguayensis* and *corderoi*) specimens, placed as subspecies by Righi (1974), and then cited as separate species once again in Righi (1999). Further examination of these species is necessary to confirm their possible differences.
[20] Collected in several locations in the north-west region of SP for sale as bait (Caballero, 1973).
[21] Collected in several locations near the river Paranapanema for sale as fish-bait. Found in low-lying marshy areas.
[22] This species is purchased from an unknown site where it is collected for reselling. Thus it is not native to the site, but has now colonized the area surrounding the house of the retailer, from specimens that escaped the soil–manure mixture in vermiculture beds where they were maintained.
[23] Originally *Thamnodrilus*, then *Martiodrilus*. Ascribed by Righi (1995a) to the new genus *Maipure*.
[24] The original name of this genus was spelled *Opistodrilus* by Rosa (1895, 1896). Many later authors, including Cognetti, Michaelsen and Righi considered it a language (latin) 'mistake' and respelled it Opisthodrilus (with 'h'). According to zoological nomenclature, article 33.2.3.1: 'when an unjustified emendation is in prevailing usage and is attributed to the original author and date it is deemed to be a justified emendation'.
[25] Length 53–58 cm, but diameter is 5 mm.
[26] This is the only species of the *Pontoscolex* (*Meroscolex*) subgenus found in Brazil. The other species are in the *Pontoscolex* (*Pontoscolex*) subgenus.
[27] Species widely collected for the bait industry. Many families derive the main or sole income strictly from the sale of this species, contributing to the reduction of its abundance in the areas of the state of Minas Gerais where it occurs. Due to overharvesting, presently considered in danger of extinction (endangered status) by Righi (1998b).
[28] See note number 11.
[29] This extremely disjunct distribution seems unlikely. With further sampling, evidence may show either that one of the two species is different or that this species is more widespread than previously thought.
[30] This is the largest earthworm in Brazil (2.1 m in length), although unfortunately its identity cannot be properly assessed, and it has apparently not been found again. Known from only one locality and one specimen, this was ill preserved and the internal organs had 'gelatinized' according to Michaelsen (1918). It is similar to *R. horsti*, although much larger in length. Considered for many years as endangered (Righi, 1998b), it was considered as extinct in the last meeting (2002) on endangered species of Brazil (MMA, 2003).
[31] According to Righi (1985), this species is very similar to *R. motucu*, and the types need further evaluation to confirm the validity of this species (i.e. if they are truly different from each other).
[32] This very large earthworm was described originally as *Anteus gigas*, syn. *A. horsti*, and finally transferred to *Rhinodrilus* by Michaelsen (1900).

[33] Synonyms: *R. motucu unais* and *R. garbei cuiabanus*. Appears to have a very large home range. Is collected and sold for bait.
[34] Species very close to *R. jucundus*. Originally ascribed to *Aicodrilus* gen. nov., but invagination of first segments was not properly accounted for, creating synonymy with *Rhinodrilus* (Righi, 1995a).
[35] Synonym: *Tuiba tipema*. This earthworm species has been found to migrate horizontally from wetter to drier soil (and vice-versa) with the flooding of the forest in the lower river Negro region (at confluence with river Solimões), near Manaus (Adis and Righi, 1989).
[36] Originally *Martiodrilus*. Righi (1995a) erected the genus *Tupinaki* to accomodate the two known species.
[37] Synonyms: *Rhinodrilus papillifer* (*papillifer* and var. *teres*) and *R. brasiliensis*. This is the second most widespread native earthworm in Brazil, after *P. corethrurus*. It inhabits forest litter and areas rich in organic materials. It easily autotomizes (fragments) when stressed, losing tails and breaking into several pieces.
[38] Species very similar to *Urobenus brasiliensis*, but with some differences. More detailed examination of internal and external features of additional adult specimens is necessary to determine if new or not.
[39] Moderately sized (20 to 30 cm long) earthworms found in irrigated low-land rice production systems. The identity of both species from Camaquã is still uncertain but they appear to belong to the genus *Criodrilus*, according to Carlos Rodriguez and Ana Cláudia Rodriguez de Lima (personal communication).
[40] Synonym: *Criodrilus iheringi*.
[41] Synonym: *Drilocrius ehrhardti*.
[42] According to Gavrilov (1981), considered a specialist on the genus *Eukerria*, this species, although placed in synonymy with *E. kukenthali* by Jamieson (1970) should remain separate until more specimens have been examined in detail.
[43] Synonym: *Eukerria hortensis*.
[44] This is a widespread ocnerodrilid, found in many locations on several other continents. Although this species is considered exotic, its origin may well have been in central South America (Argentina–Paraguay region).
[45] Synonym: *Eukerria borelli*.
[46] Synonym: *Ocnerodrilus michaelseni*.
[47] Found in bromeliad.
[48] Found in bromeliad.
[49] Found in bromeliads, growing both on trees and on the forest floor.
[50] Found in bromeliads growing on trees.
[51] Gates (1973) considered *Ocnerodrilus hendriei* as a junior synonym of *O. occidentalis*. It is likely that the subspecies *Ocnerodrilus hendriei paulistus* described by Righi (1968b) is also a junior synonym.
[52] Such a large (48 cm long, 0.9 to 1.2 cm diameter) Ocnerodrilid is extraordinary. Most of the species of this family are medium- to small-sized earthworms.
[53] Also known as 'gigante africana'. Commonly used for vermiculture in the warmer areas of Brazil. Its distribution is generally restricted to areas close to human habitations, and to activities of vermicomposting, although it can sometimes leave the beds and invade neighboring soils that have vegetation cover, particularly home gardens with fruit trees or vegetable production.
[54] May be any of three species (*A. trapezoides*, *A. tuberculata* or *A. turgida*), but most likely to be *A. turgida*.
[55] Synonym: *Eisenia rosea* (Blakemore, 2002).
[56] Also known as 'vermelha da califórnia'. *E. andrei* seems to be much more widely distributed than *E. fetida*. Both are separate species and although they can mate, the cocoons produced are sterile. Although often mentioned by producers (as a marketing strategy), it is impossible to produce what is commonly called 'hybrid-worms' in the vermicomposting process, even if several earthworm species are found together in the composting beds. Although very widespread throughout Brazil, the species is always restricted to areas close to human habitations and does not appear to survive well outside of its food substrate (the compost). *E. fetida* and *E. andrei* have been recommended for inoculation into the field by some producers, although there is little evidence that they can survive, reproduce and make any important contribution to soils when introduced.
[57] Also known as 'vermelha da califórnia'. *E. fetida* is almost always quoted as the earthworm species used by vermicompost producers, although it appears that in most cases, the species is actually *E. andrei*. Species is also restricted to areas close to human habitations with abundant organic matter (substrate).
[58] In the meantime, this species has been placed in the family Sparganophilidae, although no other species in this family is known from South America.

Continued

Table 4.1. List of earthworm genera and species found in Brazil, together with their distribution and origin. Large-bodied earthworms (minhocuçu) are identified with an asterisk.[1] – cont'd

[59] May actually be *A. gracilis*. *Pheretima aspergillum* Perrier 1872, quoted by Righi (1967d), is synonymous with *A. gracilis* (Blakemore, 2002). The specimens at the MZUSP must be looked at for confirmation.
[60] Synonym: *Amynthas* or *Pheretima diffringens*.
[61] Synonym: *Amynthas* or *Pheretima hawayana*.
[62] Synonym: *Amynthas schmardae*.
[63] Synonym: *Pheretima indica*.
[64] Synonym: *Pontodrilus bermudensis*. Some authors place this species in the Acanthodrilidae.
[65] This species was found in an ant (*Camponotus rufipes*) nest, and is in the collection of G. Righi at the MZUSP. It was not identified to the species level.
[66] In the lower river Negro region close to Manaus (at confluence with river Solimões), adults observed to ascend and descend tree trunks when the forest floor floods (Adis and Righi, 1989).
[67] Includes *D. servi, D. parva* and *D. silvestris cacaois* that were considered by Righi as separate species. Synonymized by Csuzdi (1995).
[68] Found mostly in agricultural areas or disturbed soils close to human habitation. Also found in bromeliad in coastal region (Santos) (Zicsi and Csuzdi 1999).
[69] Originally named as *N. americanus*. Csuzdi (1995) changed the valid name of the species to *N. americana*.
[70] Previously *Wegeneriella*.

and in Brazil, it is imperative that taxonomic training and capacity building be put on the forefront of the funding agencies' and governments' agendas. Without the persistent, detailed work of taxonomists, discussions on biodiversity increasingly become merely releases of hot air and CO_2.

In a survey on Brazilian biodiversity submitted to various specialists (Lewinsohn and Prado, 2002), Righi responded to a questionnaire on various aspects of the study of Oligochaete biodiversity in Brazil, including the assessment of: (i) national capacity and need for training; (ii) foreign and Brazilian collections; (iii) availability of identification keys; (iv) importance of the taxon; (v) total number of species known and estimated in Brazil, Latin America and the Neotropics; (vi) knowledge on distribution and diversity according to Brazilian biomes and regions; (vii) number of endangered species; (viii) availability of geographically based species biodiversity surveys; and (ix) genetic diversity. Unfortunately, we do not have access to his original replies, but some of them were partly published in Brandão *et al.* (2005) and summarized by Lewinsohn and Prado (2002). In the remainder of this chapter we attempt to deal with points (iv)–(ix), with the presently available data and recently gained experience.

Brazilian Earthworm Biodiversity

The annelids appeared early in the history of animal evolution, with probable representatives in the Ediacaran period of the Neoproterozoic, about 600 million years ago. Given the long time that they have had to evolve, and the adaptations needed to live in the soil, an opaque (dark), compact medium with few food resources, and generally of poor quality, it is not surprising that the number of estimated species may be as high as 8000 (Fragoso *et al.*, 1997). Of these, however, only about 50% (approximately 3800 species) are known (Reynolds, 1994).

Almost 70 years after Michaelsen's last estimate of earthworm biodiversity in Brazil, Righi's reply to the questionaire of Lewinsohn and Prado (2002) stated that 240–260 species of terrestrial oligochaetes (this included not only megadriles but also microdriles) were known from Brazil,[5] although he estimated a much higher number (800 species). The updated list,[6] including only the megadrile earthworms, contains 295 species in 64 genera (Table 4.1 and Fig. 4.1). Of these, 253 are native species (86%) and 42 are exotics (14%). The most diverse families are the Glossoscolecidae, with 193 species (all native to Brazil) in 26 genera; the Ocnerodrilidae, with 47 species (40 native) in 15 genera and the Acanthodrilidae, with 22 species (about 50% native) in 6 genera (Fig. 4.1). Within the Glossoscolecidae, the most diverse genera are *Glossoscolex* (50 spp.), *Rhinodrilus* (29 spp.) and *Righiodrilus* (20 spp.), an offshoot of *Glossodrilus*.

Although exotic species constitute only a small percentage of the total species of Brazil (14%), their distribution is relatively widespread (Table 4.1). The first confirmed report of the exotic *Amynthas* spp. being found in Brazil was by Kinberg (1867). None the less, these earthworms may have arrived centuries earlier, when trade routes with the Pacific were first established, leading to the exchange of various plants and soil between Asian countries and Brazil (Chang, 1997). The *Amynthas* spp. are widely known in Brazil and have several common names, including crazy-worm, dancing-worm, angry-worm and jumping-worm, due to their slashing, active behaviour when disturbed. Exotic invasive Megascolecidae (e.g. *Amynthas*, *Metaphire*) and some Acanthodrilidae (mainly Dichogastrini) are found throughout the country, from north to south (Table 4.1), while some Acanthodrilidae (mainly *Microscolex* spp.) and the Lumbricidae have a more restricted distribution (with the

[5]Unfortunately, Righi did not produce a complete list of the species. This is why the number provided was a range.

[6]List complete as of 15 March 2005. Many specimens in several collections, deemed to be new species, must still be examined.

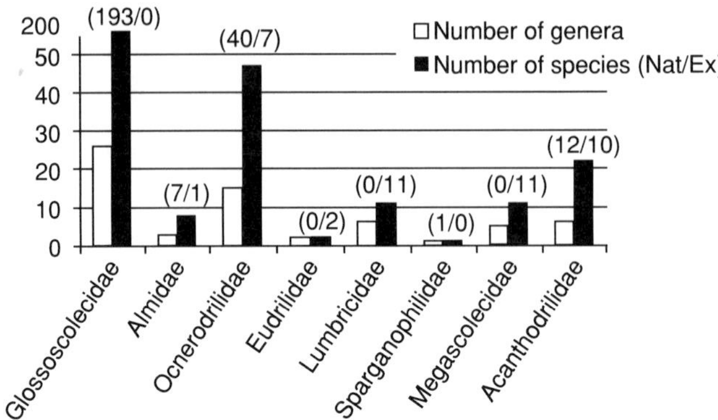

Fig. 4.1. Generic (white bars) and species (black bars) diversity of the major families of megadrile earthworms found in Brazil. Note: The number of native (Nat) and exotic (Ex) species, given in parentheses above the black column, is approximate, as there are several species from different families (particularly Ocnerodrilidae and Acanthodrilidae) whose origin is still not clearly established.

exception of *Eisenia andrei* and *Eisenia fetida*, species used in vermiculture) in the southern part of the country, particularly in Rio Grande do Sul (Table 4.1), where the cooler subtropical climate is more like their native homelands in the northern hemisphere.

According to Righi's estimates, the Neotropical oligochaete fauna (micro- and megadriles), estimated at 2000 species, represented approximately 40% of the world's total number of species (which he estimated at only 5000 species), but only 18% of them were known, despite the taxonomic effort undertaken up to his death (Righi described more than 220 species; Fragoso *et al.*, 2003). If we consider the same proportion and use the higher estimates of Reynolds (1994) and Fragoso *et al.* (2003), reaching to over 8000 species, the total number of species in the Neotropics may easily be well in excess of 3000.

Recent evidence collected by the authors and Lavelle and Lapied (2003) sheds some more light on the actual and potential diversity of earthworms in both Brazil and the Neotropics. These are related mostly to the frequency and geographic distribution of sampling and the 'endemic' nature of many species of Brazilian earthworms.

Sampling frequency and geographic distribution

Large areas of Brazil are still unexplored and have never been sampled for earthworms (see 'Biogeography' section). In fact, three states (Rio Grande do Norte, Alagoas and Piauí) have no earthworm records at all (Fig. 4.2). Furthermore, 11 states all have less than ten sample sites, or samples concentrated in only a very limited area. This means that the samples actually taken probably greatly underrepresent the actual variability of habitats and situations where earthworms may be found, and that the known number of species for these states is greatly underestimated (Fig. 4.2). The states with the largest number of samples (SP, RO, RS, AM, PR) are generally, but not necessarily, the states with the largest number of species, implying that there is certainly a lot to be gained by increased sampling intensity, in both underrepresented states and the states with larger number of samples.

For instance, Brown *et al.* (2004) and Brown and James (2006) and performed scattered sampling in about 50 sites in the states of Paraná and São Paulo (SP). These samples turned up a large number of new species (>30; Table 4.1). Despite the fact that SP is

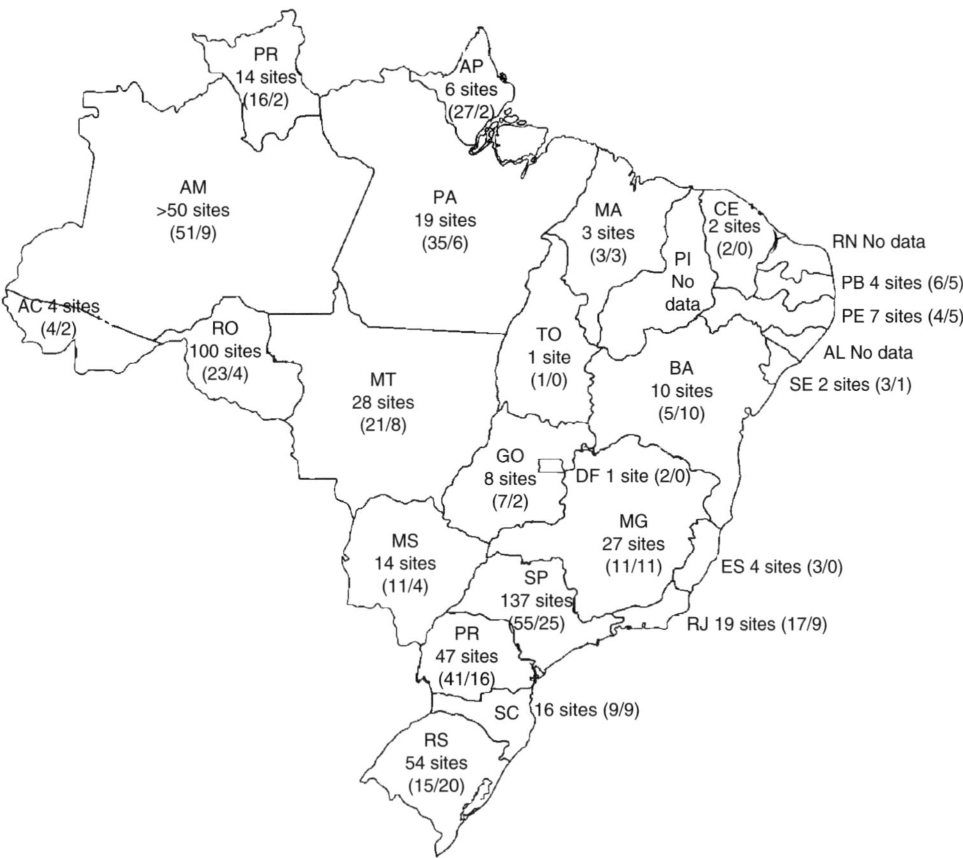

Fig. 4.2. Earthworm species diversity (number of native species/number of exotic species) and approximate number of collection sites in each Brazilian state.

the best-known state in terms of earthworm diversity, having been studied by Righi for over 30 years, visits to formerly sampled and unsampled regions/counties revealed at least ten new species (Brown and James, 2006).

When we mapped the geographical coverage of the samples taken in SP by Righi, his colleagues/students and other previous taxonomists, we observed that it was, in fact, very sparse, including only 19% (120) of the 645 counties, largely confined to regions close to São Paulo city, the north-west part of the state (where Caballero, 1973, performed her dissertation work) and the Atlantic coast. The extensive interior, particularly the south and south-west parts of SP and some more remote Atlantic forest areas (particularly in the south and east) remained and still remain, little explored.

Therefore, although Righi (1999), in his summary of the earthworm biodiversity of SP state, listed 50 species of megadrile earthworms (excluding the synonymies), divided into 7 families and 23 genera, when we considered the aforementioned field collections (made by the authors), a review of the literature, an update on the species synonyms and a visit to the Museum of Zoology of the University of São Paulo, a further 30 species (most of them native) were found, totalling 80 species plus 2 subspecies of earthworms for the state (Brown and James, 2006). This did not substantially increase the geographic coverage of the state, adding only an additional ten counties to the total studied (120 counties). Of the 80 total known species, native species (55 species) dominated over exotics (25 species). Of the native earthworms, 45 species were in the Glossoscolecidae family and 8 species in the Ocnerodrilidae. Of the exotic species, ten were in the Megascolecidae and six in the Acanthodrilidae.

In contrast, in Paraná state, from which only ten species were known prior to 1997 (Brown *et al.*, 2004), collections in northern and eastern Paraná, in only 10% of the state's counties (40 out of 399) revealed 57 species, more than 25 of which were new to science (Sautter *et al.*, 2006). Of the total found, 16 were exotic and 41 were native species. In almost each new site sampled, at least one new earthworm species was encountered.

In fact, if we consider the number of earthworm species and the number of sample sites, particularly in the Brazilian states where fewer samples have been taken (Fig. 4.2), we can see that the ratio is of approximately one species or more per sample site, meaning that, even with small number of additional samples, the total number of known species for the state (and the country, when new species are found) could easily and greatly increase.

Simple estimates of the number of species in an area can also be made, based on species/area relationships. For instance, SP state, with 250,000 km^2, has 80 known species. Thus, for each 100,000 km^2, we could expect to find 32 earthworm species (compared with an estimate of 20 species per 100,000 km^2; Fragoso, 2001). By using this estimate to calculate the earthworm species diversity in Paraná state (200,000 km^2) and by considering similar diverse habitats/vegetation and climate/soil conditions as found in SP, we find that at least 64 species should be present. However, considering the previous experience in SP, where the knowledge of the earthworm diversity was increased by 50% with little effort, and that only a small proportion of the state has been sampled and 57 species have already been found, this is probably a gross underestimate.

Using these values to calculate the total species diversity of earthworms in the 8.5 $\times 10^6$ km^2 of Brazil results in an estimate of as many as 2720 species. This is more than three times the 800 species estimate of Righi, but not completely unreasonable, given the known and potential endemicity of many species (see below). However, complicating the estimation problem is the very low density, or outright absence of earthworms, from some dystrophic soil types in Paraná and other states (more on this below). If this holds up to more extensive sampling, the estimates based on land area may have to be modified by subtraction of certain soil types. Furthermore, these estimates homogenize many of the intrinsic differences in habitat types, distribution and climate/soil variations encountered among the different regions of Brazil.

At a more regional level, calculations of the number of species and sample sites in various Brazilian regions (Fig. 4.2) continue to show the great underrepresentation of the north-east, with only 30 species found in 18 sample sites. Most likely, many new species will be found with further sampling efforts, particularly in the Atlantic forest domain, but perhaps even in the drier 'caatinga' areas, if taken in the rainy season, and concentrated in areas close to watercourses and in protected areas with native vegetation. The central west region, the main domain of the cerrado vegetation (a global biodiversity hot spot, also highly endangered by agricultural expansion activities), continues to be greatly unexplored, despite the work of Righi (1990a). Only about 50 sites have been sampled, revealing 42 species.

Endemicity of Brazilian earthworm species

Apart from a few species with fairly wide natural distributions (e.g. *P. corethrurus*, *U. brasiliensis*), most native Brazilian earthworms are known from one locality or a few closely spaced localities (Table 4.1). A brief look at the location records in Table 4.1 reveals that, of the 253 native species,

171 species are found in only one location, and 29 in only two sites. Therefore, close to 80% of all Brazilian species are found in two sites or less. This restricted distribution could reflect high degrees of endemicity or simply a lack of geographic coverage by collectors. The present coverage of the country is not sufficient to answer the endemicity question for most species.

On the other hand, our recent experience is that, within a physiographic province (e.g. PR state), localities separated by about 100 km but with comparable soils can have different species. The only earthworm species in common between the two sites are generally the invasive exotic species and the unusual species with broad natural distributions. For instance, the glossoscolecid species *U. brasiliensis* is widely distributed in southern and south-eastern Brazil, showing little morphological variation and the structural correlates of an epigeic lifestyle. In contrast, the endogeic species found in many wetlands (see section titled 'Minhocuçus in Brazil') are quite localized.

Similar results were obtained in the Amazon Basin by Lavelle and Lapied (2003): of the 106 earthworm species found in the five main regions for which data were available in the basin, 86 species were known from only one of the five regions and 14 occurred in only two regions. Thus, the ratio of local to regional species richness was the lowest of all of the invertebrates for which data were available, indicating an extremely high rate of endemism for the Amazonian earthworms. Their calculations led them to believe that the regional earthworm diversity (for Amazonia) is probably in excess of 2000 species.

Therefore, a critical factor in estimating the number of species in Brazilian states and regions is the level of endemicity. Unfortunately, these data are unavailable for the majority of species known from Brazil, mostly because of the limited number of sites studied in the country. Much more work and sampling need to be done, not only to determine total earthworm diversity in the country, but also to determine those species that may actually have a restricted distribution and/or be endangered due to their particular habitat requirements, behaviour and/or human pressure on their populations.

Biogeography of Brazilian Earthworms

The traditional southern boundary of the distribution of the predominant South American family, the Glossoscolecidae, is the Juramento-Salado River – Rio La Plata system in northern Argentina, and extending across the Peru–Chile border to the Pacific (Righi, 1972c). Righi's (1972c) paper integrated his earlier work and that of other contributors to the study of glossoscolecid worms (Michaelsen, Cognetti, Rosa and others) by mapping the natural distributions of genera in the Neotropics. Regrettably, he did not update these conclusions later in life, for after the passage of another 25 years, sufficient additional information came from his laboratory and others to justify another such paper.

The broad outlines of distributions remain unchanged, but many new genera and new records of other genera, mainly based on the discovery of new species in the latter case, fill some of the gaps in the 1972 publication. As of that time, he commented that vast areas of South America were then unexplored with respect to oligochaete worms, so that his broad outlines of genera were necessarily premature. The situation has changed but not greatly; the bulk of exploration since then has filled in some of the more accessible uncollected areas easily apparent in his 1972 range maps. Looking over his post-1972 papers, a large area south of the Amazon River to northern Paraguay, a rectangle approximately bounded by the corners 5°S, 65°W; 5°S, 45°W; 20°S, 65°W and 20°S, 45°W, except for the south-east corner of this rectangle, remains very sparsely collected, as does the Orinoco River drainage, and to a lesser extent, the northern interior Amazon Basin. North of Rio de Janeiro all the way up to Maranhão, a few coastal locations have been collected, but the interior has not. Some of this region is at least seasonally very dry, and parts may be too dry to support earthworms. Further south, the Brazilian states of Paraná and especially Santa Catarina still have very few collection records. In Rio Grande do Sul, most of the collections have been made in the vicinity of Porto Alegre.

Other valuable contributions of Righi came in later years. His work showed the centre of *Pontoscolex* diversity to be in the Guyanan highlands region of north-east South America (Righi, 1984c), pinning down the likely homeland of *P. corethrurus*, now arguably the world's most abundant earthworm. This is important in the context of determining the extent of invasions by this species, which is now so ubiquitous that people are reluctant to believe that it could be an exotic almost everywhere it occurs. This is especially important in regions with endemic Glossoscolecidae, such as Costa Rica (Lapied and Lavelle, 2003). In a famous tropical forest research area, Finca La Selva, the soils are dominated by *P. corethrurus*, which might otherwise be considered a 'native' element of the soil biota. Righi's work establishes that this is not true, and puts a different spin on the study of soil processes in that well-studied benchmark tropical forest.

Righi's work on the Glossoscolecidae is the most comprehensive of recent times, but it was not the only family to benefit from his attention. Almost of greater global significance is his work on the Ocnerodrilidae, an interesting and neglected group with a curious distribution and many unresolved phylogenetic issues. His discoveries give support to the hypothesis that the Ocnerodrilidae have an ancient presence in South America, and may have originated there. Of the 47 species of Ocnerodrilidae found in Brazil, Righi described 24 (all native). Further attention should be given to this issue, particularly using molecular genetics, to help clarify ocnerodrile phylogeny.

Another intercontinental impact is his discovery of many more South American acanthodrilid species, within the genera *Dichogaster*, *Wegeneriona* and *Neogaster*, the latter two genera possibly allied to *Dichogaster* (s.l.). These discoveries also help remove lingering doubts about the transport history of worms of these genera, which, like many other biota, show strong South America–Africa links, as would be predicted by the tectonic history of the region.

The next phase of biogeographical investigation of South American Oligochaeta should be to analyse the phylogenetic relationships of the taxa and to define areas of endemism, processes which are somewhat interdependent (e.g. Hausdorf, 2002). One should also take advantage of such paleogeographical and paleoclimatic data as exist for the continent. Such data are available for a variety of factors, including, for example, expansions and contractions of vegetation zones during the Pleistocene (Brown, 1987). These changes are potential vicariance events for earthworms. The hypothesis that the Amazon rainforest contracted to small isolated patches during glacial maxima was seriously questioned by Colinvaux and De Oliveira (2001). Nevertheless, their data still indicated a significant cooling and invasion of the lowlands by vegetation now characteristic of higher elevations. Given the greater abundance of earthworms in higher-elevation South American forests in modern times (Righi, 1972c), the glacial maxima may have been the time of advance of earthworms into lowlands, and the interglacial periods times of retreat to higher elevations and genetic isolation.

Aquatic and Semi-aquatic versus Terrestrial Earthworms

A large number of Brazilian earthworm species inhabit wet soils, seasonally submerged soils or even aquatic habitats. The first two categories we consider to be terrestrial in the main, and differ by degree rather than being sharply distinct. Wet soils along riparian areas may be flooded for brief intervals, but we are concerned here with chronically wet soils, those that are subirrigated and support characteristic wetland vegetation. In southern Brazil these places may be found in forests, as marshlands within deforested pastures and croplands, or in naturally unforested sites along river margins with impermeable rock or soil layers causing a perched water table. These chronically wet soils are important refugia for native earthworm species, because after deforestation no other native species may survive or be found in the area.

Seasonally submerged soils in *várzeas*, the forested floodplains of high-order streams, pose special environmental problems for earthworms, most of whom cannot endure flooded soils for long periods. In places such as the inundation forests of the lower river Negro (AM) and on the island of Maracá, earthworms (e.g. *Andiorrhinus tarumanis*, *Dichogaster andina*) ascend trees and inhabit the forest canopy, living in epiphytes during the rainy season (Adis and Righi, 1989; Righi, 1997). The reasonably large (up to 19-cm adults) *A. tarumanis* even developed a special means of 'scaling up' the trees (Adis and Righi, 1989): they extend the anterior part of their body and remain still until the secreted mucus binds firmly to the trunk, after which they retract the posterior part, which subsequently adheres to the trunk and enables the following upward movement of the anterior portion. Keeping the body in an 'S-shaped' form seemed to facilitate upward movement. If the individuals were mechanically disturbed, they fell to the forest floor. The animals move upwards only at night, when the high noctural humidity prevents desiccation. During the day, the animals hide in moist places along the trunk, such as under the bark. When the waters recede they descend to the ground and resume life in the soil–litter interface.

Therefore *várzeas* (wetlands) and similar habitats should be expected to harbour earthworms, though the body size limitations imposed by the necessity of climbing could affect the composition of *várzea* earthworm communities. Could different kinds of soil organic material resources be present in *várzeas*? Does the litter float away and/or accumulate in big leaf packs? Does litter become stranded in the canopies and colonized by earthworms there? Furthermore, it might be worth investigating the question of how earthworm burrowing affects soil drainage in various *várzea* soil types. According to Victor Del Mazzo (G. Righi, 1997, personal communication), the 2- to 3-cm diameter galleries of large earthworms act as major drainage channels in the *várzeas* of the river Paraná in Mato Grosso do Sul.

Another Amazonian species, *Tuiba dianae* (synonym *Tuiba tipema*) has developed a different mode of adaptation to forest floor inundation: horizontal migration (Adis and Bogen, 1982; Adis and Righi, 1989). These earthworms follow the inundation and receding fronts of the river Negro and its tributaries, escaping the waterlogged soil of the blackwater inundation forest, moving towards the dryland forest. During the inundation period, the earthworms (all juveniles) always stayed within 16–26 m beyond the water margin to avoid being drowned. At the peak of the inundation, they reached the edge of the dryland forest, at a distance of about 450 m from the river. With the receding of the waters, the earthworms (now adults) followed the front at a closer distance (5–10 m) back towards the blackwater inundation forest. A similar phenomenon at a smaller scale (a few metres) was observed by Righi (1997) for the exotic species *A. gracilis*, which moved to, and concentrated in wet areas to escape drought in soils around Itu, SP state.

But there are also earthworm species that live in truly aquatic habitats, where they can be found in decaying vegetation in marshes and swamps, in the soils of flooded rice paddies and along river margins. The current state of knowledge indicates that the few species of Almidae, in the genus *Drilocrius*, *Criodrilus* (several spp. known) and *Glyphydrilocrius* (only *ehrhardti* is known) are most often found in such places, as are several members of the Ocnerodrilidae (Gavrilov, 1981). Righi *et al.* (1978) described the single known native member of the Sparganophilidae family in South America (*Areco reco*) from specimens collected at the Reserva Ducke, near Manaus, AM. The authors themselves, however, questioned whether the species truly belongs to this family, since its organization is similar to *Drilocrius* and *Glyphydrilocrius*, but also to *Sparganophilus*, though it is distinguished from all of these by being metandric (testes in segment 11) rather than holandric (two pairs of testes in segments 10 and 11).

Bromeliad leaf tanks are another 'aquatic' habitat in which Brazilian Ocnerodrilidae (e.g. *Kerriona limae* and *Kerriona luederwaldti*, *Kerriona* n. sp. 2 and 3) and Glossoscolecidae (*Andiorrhinus planaria*, *A. tarumanis*, *Fimoscolex ohausi*) can be found, both growing on the ground and on trees (Table 4.1). However, even when many bromeliads are present in the habitat, we have observed that the proportion of inhabited plants is generally small. The exotic *Dichogaster bolaui* has also been found in a bromeliad near Santos (Zicsi and Csuzdi, 1999) and elsewhere, in both Latin America and the rest of the world, various species of Acanthodrilidae, Megascolecidae and Glossoscolecidae are known from bromeliads or their structural analogues (e.g. Pandanaceae in the Old World tropics).

The availability of water in the soil plays a key role in earthworm activity (e.g. casting, aestivation), growth, reproduction, survival and abundance (see more on this below) and therefore habitats with particular soil moisture regimes can exert particular species selection pressures. For instance, Ayres and Guerra (1981) found that 33 of the 40 species collected in the vicinity of Manaus (AM) occurred strictly in the proximity of water. Four species were found in habitats with great variation in soil moisture (*P. corethrurus*, *Rhinodrilus priollii* and two exotic *Dichogaster* spp.) and the remaining three (two native *Dichogaster* spp. and *P. eudoxiae*) were found in decomposing tree logs. If this trend is also true for most of the country, this means that the greater part of the earthworm species (particularly native) still to be encountered in Brazil is likely to be associated with aquatic or semiaquatic habitats, while the number associated with strictly terrestrial habitats may be much lower.

Minhocuçus in Brazil

One of Fritz Müller's favorite stories was 'Der Minhocão', or 'the big earthworm' (Müller, 1877). The story actually consists of a compilation of eyewitness reports of the feats of very large 'earthworms', from 1 to 3 m diameter and up to 30 m length. These 'earthworms' produced huge trenches and holes in the ground, which led to, among other things, creation of large canals in swamps and flooded areas, tree falls, drying up of lakes and the muddying of rivers. Their activities were always observed after long rainfall events.

As we pointed out in the introduction to the history of earthworm collecting in Brazil, there are some very large species in the country. However, none reach the dimensions of 'der minhocão'. The largest earthworm species collected in Brazil is *Rhinodrilus fafner*, a giant earthworm 2.1 m in length, described from a single specimen taken from somewhere near Belo Horizonte, MG (Michaelsen, 1918). However, this earthworm, described from ill-preserved material, has not been found again, and was thus declared officially extinct by scientists participating in the last meeting on endangered species of Brazil (MMA, 2003). However, considering the relatively few localities sampled in MG, we believe that this conclusion may be premature, despite the rather large endemicity of these earthworms (more on this above). Only with more intense sampling will a proper assessment of the status of this species be possible, as several cases of supposedly extinct invertebrates being found once again are known, particularly in the Atlantic rainforest region (Brown and Brown, 1992).

Presently, the largest known species come from forested areas, particularly the Atlantic rainforest at middle and upper elevations. However, large worms (>30 cm in length and ~1 cm in diameter or more) seem to be available almost everywhere. There is a Brazilian word for large-bodied earthworms, 'minhocuçu', derived from the general word for earthworm, 'minhoca', and 'açu', the word in Tupí-Guaraní (native Indian Brazilian language) for big or large.

The present list of minhocuçus in Brazil includes at least 41 species, all except one (*Paulistus taunayi*, Ocnerodrilidae) in the Glossoscolecidae family, primarily in the *Glossoscolex* and *Rhinodrilus* genera (Table 4.1). There is a reliable report of worms reaching > 2.5 m length (K.S. Brown, personal observation, 27 June 1978) from Amapá, but metre-long worms are more common (at least five species), and worms of 30–40 cm are rather ordinary. Giant earthworms occur on all continents but nowhere are they apparently so numerous as in Brazil.

This leads us to ask why worms evolve giant body size, under what conditions and whether or not these conditions are frequently met in Brazil. Classical models (r and K selection, among other terms) of life-history strategy evolution generally agree that large-bodied, long-lived organisms with repeated reproduction of few well-provisioned offspring are expected where the primary mortality is in the juveniles, environmental factors are predictable and there is robust competition for resources. Where the cost of producing an amply provisioned offspring is high, one would expect delayed reproduction, given the trade-off between allocating resources to growth or to reproduction. Allocation to viable reproduction at an early age may cause greater loss of future reproductive value (current reproduction lowers residual reproductive value) than would be gained by waiting and growing larger (at which point current reproduction has negligible effect on residual reproductive value).

Is earthworm life-history evolution comparable with other animals? Epigeic earthworms are more exposed to predation at all stages of development. Their generally smaller body sizes and more rapid achievement of reproductive maturity compared with endogeic and anecic species is consistent with life-history theory. If we assume this to be a good indicator, the following factors would select for large body size in endogeic and anecic earthworms: low adult mortality, uncommon disturbance, and unfavourable conditions for growth (low soil fertility, low resource availability, competition). Increased provisioning of an earthworm embryo can only be accomplished by putting more food into the cocoon. Generally, this means a larger cocoon is needed, along with the parental bodily reserves required for making the food. Both come from larger body size. So it appears that the above conditions should cause the evolution of large body size in earthworms.

The next issue is whether or not the conditions occur in Brazil, and in the places where giant worms exist. This is much harder to determine. Unknown at this point are 'minhocuçu' mortality curves, age-specific fecundity, intensity of competition, availability of resources and the predictability of the environment from an earthworm perspective. For the time being, we offer this as an open research topic.

The Brazilian Worm-Bait 'Industry'

The native earthworm fishing bait market in central west, south-east and south Brazil appears to be almost entirely based on collecting from natural populations, with little contribution of vermicultured species or maintenance of managed earthworm habitat for the target species. At present, there is no incentive for investing in anything other than digging tools and labour costs. Collecting from the commons or from private land without much financial loss is the most economical choice.

In our limited experience with bait dealers and collectors in SP and PR states, the main target species are those of wetlands. In some cases, particularly near larger cities (e.g. Londrina, PR), the exotic invading *Amynthas* are the main species collected, although in most cases the species extracted appear to be natives of the genus *Glossoscolex* or *Andiorrhinus*, of sizes ranging from 15 to 40 cm length. In Mato Grosso, Goiás and Minas Gerais, the species extracted are taken from both dry- and wetlands, and are mostly of the *Rhinodrilus* and *Goiascolex* genera, ranging in size from 20 to 50 cm. They are manually removed from the soil and sold to fishermen in the region and, in the case of the minhocuçus from MG, shipped to other areas for sale, particularly for fishermen going to the Pantanal. In one case, we even found a Paraguayan minhocuçu (*Glossoscolex* sp.) for sale in Foz do Iguaçú, Brazil.

The collection of these animals without due permit/authorization is illegal in Brazil according to law number 9605 of 12 February 1998, Article 29 (Guimarães, 2003). To collect these animals, proper permits must be obtained from the Instituto Brasileiro de Meio Ambiente e Recursos Naturais (IBAMA). Furthermore, the transport of these live animals (as when they are sent to the Pantanal or to retailers for resale) is also illegal, without the Guia de Trânsito Animal (GTA); a certificate issued by the Federal Government's Animal Defense Secretariat. However, this is no deterrent to the many families who collect and sell these earthworms for up to $7 per dozen, depending on the size, species, location and time of the year. A dealer (reseller) of minhocuçús may earn over $1000 per week selling more than 10,000 earthworms (Guimarães, 2003). Unfortunately, many of these families and dealers are either unaware of or cynically disregard the above laws, contributing to the destruction of many native earthworms.

Little is known of the effects of the removal of these large earthworms on the soil and ecosystem services. The digging of earthworms often leaves the habitat in disarray: frequently, no attempt is made to replace the soil in its original orientation with the vegetation on top. Gradually, these habitats are degraded and erosion is likely to increase. Some habitats are already extensively modified from the original condition after deforestation, but others have diverse natural herbaceous vegetation. Close to Sete Lagoas and Paraopeba, Minas Gerais, *Rhinodrilus alatus* is widely collected from cerrado areas, many of which are even intentionally burned to facilitate entry and extraction of the worms. Many families live off this predatory extractivism (Righi, 1977; Guimarães, 2003), which has led to serious decline in the populations of *R. alatus*, now considered endangered (Righi, 1998b; MMA, 2003).

In other cases, bait collecting may just be an additional income source, rather than a complete livelihood. However, the income can be the difference between affording and not affording the basic necessities. We suspect that the current rate of extraction of most species collected is unsustainable and that the bait 'industry' will suffer numerous local collapses. When the natural resource collapses below the point of economic viability, the people will be deprived of the income until the resource (the environment and the earthworm populations it supports) recovers, if ever. The economic, ecological and social importance of the bait collection industry must be evaluated, in order to formulate appropriate regulatory action to ensure the future of the industry, not to mention of the earthworm species involved.

Biological/Ecological Studies on Brazilian Earthworms

Over the last 10 years, the number of people working on soil fauna, both in universities and research centres, has greatly increased. The growing number of contributions on the topic in various national congresses (Zoology, Soil Science, Fertbio) attests to this increased interest. Nevertheless, in most of these studies (the majority published as abstracts or short papers in the conference proceedings), earthworms are only considered briefly, as part of the soil macrofauna; rarely are data on numerical abundance and biomass values actually provided, and almost never are the species present mentioned.

A topic that has received much attention since the early 1980s is the practical aspect of vermiculture production, a popular practice in Brazil. Nevertheless, as these exotic earthworms are taken out of their natural habitat and grown artificially by human beings, we will not consider this topic further in the present chapter. Further information on this topic can be found in several books and papers (e.g. Aquino *et al.*, 1994; Ricci, 1996; Martinez, 1998) and on the Web.[7]

Earthworm biology

Righi (1972c, 1997, 1999) mentions the burrowing and casting habits of some native earthworms. For instance, in the *Mata Atlantica* forests of the coastal range of SP, the minhocuçú *Fimoscolex sacii* produces large (up to 4 cm diameter) and deep (up to 5 m) somewhat vertical burrows with several ramifications. In central SP along the Rio Claro River floodplain, *Glossoscolex paulistus* creates U-shaped burrows of about 30 cm depth and open to the soil surface, but then in the dry season burrows deeper following the water table and, if the dry season is long, aestivates in a chamber at ~50 cm depth (Abe and Buck, 1985). In the cerrado region near Paraopeba valley, MG, *R. alatus* produces two main galleries that emanate from their annual diapause chamber towards the soil surface, one is bent from 30° to 60° and the other is perpendicular. Both are plugged at the soil surface, but only one is easily recognizable by the castings raised 2–3 cm above the soil surface, occupying an area of about 10–15 cm diameter. Three species of earthworms were reported to produce tower castings: *Rhinodrilus motucu*, *G. paulistus* and *F. sacii*. These towers reached a height of 20–30 cm, and in the latter two species, they had holes in the middle where the earthworms inserted their hind ends to further build up the towers. The production of tower castings by the minhocuçú *Chibui bari* was also observed and measured over a one year period by Guerra (1988a; see later). Each individual constructed one cast tower 'group' only during the 6 months of higher rainfall. In the dry season, the species entered into aestivation and no castings were produced.

[7]See, for example, the Minhobox site at http://www.minhobox.com.br

The role of seasonal changes in soil moisture and precipitation on the activity of the minhocuçus *C. bari*, *Andiorrhinus samuelensis* and *G. paulistus* were studied by Guerra (1985, 1988a), Buck and Abe (1990), and Abe and Buck (1985), respectively. *C. bari* was active for 6 months when the soil moisture was above 20% in the wet season and inactive the rest of the year, aestivating at a depth of around 1 m. The latter two species followed the retreat of the water table with the onset of the dry season; *A. samuelensis* burrowed to a depth of more than 9 m (Righi, 1990a). This species did not enter diapause, but remained inactive in the burrow until suitable soil moisture conditions were re-established in the upper soil horizons. The energy cost of burrowing to feed on the poor soil at lower depths is probably too large to induce feeding, so the worms remain inactive (Buck and Abe, 1990). *G. paulistus*, on the other hand, aestivated in a chamber, rolled up into a ball (this posture reduces water loss by decreasing surface area for desiccation), as is typical in many glossoscolecid earthworms (Jiménez *et al.*, 2000). Some earthworm species may also increase osmotic concentration of the body fluids, allowing for greater resistance to desiccation and perhaps even permitting re-absorption of soil moisture in a manner similar to amphibians (Buck and Abe, 1990).

Seasonal differences in the body water contents of *P. corethrurus* and *A. gracilis* according to changes in soil moisture were also observed by Caballero (1979). Both species had lowest body water content in August, at the height of the dry season, when they were found quiescent in the soil, at greater depths than during other sampling periods. Studying two *Pontoscolex* species (*corethrurus* and *marcusi*) and *Andiorrhinus caudatus*, Ayres and Guerra (1981) showed that the lethal percent moisture loss in the tissues ranged from 56% to 64%, with *P. marcusi* being the most susceptible to water loss, and *P. corethrurus* the most resistant. Of the 40 species encountered in their survey (Ayres and Guerra, 1981), *Andiorrhinus amazonius* appeared to be most euryhydric (wide-ranging tolerance to different soil moisture conditions), being found in habitats very unfavourable in terms of soil moisture (e.g. 'campinas'; open, short, scleromorphic forests on very sandy, nutrient-poor soils), and surviving the dry season by aestivation. On the other hand, *A. caudatus*, which inhabits saturated soils, was found to be highly resistant to anoxic conditions, surviving up to 24 h immersed in water. Under these conditions, the worms became pale and autotomized the caudal zone. When aerobic conditions were re-established, they rapidly regained an intense reddish colour. *P. corethrurus*, although resistant to dessication, was susceptible to anoxic conditions and, contrary to *A. caudatus*, did not autotomize their tails.

In what may be the first paper dealing strictly with the ubiquitous *P. corethrurus* (the 'tame-worm' as it is commonly called in Brazil), Vanucci (1953) provided several interesting notes on the distribution, habits and biology of this species, making some additional comments on *A. gracilis*. Based on empirical observations, she concluded that *A. gracilis* substituted *P. corethrurus* in the urban areas, while the latter species remained more abundant in the city outskirts and rural areas. Little evidence other than this has been found in the literature on this topic, which deserves further verification. In laboratory cultures of *P. corethrurus*, she 'never found this species copulating', but she measured cocoon production and made several interesting observations on the cocoons, including their parasitism by small enchytraeids (also observed by Hamoui, 1991), as well as their placement 'suspended to the chamber by means of a thread made of the same material as the substance of the egg capsule'. As Müller (1857a,b) observed more than 100 years previously in adult and juvenile specimens, she also frequently found the 'caudal zone' in newborn specimens. The function of this caudal zone is still not certain, but may be related to growth, regeneration or autotomy, sensorial functions, anchorage in the galleries or respiration (Righi, 1990a).

P. corethrurus has been the topic of several laboratory incubation experiments: Hamoui (1991), Guerra and Bezerra (1989), Bernardes and Kiehl (1992, 1993, 1994, 1995a,b, 1997), Bernardes *et al.* (1998), Ferraz and Guerra (1983) and Soares *et al.* (1997) studied various aspects of the life cycle of this species. Their experiments demonstrated the great versatility of this species to human manipulation and to living in different substrates, moisture and temperature conditions. Cast and cocoon production and growth rates were highest when a mixture of soil and added organic matter (composted manure) was used. Furthermore, the ideal temperature for activity was 25°C and the ideal moisture for cocoon and cast production was 55% and 70–80%, respectively, of the field moisture capacity.

Species distribution according to various habitats

Of all the Brazilian earthworms, *P. corethrurus* is the most well known, both in the country and internationally. It is the most common and widespread earthworm in Brazil (arguably the world), and probably dispersed from its supposed place of origin (the Guyanan Shield area; Righi, 1984c) both naturally and aided by indigenous groups, who transported various materials that may have contained either cocoons or small individuals (Righi, 1990a; P. Lavelle, personal communication). This species, although native to Brazil, must therefore be considered a euryecious (wide-ranging tolerance for different habitats) peregrine invader in most of Brazil. Nevertheless, it has not been treated as such, and little has been done to reduce its spread to new areas. This phenomenon has occurred regularly with deforestation and other land use transformations (Barros *et al.*, 2001; Lavelle and Lapied, 2003), and has been associated with negative effects on soil structure (Chauvel *et al.*, 1999; Barros *et al.*, 2004;), plant production (Brown *et al.*, 1999) and native earthworm communities (Lapied and

Lavelle, 2003), although there is little solid evidence for the latter, as of yet.

The first earthworm ecological surveys in Brazil were performed by Christa Knäpper in the late 1960s in Rio Grande do Sul, in collaboration with Josef Hauser and with the taxonomic help of Righi. These studies addressed earthworm distribution in various different habitat types of 36 counties in RS (Righi and Knäpper, 1965; Righi, 1967c; Knäpper 1972a,b, 1977; Knäpper and Porto, 1979). Eighteen earthworm species were found, 15 of them exotic, as the samples were taken primarily in disturbed habitats. In many cases, earthworm abundance was very low, often less than 1 individual/m^2 (Table 4.2). Similarly low densities have been found in natural environments as well (Fragoso and Lavelle, 1992) (Table 4.2; see later discussion): in various vegetation types of the cerrado region by Dias *et al.* (1997), in central Amazonia by Römbke *et al.* (1999) and in high-altitude grasslands and forests by Castro and d'Agosto (1999). These data seem to support the notion that native earthworm abundance is generally low in well-preserved natural ecosystems, except in swampy areas and where *P. corethrurus* has invaded (Table 4.2, S.W. James and G.G. Brown, personal observation).

From 1969 to 1972, under the direction of Righi and as part of her doctorate dissertation, Caballero (1973) performed a large-scale study of earthworm species diversity and distribution in north-west São Paulo, including 52 sites in 48 counties, covering an area of approximately 31,000 km^2. At each site, earthworms were collected in five 60 × 60 cm holes, to a depth of 60 cm in both the rainy and dry seasons. The dominant vegetation in the area was originally cerrado, although most of the samples were taken in secondary vegetation, grasslands, pastures, riverbanks and swamps. The mean number of earthworms found at each of the 52 sites ranged from 58 to 188 individuals/m^2. She found eight earthworm species, four native and four exotic. *P. corethrurus* was found at every site, and a native minhocuçu, *Glossoscolex vizottoi*, was found in slightly over one-half the sites. *Drilocrius iheringi* (wetland species) was found at four sites and *Glossoscolex grandis ibirai* was found at only one site.

Caballero (1975) also tested the best size for sampling earthworms in the region, and concluded that 30 × 30 cm holes at a depth of 30 cm were not large enough to properly estimate the number of large earthworm species present at her sample sites. The dimensions that recovered the greatest number of individuals per unit area were 60 × 50 and 60 × 60 cm to a depth of 50 or 60 cm. In addition, she evaluated the effect of pasture transformation to annual cropping on earthworm populations in Votuporanga (Table 4.2), and mapped the distribution of *P. corethrurus* in a grassland, correlating their abundance with soil moisture and vegetation characteristics (Caballero, 1973). Finally, both Caballero (1973) and Knäpper and Porto (1979) related species presence with soil types and their properties and main vegetation/landscape physionomies, pointing out the preference of particular earthworm species to specific soil types studied. For instance, the two *Glossoscolex* spp. collected in São Paulo were found in only one of the seven soil types studied (Caballero, 1973), while *D. iheringi* was found only in the inundated areas with hydromorphic soils.

In an extensive survey of the earthworms of Mato Grosso and Rondônia, Righi (1990a) found 45 earthworm species, 37 of them native and 8 exotic. Of the natives, 26 were glossoscolecids, 10 ocnerodrilids and 1 acanthodrilid. Of the exotics, six were *Dichogaster* spp. and the other two ocnerodrilids. The majority of the species were associated with hygrophylous habitats; 27 species were found strictly next to or close to watercourses. The remaining 18 species were associated with mostly terrestrial habitats, being found in forests, gardens and urban and agricultural areas. Exotic species were found mainly in the disturbed habitats, while several native species (various ocnerodrilids, *P. corethrurus*, *Goiascolex pepus*) were present in both natural and disturbed habitats.

Table 4.2. Quantitative estimates of earthworm populations in various Brazilian ecosystems (from various sources).

State county	Ecosystem/management[71]	Abundance[72] (individuals/m^2)	Biomass[73] (g/m^2)	Species/families	References
Acre					
Rio Branco	Rubber plantation	23	39.0	*Rhinodrilus curiosus*, *P. corethrurus*, *Chibui bari*	Guerra (1988b)
	Pasture	3*–45	0.8*–27.6	*Rhinodrilus curiosus*, *P. corethrurus*, *Chibui bari*	Guerra (1994)
	Secondary forest	3–30	0.3*–19.9	*Rhinodrilus curiosus*, *P. corethrurus*, *Chibui bari*	
Amazonas					
Near Manaus	Agrosilvicultural systems	232–323	13.1–39.8	*P. corethrurus*	Barros *et al.* (2003)
	High-input agrosilviculture	205	16.7	*P. corethrurus*	
	Low-input agrosilviculture	107	5.7	*P. corethrurus*	
	Fallow (secondary forest)	43	4.9	*P. corethrurus* and native spp.	
	Primary forest (next to river)	635–1300	ND[74]	*Tuiba dianae*[75]	Adis and Bogen (1982)
	Forestry polycultures	0–5.5†	0–33†	*Andiorrhinus amazonius*, *P. corethrurus*, *R. contortus*, *R. priollii*, *U. brasiliensis*, *T. dianae*	Römbke *et al.* (1999)
	Secondary forest	1–4†	1.1–5.6†	*Andiorrhinus amazonius*, *P. corethrurus*, *R. contortus*, *R. priollii*, *U. brasiliensis*, *T. dianae*	
	Primary forest	1–9†	2–35†	*Andiorrhinus amazonius*, *P. corethrurus*, *R. contortus*, *R. priollii*, *U. brasiliensis*, *T. dianae*	
	Pastures (*Brachiaria* sp.), 2–15 years	0*–602	0*–50.2	*P. corethrurus*	Blanchart and Antony (1996)
	Agrosilviculture	61	20.9	*P. corethrurus*	
	Primary forest	202	73.2	*P. corethrurus*	
	Tree plantations	14–21*	ND	ND	Harada and Bandeira (1994)
	Primary forest	10*	ND	ND	

Distrito Federal					
Brasília	Native grassland	0	0	ND	Dias *et al.* (1997)
	Open cerrado	0–3	ND	ND	
	Closed cerrado	0	0	ND	
	Gallery forest	10	4.7	ND	
Goiás					
Santa Helena	Annual crops, NT	288–340	5.1–27.0	*Dichogaster* sp., *P. corethrurus*	Minette (2000)
	Annual crops, CT	0–52	0–0.3	*Dichogaster* sp., *P. corethrurus*	
	Pasture (*Brachiaria* sp.)	36	0.3	*Dichogaster* sp., *P. corethrurus*	
	Cerrado	16	0.1	*Dichogaster* sp., *P. corethrurus*	
Mato Grosso do Sul					
	Annual crops, NT	6–435	ND	ND	Aquino *et al.* (2000b)
	Annual crops, CT	10	ND	ND	
	Integrated pasture cropping	30–346	ND	ND	
	Pasture (*Brachiaria* sp.)	195	ND	ND	
	Cerrado	13	ND	ND	
Minas Gerais					
Uberlândia,	Maize, NT	19.2	0.1	Two unidentified species	Pasini *et al.* (2003)
	Recovered pastures (*Brachiaria* sp.)	26–147	0.1–2.0	Two unidentified species	
	Cerrado	0	0	NA	
São Sebastião do Paraíso	Organic coffee	145–640	ND	*P. corethrurus*	Aquino *et al.* (1998a, 2000a)
	Conventional coffee	3–112	ND	*P. corethrurus*	
Ibitipoca State Park	High altitude grassland	<1	ND	*R. garbei*	Castro and d'Agosto (1999)
	High altitude forest	<1	ND	*R. garbei*	
	Gallery forest	<1	ND	*R. garbei, R. hoeflingae, R. senchenbergi* (?)	
Viçosa	Pasture (*Panicum maximum*)	~235	ND	ND	Resende *et al.* (2002)
	Forest	~55	ND	ND	

Continued

Table 4.2. Quantitative estimates of earthworm populations in various Brazilian ecosystems (from various sources). – cont'd

State county	Ecosystem/management[71]	Abundance[72] (individuals/m^2)	Biomass[73] (g/m^2)	Species/families	References
Paraíba					
Areia	Polyculture	14*–152	3.9*–57.7	Mostly *P. corethrurus*, also *A. gracilis*, *D. affinis*, *E. eugeniae*	Guerra and Silva (1994)
João Pessoa	Pasture	10*–31	2.3*–9.8	Mostly *P. corethurus* and *R. tocantinensis*; also *D. gracilis*, *H. africanus*	
	Secondary forest	0*–4	0*–2.4	Mostly *L. paraibaensis*, some *P. corethrurus*	
Paraná					
Guarapuava	Annual crops, NT	3–12	1.4–2.4	ND	Mafra *et al.* (2002)
	Annual crops, CT	0	0	ND	
	Native forest	6	1.5	ND	
São Mateus do Sul	Recently reclaimed (1 year)	0†	ND	ND	Sautter *et al.* (1995)
	Wheat/clover (6 year)	15†	ND	ND	
	Eucalyptus plantations (8–16 year)	0–2†	ND	ND	Sautter *et al.* (1995), Dionísio *et al.* (1995)
	Pasture (6 year)	1–6†	ND	ND	Dionísio *et al.* (1995)
	Secondary forest	39	ND	ND	
Londrina	Annual crops, NT	40*–100	0.4*–0.8	*Dichogaster* sp., *P. corethrurus*, *Belladrilus* n. sp.	Brown *et al.* (2003, 2004)
	Annual crops, CT	0–24*	0–0.2*	*Dichogaster* sp., *P. corethrurus*, *Belladrilus* n. sp.	
	Annual crops, MT	8*–80	0.3–1.0	*Dichogaster* sp., *P. corethrurus*, *Belladrilus* n. sp.	
	Cover crops	56	ND	ND	
	Native forest	40	1.6	*Glossoscolex* n. sp., *U. brasiliensis*, *Urobenus* sp.	
Sertanópolis	Native forest	18*–54	0.2*–2.0	*Glossoscolex* n. sp., *U. brasiliensis*, *Urobenus* sp.	
Cornélio Procópio	Native forest	16	0.2	ND	Brown *et al.* (2004)
	Soybean, NT	176	2.2	ND	

Jaguapitã	Sugar cane	0–20	0–0.2	*D. saliens*, *D. affinis*, *P. corethrurus*, *Fimoscolex* n. sp.	Nunes *et al.* (2004), Pasini *et al.* (2004)
	Soybean, CT	2–13	0.1–0.2	*D. affinis*, *Fimoscolex* n. sp.	
	Pastures	15*–189	up to 11.4	*D. affinis*, *D. bolaui*, *D. saliens*, *P. corethrurus*, *Eukerria stagnalis*, Ocnerodrilidae n. sp., *Haplodrilus michaelseni*, *Ocnerodrilus occidentalis*, *Fimoscolex* n. sp., *Glossoscolex* n. sp., *Belladrilus* n. sp.	
	Secondary forest	0–2	ND	ND	
São Jerônimo da Serra	Organic soybean, CT	42	1.6	*P. corethrurus*	Brown *et al.* (2004)
	Organic soybean, NT	142	10.9	*P. corethrurus*	
Lerroville	Soybean, NT	48–240	0.3–12.2	*Dichogaster* sp., *Andiorrhinus* n. sp.	
Cafeara	Annual crops, NT	13–35	0.2–0.9	*Dichogaster* spp., *P. corethrurus*, *Glossoscolex* n. sp., *Fimoscolex* n. sp., Ocnerodrilidae spp.	
	Pasture (*Brachiaria*)	90	10.7	*Dichogaster* spp., *P. corethrurus*, *Glossoscolex* n. sp., *Fimoscolex* n. sp., Ocnerodrilidae spp.	
Campo Mourão	Soybean, NT	12–144	0.1–1.4	*Dichogaster* spp.	
	Soybean, MT	36	0.3	*Dichogaster* spp.	
	Soybean, CT	24	0.1	*Dichogaster* spp.	
Carambeí	Annual crops, NT	38–170	5.1–50	*Amynthas corticis*, *A. gracilis*, *Dichogaster* spp.	Tanck *et al.* (2000), G. Brown (personal observation, 2004)
	Annual crops, CT	0–6	0–0.6	*Amynthas corticis*, *A. gracilis*, *Dichogaster* spp.	
	Native grassland	0	0	NA	
	Secondary *Araucaria* forest	30–95	16.5–65.1	*Amynthas corticis*, *A. gracilis*, *Dichogaster* spp.	

Continued

Table 4.2. Quantitative estimates of earthworm populations in various Brazilian ecosystems (from various sources). – cont'd

State county	Ecosystem/management[71]	Abundance[72] (individuals/m^2)	Biomass[73] (g/m^2)	Species/families	References
Rolândia	Annual crops, NT	3*–13	ND	ND	Brown *et al.* (2003)
	Annual crops, CT	6	ND	ND	
	Eucalyptus sp.	12*	0.7	ND	
	Pasture (*Brachiaria* sp.)	2*	0.4	ND	
Arapoti	Annual crops, NT	72–168	ND	*A. corticis, A. gracilis, P. corethrurus, Dichogaster* spp.	Peixoto and Marochi (1996) G. Brown (personal observation, 2004)
Ponta Grossa	Annual crops, NT	44–117	ND	*A. corticis, A. gracilis*	Voss (1986)
	Annual crops, CT	0	0	NA	
Bela Vista do Paraíso	Annual crops, NT	10*–291	0.2*–1.5	ND	Brown *et al.* (2002)
	Annual crops, MT	86*–122	0.4*–1.7	ND	
	Secondary forest	13*–51	0.7	ND	
	Pasture	48*–182	0.8	ND	
Castro	Pasture	270†	83.4†	*D. bolaui, A. corticis, A. gracilis, A. morrisi, M. californica, Lumbricidae sp.*	Ressetti (2004)[76]
	Annual crops, NT	123†	21.7†	*Lumbricidae sp., A. corticis, A. gracilis, A. morrisi, M. californica*	
	Secondary forest	39†	4.3†	*D. bolaui, A. corticis, A. morrisi*	
Curitiba	Pasture	8–123†	0.3–45.4†	Mostly *Amynthas* sp.	
	Orchard	8–92†	0.1–20.3†	Mostly *Amynthas* sp.	
	Secondary forest	0–54†	0–25.5†	Mostly *Amynthas* sp.	
	Grassland	0–31†	0–25.1†	Mostly *Amynthas* sp.	
Rio de Janeiro					
Seropédica, RJ	*Passiflora* + green manure	640	17	Mostly *P. corethrurus*	Aquino *et al.* (1998b)
	Banana plantation	85	1.2	Mostly *P. corethrurus*	
	Annual crops, organic	0–90	ND	Mostly *P. corethrurus*	Aquino (2001)
	Annual crops, NT	67–320	37	Mostly *P. corethrurus*	Aquino (2001), Rodrigues *et al.* (2004)

	Annual crops, CT	140–180	19–23	Mostly *P. corethrurus*	Rodrigues *et al.* (2004)
	Fallow	155	14	Mostly *P. corethrurus*	
	Secondary forest	64	ND	Mostly *P. corethrurus*	Aquino (2001)
Juparaná	Coffee	18–34	ND	ND	Pimentel *et al.* (2002)
	Pasture	0	ND	ND	
	Secondary forest	88	ND	ND	
Roraima					
Boa Vista	Savannah	0*–23	0*–8.8	Mostly *P. corethrurus*, few *P. roraimensis*	Guerra (1994b)
	Transition savannah forest	0*–43	0*–21.1	Mostly *P. roraimensis*, also *P. corethrurus*	
	Forest	0*–27	0*–12	*P. roraimensis* and *P. corethrurus*	
Ilha de Maracá	Savannah-forest ecotone	up to 625	ND	*Pontsocolex*, *Glossodrilus* and *Righiodrilus* spp.	Righi (1998a)
Confiança	Agroforestry systems	64	ND	ND	Moreira *et al.* (1998)
	Fallow (secondary forest)	176	ND	ND	
	Primary forest	464	ND	ND	
Rio Grande do Sul					
Santa Maria	Annual cropping	8–20	ND	ND	Lasta *et al.* (2002)
	Cover crops after CT	0*–35	ND	ND	Campos *et al.* (1993)
Teutônia	Annual crops, NT	28*–299	ND	*A. morrisi*, *A. gracilis*, *M. californica*, *A. corticis*	Krabbe *et al.* (1993, 1994)
	Annual crops, CT	0*–13	ND	ND	
	Secondary forest	35*–96	ND	*A. morrisi*, *A. corticis*	
Rio Grande	Lowland, irrigated rice	>600	ND	Mostly Ocnerodrilidae (*Eukerria* spp.)	Silva *et al.* (2003), G. Brown (personal communication, 2003)
Not specified	Grassland	<1	ND	Mostly *A. caliginosa*	Knäpper (1972)
	Annual crops	1	ND	Mostly *Amynthas spp.*	
	Forest	<1	ND	Mostly *P. corethrurus*	
São Paulo					
Botucatu	N.S.	243	183	*P. corethrurus*	Miklós (1996)
Onda Verde	Grassland	3–139	ND	*P. corethrurus*	Caballero (1973)

Continued

Table 4.2. Quantitative estimates of earthworm populations in various Brazilian ecosystems (from various sources). – cont'd

State county	Ecosystem/management[71]	Abundance[72] (individuals/m^2)	Biomass[73] (g/m^2)	Species/families	References
Votuporanga	Annual crops (1st year)	105	ND	*P. corethrurus*, *G. vizottoi*	Caballero (1973)
	Pasture	120	ND	*P. corethrurus*, *G. vizottoi*	
São Roque	Organic strawberry	72	ND	*P. corethrurus*, *Amynthas* sp.	Uzêda and Garcia (2006)
	Secondary forest	84	ND	*P. corethrurus*, *Amynthas* sp.	
	Organic farms	86	ND	ND	Borges and Espíndola (1999)
	Transitional farms	51	ND	ND	
	Conventional farms	8	ND	ND	
Taciba	Soybean, NT (1st year)	138	7.4	*Glossoscolex* n. sp., *P. corethrurus*, *Dichogaster* spp.	Brown *et al.* (unpublished data)
	Soybean, CT (1st year)	74	4.3	*Glossoscolex* n. sp., *P. corethrurus*, *Dichogaster* spp.	
	Pasture (*Brachiaria* sp.)	240	8.8	*Glossoscolex* n. sp., *P. corethrurus*, *Dichogaster* spp.	
São Carlos	Pasture (*Brachiaria* sp.)	218*–278	NA	ND	Brigante (2000)
	Pasture (*Panicum* sp.)	550	NA	ND	
	Forest	70*–83	NA	ND	

[71] CT = conventional tillage, NT = no tillage, MT = minimum tillage.
[72] Most samples were taken in the wet (rainy) season using the TSBF method (Anderson and Ingram, 1993) or slightly modified version thereof. When taken with formalin, samples are identified by a dagger. When taken in the dry season, samples are identified with an asterisk.
[73] Fresh weight (directly from field or formalin preserved).
[74] Not determined.
[75] The authors state the earthworm species as being *Tairona tipema*. The correct name is *Tuiba dianae*.
[76] The study of Ressetti (2004) actually contains a comparison of two extractants: formalin and alil isothiocyanate.

Earthworm diversity and distribution according to habitat type has been evaluated in several Amazonian sites (mostly near Manaus), although only a few of these studies are published in the readily available literature. In the first study performed by researchers and students of INPA, Ayres and Guerra (1981) listed the preference of 40 earthworm species (identified by Righi *et al.*, 1976, 1978) to eight habitat types near Manaus, showing how most species were associated strictly with hygrophylous habitats, while only a few species were found in the 'terra firme' (dry upland) forest (see earlier discussion).

More recently, Römbke *et al.* (1999) and Zicsi *et al.* (2001) showed the differences in species composition, total biomass and density of earthworm communities found in various land use management systems (forestry polycultures, primary and secondary forest) at the Embrapa research station north of Manaus (Table 4.2). Of the ten species found, only one was exotic (*D. bolaui*). The natives *Rhinodrilus contortus* and *R. priollii* were the most conspicuous due to their large size (*R. priollii* can be >1 m in length), while *U. brasiliensis* and *T. dianae* were the most common. The smaller species, *Pontoscolex vandersleeni* and *C. righii*, were very rare and encountered only in Berlese samples (Höfer *et al.*, 2001).

In Rio Branco, AC, Guerra (1988b, 1994a) found three species (*C. bari*, *P. corethrurus*, *Rhinodrilus curiosus*) living in a secondary forest, rubber plantation and a neighbouring pasture (Table 4.2). *C. bari* dominated in terms of biomass, followed by *R. curiosus* (both are large earthworms). *P. corethrurus* was dominant in abundance, remaining active throughout the year, even during the dry season. The other two species both aestivated during the dry season, below a depth of 30 cm (maximum depth of sampling).

In a forest–savannah ecotone near Boa Vista, RR, Guerra (1994b) found only two earthworm species (*P. corethrurus* and *Pontoscolex roraimensis*). *P. corethrurus* was the most abundant earthworm species in all habitats, while *P. roraimensis* seemed to prefer the transition zone rather than the other two habitats (Table 4.2), concentrating their surface casting activities in this zone. In contrast, not far from the above site, Righi (1990b, 1998a) found 12 species inhabiting the Ilha de Maracá (ten native, two exotics), and related this higher diversity to the greater diversity of sample sites (Righi, 1997), although the larger collection area probably also plays a role.

Near João Pessoa, PB, Guerra and Silva (1994) found a mixture of native and exotic earthworm species in a polyculture (*P. corethrurus* + three peregrines/exotics) and a pasture (*P. corethrurus*, *G. tocantinensis* + two exotics). Earthworms, particularly *P. corethrurus*, were more abundant (up to 152 individuals/m^2) in the polyculture than in the pasture (up to 31 individuals/m^2). In contrast, only two native species (mostly *Lourdesia paraibaensis*, some *P. corethrurus*) were found, in very low densities, in a nearby secondary forest (Table 4.2).

In a variety of different highland (1050–1650 m) ecosystems of the Parque Estadual do Ibitipoca, MG, Castro and d'Agosto (1999) found five species of *Rhinodrilus*, all of them very rare (Table 4.2). On 38 sampling dates from April 1993 to January 1997, five 4 m^2 samples were taken (to a depth of 25 cm) and hand-sorted in each of the four vegetation types studied (total 80 m^2 sampled per sample date). In all these samples, the authors encountered only 64 individuals. In most samples the authors encountered no earthworms. Most researchers, even the most persistent, would have given up after the first year, but the authors continued their work for almost 3 more years. The most commonly encountered species was *Rhinodrilus garbei* (78% of all individuals). However, two of the species found are probably misidentified and must be restudied: *R. curtus* (four individuals found), known from Trinidad, is unlikely to be the species found in this remote and well-conserved park; the description of *R. fafner*, a giant minhocuçu that may be extinct, did not correspond to the photos of a moderately sized earthworm taken by the author (S.W. James and G.G. Brown, personal observation). These two are probably new species. Another species, *Rhinodrilus senckenbergi*, known from the Rio Doce region in the neighbouring state (ES), is also unlikely to be the same species as that found by the author in this highland habitat. All these earthworms must be re-evaluated to confirm their identification.

In the state of Paraná, Brown *et al.* (2004) studied approximately 50 sites, of which 18 were annual cropping systems (see earlier discussion). Exotic species were found only in secondary vegetation (forests and disturbed grasslands) and agricultural areas. Under well-conserved native vegetation, exotics were absent and native species predominated. Only a few native species (particularly of the Glossoscolecidae and Ocnerodrilidae families) were found in disturbed areas, such as pine forests, grazed grasslands, introduced pastures and gardens/orchards.

In much of Brazil, *Amynthas* spp. have extensively colonized gardens, orchards and croplands, especially in sites close to human habitations (Brown *et al.*, 2004). Under no-tillage (NT) planting, which now occupies 20 million ha in Brazil, both *Dichogaster* and *Amynthas* spp. have spread rapidly, and are quite commonly found in Paraná, where NT covers 5.5 million ha (25% of the state's surface). With NT, organic matter contents in the topsoil recover gradually (at rates from 0.5 to 1 t ha/year (Sá *et al.*, 2001; Franchini *et al.*, 2004)), due to the abandonment of tillage and lower residue decomposition rates. Consequently, soil macrofauna (Brown *et al.*, 2001) and earthworm populations increase as the number of years of NT adoption also increases (Brown *et al.*, 2003). Furthermore, the earthworms also concentrate in higher numbers where the soil has higher organic matter content (Brown *et al.*, 2004).

Considering the data presented above, it appears that native earthworms are ill suited for survival in disturbed habitats, particularly annual cropping systems, while exotic or peregrine species are better suited to the conditions created by cropping, and/or may be just opportunistic invaders that occupy empty niches left by the native species that disappeared after transformation of the native or former vegetation (e.g. tropical forests) for agricultural uses. The only exception to this rule appears to be *P. corethrurus*.

Quantitative estimates of earthworm populations

A large number of Brazilian sites have been sampled for quantitative estimates of earthworm populations. These results are summarized in Table 4.2 and show that:

- very few authors measured earthworm biomass and many did not identify the species they found;
- native earthworms were found in low abundance in natural ecosystems (see above), with the notable exception of the horizontally migrating earthworms of the Negro River valley (Adis and Bogen, 1982);
- native earthworms were rarely found in transformed or more intensively managed ecosystems;
- agrosilviculture, fruit production systems (coffee, passion fruit), improved pastures, organic and NT crop production favoured earthworm populations, generally of peregrine species (e.g. *P. corethrurus*) or exotics (*Dichogaster* and *Amynthas* spp.);
- conventional agricultural practices (including tillage) were detrimental to earthworm populations;
- in most sites with seasonal rainfall regimes, earthworm populations essentially followed the patterns of monthly rainfall and soil moisture contents, being less abundant in the dry season and more in the wet season.

The highest earthworm abundances were generally observed in pastures and other areas with permanent cover, where *P. corethrurus* often dominated, and in hygrophylous habitats (rice, *várzea* soils), where water is not a major limitation to earthworm activities, or they are able to overcome those activities by adaptation (see earlier discussions). Both the native habitats (forests, cerrado, grasslands) and annual cropping systems (particularly those with frequent tillage) generally had very low earthworm abundance values. Nevertheless, the fact that certain management practices promoted or maintained reasonably high earthworm abundances is a positive aspect of earthworm management that needs further investigation. Techniques for increasing earthworm populations where they are low and for maintaining large and active populations must be found and promoted, particularly considering the potentially important effects of earthworms on soils and plants (Brown *et al.*, 1999; see later). For instance, Voss (1986), reported a large increase in the population of *A. corticis* and *A. gracilis* (from nearly 0 to 108 individuals/m^2) after only 4 years of adoption of NT, and the abandonment of conventional disc tillage in the region near Ponta Grossa, PR. Nearby, in Arapotí, after 6.5 years of NT, the *Amynthas* invasion front stabilized at over 200 individuals/m^2 (Peixoto and Marochi, 1996). In fact, farmers in this county even developed a method of field inoculation consisting of spreading batches of composted manures containing high populations of *Amynthas* spp. in selected sites of their fields (e.g. close to the bunds for erosion control), and at intervals of a certain distance, to promote and speed colonization by these earthworm species.

Effects on soil properties

Few estimates of the annual surface cast production by earthworms have been made in Brazil: the native minhocuçu *C. bari* produced 24 mg/ha/year in a grassy (treeless) area and 88 mg/ha/year under arboreal cover in Rio Branco (AC) (Guerra 1988a); *P. roraimensis* only produced from 0.5 to ~1.3 mg/ha/year in a forest–savannah ecotone north of Boa Vista (RR) (Guerra, 1994b). The casts of *C. bari* had a higher pH and had higher levels of in C, Ca, Mg, K and P compared with the surrounding soils at various depths (Guerra, 1994a). Other authors (Dadalto and Costa, 1990; Santos *et al.*, 1996; Demattê *et al.*, 1998; Quadros *et al.*, 1998; Peneireiro, 1999) collected surface castings of various species (most not identified) only at particular points in time (not seasonally) to measure their nutrient contents. They found the typical changes in soil texture and increases in organic matter and plant-available nutrients (e.g. P, Ca, Mg) in the castings (compared with the surrounding topsoil), observed in many earthworm species' casts in the tropics (Barois *et al.*, 1999). Bernardes and Kiehl (1992) believed that the changes in nutrient contents of castings of *P. corethrurus* were mainly attributed to preferential selection of smaller (clay + silt) particles by this species, and not to contributions by earthworm metabolism. Nevertheless, we know that the nephridial (N) excretions and $CaCO_3$ secretions of earthworms play an important role in Ca and inorganic N availability in castings and their pH.

When casts are deposited within the soil and back-fill the burrows, they are called crotovinas or pedotubules. These structures often contain soil of a slightly or very different colour than the surrounding soil, depending on the differences in soil colour within the profile and the food source of the earthworm. If the earthworms are feeding at the surface and burying litter (anecics), the crotovinas are generally much darker in colour than the surrounding soil, due to the organic matter in the casts. These casts can be important sources of plant nutrients and act as hot spots of plant root growth and/or microbial activity (Brown *et al.*, 1999; Resende *et al.*, 2002), particularly when the difference between their nutrient content and the surrounding soil is high (e.g. casts deposited deeper in the soil). Crotovinas at depths of 2 m were observed at the Ilha do Maracá, RR (Righi, 1997), and galleries of extinct earthworms at depths of up to 20–30 m have been found especially in latossols of south-east Brazil but also in other regions of the country (Resende *et al.*, 2002).

The short- and long-term casting activities of earthworms can have significant cumulative effects on populations and activity of other soil organisms (Brown and

Doube, 2004), as well as on soil chemical characteristics (especially organic matter, nutrient availability and pH), structure and pedogenesis (Darwin, 1881; Lee, 1985).

Hence, various authors (e.g. Primavesi and Covolo, 1968; Guerra, 1982; Langenbach *et al.*, 2002) found higher pH and nutrient contents (e.g. Ca, Mg, P, K, NO_3) in worm-worked soils although these differences were probably due in part to changes in the surface organic matter decomposition and nutrient mineralization rates induced by earthworm presence. In greenhouse cultures, Guerra and Asakawa (1981) found two to five times greater microbial populations (measured by plate counts) in soils with *P. corethrurus* than in soils without earthworms. This increase was attributed to the intense soil + organic matter mixing activities of the earthworms, and the deposition of castings throughout the soils of the experimental pots. On the other hand, Bernardes *et al.* (1998) observed no effect of *P. corethrurus* on soil respiration rates in laboratory cultures, and Sparovek *et al.* (1999) found no effects of these earthworms on various soil microbial and chemical parameters when they were inoculated under field conditions.

The continual deposition of surface castings may be an important factor in the alteration of topsoil structure and even its texture, if the earthworms are ingesting soil particle sizes preferentially (Guerra, 1994b; Bernardes and Kiehl, 1995a; Nooren *et al.*, 1995). In Botucatu, SP, Miklós (1992) observed an abundant grumous (well-aggregated) structure in the topsoil of a toposequence, and related this to the deposition of castings (mainly *P. corethrurus*). Furthermore, he (Miklós, 1996) estimated that the earthworms present (numerical abundance and biomass in Table 4.2) could be ingesting as much as 3100 mg of soil per year.

In hydromorphic soils of the upper river Negro basin, Mafra *et al.* (2006) observed the mound-building activities of two species of minhocuçus (not identified). More than three mounds per square metre were counted, with an average height of 14 cm and a basal diameter of 17 cm. They calculated a surface cast deposition rate of 30 mg/ha. In the topsoil, evidence of intense mixing was found, with an abundance of large castings of 6–12 mm in diameter. In the subsurface horizons (below 20 cm), the evidence of earthworm activities was visible mostly as pedotubules, frequently enriched with organic matter brought from the surface by the earthworms. Similar mounds are produced by minhocuçus in hydromorphic soils at various sites throughout Brazil, in both the cerrados and Amazonian regions (Assad, 1997; G.G. Brown, personal observation, 1999; F. Bernardes, personal communication, 2003). The species involved are generally of the *Rhinodrilus* and *Glossoscolex* genera.

With deforestation and the introduction of pastures (*Brachiaria* spp.) near Manaus, native earthworms from the forest quickly disappear, and *P. corethrurus* invade, taking over completely the initial population (Barros *et al.*, 1998). Under these conditions, the invading *P. corethrurus* reach numerical abundance and biomass values up to ~365 individuals and 45 g/m^2 (Barros *et al.*, 1996) and produce more than 100 mg/ha of surface castings, which coalesce with the abundant rainfall and dramatically decrease soil macroporosity down to a level equivalent to that produced by the action of heavy machinery on the soil (Chauvel *et al.*, 1999; Barros *et al.*, 2001). During the rainy season these casts plug up the soil surface, saturating the soil and producing a thick muddy layer, where anaerobic conditions prevail (simultaneously increasing both methane emission and denitrification). In the dry season, desiccation cracks the surface, blocking root growth and hindering their ability to extract water from the soil. The plants then wilt and die, leaving bare patches in the pasture.

Barros *et al.* (2001) demonstrated the role of a diverse assemblage of soil macrofauna, including decompacting species, in the recovery of these compacted soils. Soil monoliths measuring 25 × 25 cm^2 were removed from the pasture and placed in the forest; similar blocks were also taken from the forest and placed in the pasture. After 1 year, the structure of the compacted pasture soil was completely restored to levels of those typical in native forest soils by the action of the diverse community of forest soil invertebrates. Meanwhile, the macroaggregate structure of the forest soil was completely destroyed by *P. corethrurus* in the pasture, reaching compaction and porosity levels similar to those of the degraded pasture. This research highlights not only the important role of a diverse macroinvertebrate community in soil structure maintenance (especially in these kaolinitic soils), but also the problems associated with management practices that are not well adapted to the environment (extensive pastures on problem soils after deforestation), and the role of invasive earthworm species on ecosystem properties and processes.

Effects on plant growth

Besides the observations of Chauvel *et al.* (1999), only a few experiments have been performed on the effects of earthworms on plants in Brazil. Guerra (1982) grew maize in large containers in the greenhouse in the presence and absence of incorporated maize residues and earthworms (*P. corethrurus*). One half of the treatments received a liquid solution of radioactive ^{32}P fertilizer and the other half received ^{32}P-labelled maize residues. After 30 days, no differences were found in maize biomass between the worm and no-worm treatments, although the presence of earthworms significantly increased total plant shoot P content by 50%, ^{32}P content by three times when the P fertilizer was applied and by 30% when the source of ^{32}P was the maize residues. He attributed these results to the enhanced liberation and availability of P from the soil, residues and the fertilizer where earthworms were inoculated.

In another greenhouse experiment, Soares and Lambais (1998) evaluated the effect of *P. corethrurus* with or without added organic matter (composted manure) on *Brachiaria decumbens* growth and AMF root colonization. After 60 days, pots with earthworms produced about two and three times more shoot biomass than no-worm pots when organic matter was absent and present, respectively. Although no differences in mycorrhizal colonization were observed, there were indications that the positive effects of earthworms on the plants were due to increased N availability and uptake by the plants.

In contrast, a study by Sparovek *et al.* (1999) showed a negative effect of *P. corethrurus* on soil structure (increased compaction) and no effects on the grain and/or shoots of wheat, *Crotalaria juncea*, sorghum and maize, when this species was inoculated into field mesocosms that had the top 50 cm of the soil removed (to simulate a degraded soil). Organic manure, amended in some treatments, was much more efficient than the earthworms in stimulating soil regeneration and plant production, as it helped reduce mostly the physical (compaction) limitations to plant growth.

In Arapotí, PR, Peixoto and Marochi (1996) showed how the invasion front of *Amynthas* spp. into NT cropland significantly altered soil structure and water-holding capacity. Most of the top 10 cm of the soil consisted of earthworm castings, and these had a major effect on increasing water infiltration and on the availability of several plant nutrients. In the invaded area, grain yields of wheat and soybean increased by 47% and 51%, respectively, while the dry mass of black oat increased 22%.

Similarly, Kobiyama *et al.* (1995) found enhanced growth of tree seedlings (*Mimosa scabrella*) when *Amynthas* spp. were inoculated in 1.5 × 1.8 m field mesocosms at different densities (30, 60 or 90 individuals/m^2). Earthworms affected especially the saturated hydraulic conductivity and total porosity to a depth of 30 cm, increasing the proportion of macropores >0.06 mm in diameter. Consequently, soil water-holding capacity and plant growth increased, and was greatest when 60 individuals/m^2 were inoculated (Kobiyama, 1994; Kobiyama *et al.* 1995).

However, another study performed by Santos (1995) in 1 × 1 m field mesocosms at Guarapuava, Paraná, found only few differences in soil properties (nutrients, bulk density, water infiltration and mesofauna populations) and no significant differences in yields of wheat or black beans (*Phaseolus vulgaris*) when 30, 60 or 90 *Amynthas* spp. earthworms were inoculated. Furthermore, in greenhouse trials performed in Curitiba, Paraná, Kusdra (1998) observed negative effects of *Amynthas* spp. inoculation on black bean shoot and root biomass, and nodulation by symbiotic *Rhizobia* spp. Various hypotheses were raised for the observed decreases, including alteration of soil microbial communities with a reduction in rhizobia and mycorrhiza populations, direct contact with roots, decreasing their growth and changes in soil nutrient status, especially available N. None of these, however, was adequately addressed by his research, and many questions still remain to be answered as to the possible effects of *Amynthas* spp. invasion on soil properties and plant growth, given the highly variable responses observed so far.

Unfortunately, all the experiments and observations on earthworm effects on plants have involved exotic or peregrine species. As yet, no studies have used native worms, except for *P. corethrurus*. Considering the large diversity of earthworm species in Brazil, and the presence of various native glossoscolecid, ocnerodrilid and acanthodrilid species in some human-managed ecosystems, there is a need to study their role in soil fertility and plant production. Furthermore, as NT practices continue to expand and improve, and earthworm populations in these fields increase, it is important to assess their effects on soil aggregation, porosity, nutrient cycling and plant growth. Clearly, there are still many opportunities for further applied and basic research in this area.

Considerations for the Future of Earthworm Research in Brazil

If current estimates of earthworm biodiversity in Brazil are correct, then most of the species in the country must still be found and described. We have seen that this remains a major challenge for the few Brazilian researchers active in this field. None the less, various actions have been undertaken to overcome some of the limitations to achieve this goal:

1. Two short taxonomy courses have been offered (December 2003, May–June 2004), and more than 20 Brazilians have been trained in the basics of earthworm taxonomy and identification up to generic level.
2. The first international meeting on earthworm ecology and taxonomy (ELAETAO) was held in Londrina in late 2003. Forty people attended the meeting from 12 countries. The proceedings (Brown and Fragoso, 2006) will cover the state-of-the-art knowledge of earthworm ecology and biodiversity in most Latin American countries.
3. In 2002, the CNPq group 'Biology, ecology and function of Brazilian terrestrial oligochaetes (enchytraeids, earthworms)' was created, and it now contains about 20 researchers/professors and 4 students, in various fields of expertise. The group's six main research foci are: (i) ecological functions of terrestrial oligochaetes in natural and anthropic ecosystems (agricultural and urban); (ii) biology and mode of life of earthworms and enchytraeids; (iii) oligochaete ecotoxicology; (iv) vermiculture and its uses; (v) taxonomy and distribution of Brazilian oligochaetes; and (vi) biology and conservation of minhocuçus.
4. Two major projects with wide-ranging geographic scope are presently under way, both funded by Embrapa, with partners from several states (RR, MS, DF, PR, RJ, PR, RS). These projects are collecting earthworms from both agroecosystems and neighbouring natural native vegetation of various types (e.g. cerrado, Amazonian and Atlantic forests, grasslands, swamps). Both quantitative and qualitative samples are being taken and the species found will be identified at least to genera. New species will have to await description, joining the long list of species still to be described by the first author.

The lack of identification keys, particularly for the native species, is still a major task to be overcome in the process of building knowledge on Brazil's earthworm fauna. Righi (1979, 1995a) provided some useful keys for the general study of earthworm taxonomy and the identification of the Glossoscolecidae family genera. However, more comprehensive keys, particularly to species level for the native species, are needed to help guide those entering the study of earthworm taxonomy.

Studies on earthworm biology, population structure and distribution according to the various Brazilian habitats, ecosystems and geographic regions are also urgently needed. Very few earthworm species have been the object of biological or ecological studies; most of the work so far has been done on a single species: *P. corethrurus*. It is also essential that we advance our knowledge of South American oligochaete biogeography beyond the level of dots on a map, to an understanding of the roles of various processes responsible for placement of those dots and diversification of the fauna. Furthermore, given the rather sedentary nature of most earthworms, there is the potential to see the pattern of splitting lineages most clearly, as dispersal is quite slow. This will be valuable to students of other faunal components of the South American biota, who would then see the value of Righi's work to the community of scientists.

The ecological relationships of earthworms with their habitat and the role of earthworms in the alteration of soil properties and processes have also been poorly studied in Brazil. Most studies involved the widespread *P. corethrurus*. Nothing is known (besides some empirical observations) on the role of native earthworms, particularly minhocuçus, in plant production. We can hypothesize that these large species will have important roles based on their observed effects on soil properties. In addition, little is known (besides empirical observations) of the short- and long-term effects of the removal of minhocuçus (and other bait-worms) from their habitat on soil processes and the sustainability of these practices. We also cannot underestimate the role of smaller species that are often much more abundant and frequently concentrate in the surface horizons, living within the zone of intense biological and plant rooting activities. Some of the aquatic and semi-aquatic species may also play important ecological roles, but these have not yet been studied.

What now? Will Gilberto Righi's legacy be his publications and the specimens he gathered? The small company of earthworm taxonomists often loses its members to death, career change or retirement, without anyone stepping into the vacancy and without an orderly transfer of knowledge from mentor to student. These mishaps are presently being amended, but much help will be needed from both the scientific community and research funding agencies. We can only hope that in the forthcoming years, the great enthusiasm for biodiversity and the growing interest in soil ecology and sustainability will stimulate research on the role of these fantastic organisms in nature and their importance to humankind.

Acknowledgements

We thank Embrapa, Fullbright and Prodetab (consultancy of S.W. James) and CNPq (fellowship to G.G. Brown) for their financial assistance; P.T. Martins for help in construction of the database on earthworm species and their distribution in Brazil; K.D. Sautter for help in translation of German text; the MZUSP (C.R. Brandão, E. Cancello, S. Casari and E. Gonçalves) for access to the collection, book of specimen entries and Righi's large bibliography; J.W. de Morais for information in the book of entries of the INPA collection; R. Silva for information on the collection at UNISINOS; G. de Castro and A. Guimarães for information on their personal collections; P. Lavelle for information on the collection at MPEG; C. Young and C. Ratto for opening the collection at the MN in Rio de Janeiro; the many students, colleagues, farmers and minhoqueiros who have opened our eyes to the biodiversity and habits of Brazilian earthworms.

References

Abe, A.S. and Buck, N. (1985) Oxygen uptake of active and aestivating earthworm *Glossoscolex paulistus* (Oligochaeta, Glossoscolecidae). *Comparative Biochemistry and Physiology* 81A, 63–66.

Adis, J. and Bogen, V. (1982) Reaction of Glossoscolecidae (Annelida, Oligachaeta) to flooding in a central Amazonian inundation forest. *Acta Amazônica* 12, 741–743.

Adis, J. and Righi, G. (1989) Mass migration and life cycle adaptation – a survival strategy of terrestrial earthworms in central Amazonian inundation forest. *Amazoniana* 11, 23–30.

Anderson, J.M. and Ingram, J.S.I. (1993) *Tropical Soil Biology and Fertility: A Handbook of Methods*, 2nd edn. CAB International, Wallingford, UK.

Aquino, A.M. (2001) Comunidades de minhocas (Oligochaeta) sob diferentes sistemas de produção agrícola em várias regiões do Brasil. *Embrapa Agrobiologia, Série Documentos 146*. Seropédica.

Aquino, A.M., Almeida, D.L., Freire, L.R. and De-Polli, H. (1994) Reprodução de minhocas (Oligochaeta) em esterco de bovino e bagaço de cana-de-açúcar. *Pesquisa Agropecuária Brasileira* 29, 161–168.

Aquino, A., De-Polli, H. and Ricci, M.S.F. (1998a) Estudos preliminares sobre a população de minhocas (Oligochaeta) e biomassa microbiana no solo na transição de café sob manejo convencional para orgânico. In: *Resumos da FERTBIO 1998*. SBCS and UFLA, Lavras, Brazil, pp. 403.

Aquino, A., Belloti, R., Abboud, A.C.S. and Almeida, D.L. (1998b) Monitoramento da população de minhocas (Oligochaeta) em sistema integrado de produção agroecológica (SIPA). In: *Resumos da FERTBIO 1998*. SBCS and UFLA, Lavras, Brazil, pp. 402.

Aquino, A.M., Ricci, M.S. and Pinheiro, A.S. (2000a) Avaliação da macrofauna do solo em café orgânico e convencional utilizando um método modificado do TSBF. In: *Fertbio 2000, Biodinâmica do Solo*. Universidade Federal de Santa Maria, Santa Maria, Brazil. CD-ROM.

Aquino, A.M., Merlim, A.O., Correia, M.E.F. and Mercante, F.M. (2000b) Diversidade da macrofauna do solo como indicadora de sistemas de plantio direto para a região Oeste do Brasil. In: *Fertbio 2000, Biodinâmica do Solo*. Universidade Federal de Santa Maria, Santa Maria, Brazil. CD-ROM.

Assad, M.L.L. (1997) Fauna do solo. In: Vargas, M. and Hungria, M. (eds) *Biologia dos solos dos Cerrados*. Embrapa Cerrados, Planaltina, Brazil, pp. 363–443.

Ayres, I. and Guerra, R.A.T. (1981) Água como fator limitante na distribuição das minhocas (Annelida, Oligochaeta) da Amazônia Central. *Acta Amazônica* 11, 77–86.

Barois, I., Lavelle, P., Brossard, M., Tondoh, J., Martínez, M.A., Rossi, J.P., Senapati, B.K., Angeles, A., Fragoso, C., Jiménez, J.J., Decaëns, T., Lattaud, C., Kanyonyo, J., Blanchart, E., Chapuis-Lardy, L., Brown, G.G. and Moreno, A.G. (1999) Ecology of species with large environmental tolerance and/or extended distributions. In: Lavelle, P., Brussaard, L. and Hendrix, P.F. (eds) *Earthworm Management in Tropical Agroecosystems*. CAB International, Wallingford, UK, pp. 57–85.

Barros, M.E., Blanchart, E., Neves, A., Desjardins, T., Chauvel, A., Sarrazin, M. and Lavelle, P. (1996) Relação entre a macrofauna e a agregação do solo em três sistemas na Amazônia Central. In: *Solo-Suelo 1996, Congresso Latino-Americano de Ciência do Solo*. SBCS and SLCS, Piracicaba, Brazil. CD-ROM.

Barros, E., Grimaldi, M., Desjardins, T., Sarrazin, M., Chauvel, A. and Lavelle, P. (1998) Efeito de pastagens sobre a macrofauna e o funcionamento hídrico do solo na Amazônia Central. In: *Resumos da FERTBIO 1998*. SBCS and UFLA, Lavras, Brazil, p. 798.

Barros, E., Curmi, P., Hallaire, V., Chauvel, A. and Lavelle, P. (2001) The role of macrofauna in the transformation and reversibility of soil structure of an oxisol in the process of forest to pasture conversion. *Geoderma* 100, 193–213.

Barros, E., Neves, A., Blanchart, E., Fernandes, E.C.M., Wandelli, E. and Lavelle, P. (2003) Development of the soil macrofauna community under silvopastoral and agrosilvicultural systems in Amazonia. *Pedobiologia* 47, 273–280.

Barros, E., Grimaldi, M., Sarrazin, M., Chauvel, A., Mitja, D., Desjardins, T. and Lavelle, P. (2004) Soil physical degradation and changes in macrofaunal communities in central Amazonia. *Applied Soil Ecology* 26, 157–168.

Beddard, F.E. (1891) The classification and distribution of earthworms. *Proceedings of the Royal Physical Society* 10, 235–290.

Beddard, F.E. (1892) XVII – The earthworms of the Vienna Museum. *The Annals and Magazine of Natural History* (*Series 6*) 9, 113–134.

Benham, W.B. (1887) Studies on earthworms. No II. *Quarterly Journal of Microcopic Science (n.s.)* 27, 77–108.

Benham, W.B. (1890) Oligochaeta. *Journal of the Linnean Society (London) Series Zoology* 20, 560–563.

Bernardes, F.F. and Kiehl, J.C. (1992) Alteração das propriedades químicas do solo pelas oligoquetas. In: *Anais da XX Reunião Brasileira de Fertilidade do Solo e Nutrição de Plantas*. SBCS, Piracicaba, Brazil, pp. 142–143.

Bernardes, F.F. and Kiehl, J.C. (1993) Alterações no crescimento e na atividade de minhoca do gênero *Pontoscolex* pela adição de matéria orgânica e carbonato de cálcio. In: *Resumos do XXVI Congresso Brasileiro de Ciência do Solo*. SBCS, Goiânia, pp. 257–258.

Bernardes, F.F. and Kiehl, J.C. (1994) Comportamento da minhoca *Pontoscolex corethrurus* sob diferentes condições de temperatura e umidade do solo. In: *Anais da XXI Reunião Brasileira de Fertilidade do Solo e Nutrição de Plantas*. SBCS, Petrolina, Brazil, pp. 245–246.

Bernardes, F.F. and Kiehl, J.C. (1995a) Comportamento de oito espécies de minhocas com potencial de utilização agronômica no Brasil. In: *Anais do XXV Congresso Brasileiro de Ciência do Solo*. SBCS, Viçosa, Brazil, pp. 454–456.

Bernardes, F.F. and Kiehl, J.C. (1995b) Interações entre a *Pontoscolex corethrurus* e as propriedades químicas do solo. In: *III Simpósio de Iniciação Científica da Universidade de São Paulo, v.1*. USP, São Paulo, Brazil, pp. 386.

Bernardes, F.F. and Kiehl, J.C. (1997) Reprodução da minhoca *Pontoscolex corethrurus* em dois solos e sob dois níveis de matéria orgânica. In: *Anais do XXVI Congresso Brasileiro de Ciência do Solo*. SBCS, Rio de Janeiro. CD-ROM.

Bernardes, F.F., Ribeiro, C.M. and Klein, S.I. (1998) On the interaction of *Pontoscolex corethrurus* (Müller, 1857) and the microbiology of tropical soils. In: *16th World Congress of Soil Science*. IUSS, Montpellier, France. CD-ROM.

Blakemore, R.J. (2000) *Cosmopolitan earthworms – an Eco-Taxonomic Guide to the Peregrine Species of the World* (First CD Edition). VermEcology, Kippax, Australia, 426 pp.

Blakemore, R.J. (2002) *Cosmopolitan earthworms – an Eco-taxonomic Guide to the Peregrine Species of the World*. VermEcology, Kippax, UK. CD-ROM.

Blanchart, E. and Antony, L.M.K. (1996) Structure et dynamique de la faune du sol dans des système forestiers tropicaux convertis en pâturages ou en agrosystèmes. In: *Changements dans les chaines de décomposeurs de la matière organique dus à la mise en valeur de sols forestiers en Amazonie Centrale. Relations avec les transformations de la matière organique et la structure des sols, Final Report*. Orstom and INPA, Manaus, Brazil, pp. 6–22.

Borges, M. and Espíndola, C.R. (1999) Influência do sistema de manejo na população de minhocas, na região de São Roque (SP). In: *Anais do XXVII Congresso Brasileiro de Ciência do Solo*. SBCS, Brasília. CD-ROM.

Bouché, M.B. (1972) *Lombriciens de France. Écologie et Systématique*. INRA Publications, Paris, 671 pp.

Brandão, C.R.F., Cancello, E.M. and Yamamoto, C.I. (2005) Invertebrados terrestres. In: Lewinsohn, T.M. (ed.) Avaliação do conhecimento da diversidade biológica do Brasil. *Ministério do Meio Ambiente, Secretaria de Biodiversidade e Florestas* (in press).

Brigante, J. (2000) *Comparação de algumas comunidades de macrofauna e microrganismos de solo, encontrados em áreas de mata e pastagens, em um Latossolo*. PhD thesis, Universidade Federal de São Carlos, São Carlos, Brazil, 105 pp.

Brown, G.G. and Doube, B.M. (2004) Functional interactions between earthworms, microorganisms, organic matter, and plants. In: Edwards, C.A. (ed.) *Earthworm Ecology*. CRC Press, Boca Raton, Florida, pp. 213–239.

Brown, G.G. and Fragoso, C. (2006) *Biodiversidade e ecologia das minhocas na América Latina*. Embrapa, Londrina, Brazil.

Brown, G.G. and James, S.W. (2006) Biodiversidade e biogeografia das minhocas no estado de São Paulo, Brasil. In: Brown, G.G. and Fragoso, C. (eds) *Biodiversidade e ecologia das minhocas na América Latina*. Embrapa Soja, Londrina, Brazil (in press).

Brown, G.G., Pashanasi, B., Villenave, C., Patrón, J.C., Senapati, B.K., Giri, S., Barois, I., Lavelle, P., Blanchart, E., Blakemore, R.J., Spain, A.V. and Boyer, J. (1999) Effects of earthworms on plant production in the tropics. In: Lavelle, P., Brussaard, L. and Hendrix, P.F. (eds) *Earthworm Management in Tropical Agroecosystems*. CAB International, Wallingford, UK, pp. 87–147.

Brown, G.G., Pasini, A., Benito, N.P., Aquino, A.M. and Correia, M.E.F. (2001) Diversity and functional role of soil macrofauna communities in Brazilian no-tillage agroecosystems. In: *Proceedings of the International Symposium on Managing Biodiversity in Agricultural Ecosystems*. UNU and CBD, Montreal. CD-ROM.

Brown, G.G., Benito, N.P., Pasini, A., Sautter, K.D., Guimarães, M.F. and Torres, E. (2003) No-tillage greatly increases earthworm populations in Paraná state, Brazil. *Pedobiologia* 47, 764–771.

Brown, G.G., James, S.W., Sautter, K.D., Pasini, A., Benito, N.P., Nunes, D.H., Korasaki, V., Santos, E.F., Matsumura, C., Martins, P.T., Pavão, A., Silva, S.H., Garbelini, G. and Torres, E. (2004) Avaliação das populações de minhocas como bioindicadores ambientais no Norte e Leste do Estado do Paraná (03.02.5.14.00.02 e 03.02.5.14.00.03). In: *Resultados de Pesquisa da Embrapa Soja 2003*. Manejo de solos, plantas daninhas e agricultural de precisão. Embrapa Soja, Londrina, Série Documentos, no. 253, pp. 33–46.

Brown, K.S. Jr (1987) Conclusions, synthesis and alternative hypotheses. In: Whitmore, T.C. and Prance, G.T. (eds) *Biogeography and Quaternary History in Tropical America*. Clarendon Press, Oxford, UK, pp. 175–196.

Brown, K.S. Jr and Brown, G.G. (1992) Habitat alteration and species loss in Brazilian forests. In: Whitmore, T.C. and Sayer, J.A. (eds) *Tropical Deforestation and Species Extinction*. Chapman & Hall, London, pp. 119–142.

Buck, N. and Abe, A.S. (1990) Atividade sazonal do minhococu *Andiorrhinus samuelensis* na região de Porto Velho, Rondônia (Oligochaeta, Glossoscolecidae). *Ciência e Cultura* 42, 835–838.

Caballero, M.E.S. (1973) *Bionomia dos Oligochaeta terrestres da Região Norte-Ocidental do Estado de São Paulo*. PhD thesis, Universidade Estadual Paulista Júlio de Mesquita Filho, São José do Rio Preto, Brazil.

Caballero, M.E.S. (1975) Bionomia de Oligochaeta terrestres da Região Norte Ocidental do Estado de São Paulo I- Métodos. *Ciência e Cultura* 28, 762–765.

Caballero, M.E.S. (1979) Influência dos fatores hígricos sobre a biomassa de *Pheretima hawayana* e *Pontoscolex corethrurus* (Annelida, Oligochaeta). *Zoo Intertrópica* 2, 1–11.

Campos, B.C., Pessoa, A.C.S., Figueiredo, L.G.B., Goi, C., Vogt, A.I., Neusser, C., Antoniolli, Z.I. and Giracca, E.M.N. (1993) Estudo da meso e macrofauna do solo em espécies vegetais antecedendo milho no sistema de plantio direto. In: *Resumos do XXVI Congresso Brasileiro de Ciência do Solo*. SBCS, Goiânia, Brazil, pp. 261–262.

Castro, G.A. and d'Agosto, M. (1999) Ocupação ambiental dos oligoquetos terrestres em diferentes ambientes fitofisionômicos do Parque Estadual do Ibitipoca-MG. *Revista Brasileira de Zoociências* 1, 103–114.

Cernosvitov, L. (1934a) Les Oligochètes de la Guyane Française et d'autres pays de l'Amérique du Sud. *Bulletin du Museum Nationale d'Histoire Naturelle de Paris* 6(2), 47–59.

Cernosvitov, L. (1934b) Eine neue *Glossoscolex*-art aus den sammlungen des National Museums in Prag. *Zoologicher Anziger* 105, 183–185.

Cernosvitov, L. (1935) Oligochaeten aus dem tropischen Sud Amerika. *Capita Zoologica* 6, 1–36.

Cernosvitov, L. (1938) Deux nouveaux Oligochétes Glossoscolecides du Brésil. *Bulletin de l'Association Philomathique d'Alsace et de Lorraine* 8, 401–407.

Cernosvitov, L. (1939) Résultats scientifiques des croisières du navire-école Belge 'Mercator', Vol. 2. VII. Oligochaeta. *Musée Royal d'Histoire Naturelle de Belgique, 2ème Ser* 15, 115–122.

Cernosvitov, L. (1942) Oligochaeta from various parts of the world. *Proceedings of the Zoological Society, Series B* 111, 197–236.

Chang, Y.-C. (1997) Minireview: natural history of *Amynthas hawayanus* (Rosa, 1891). *Acta Biologica Paranaense, Curitiba*, 26, 39–50.

Chauvel, A., Grimaldi, M., Barros, E., Blanchart, E., Sarrazin, M. and Lavelle, P. (1999) Pasture degradation by an Amazonian earthworm. *Nature (London)* 389, 32–33.

Cognetti de Martiis, L. (1900) Contributo alla conoscenza degli Oligocheti neotropicali. *Boletino dei Museu di Torino*, 15, 1–15.

Cognetti de Martiis, L. (1905) Gli Oligocheti della Regione Neotropicale, I. *Memoria della Reale Academia di Scienze di Torino* 55(2), 1–72.

Cognetti de Martiis, L. (1906) Eine neue *Opisthodrilus*-art aus Brasilien. *Denkschriften der Mathematisch-Naturwissenschaftlichen Klasse der Kais Akademie der Wissenschaften, Wien* 76, 41–42.

Colinvaux, P.A. and De Oliveira, P.E. (2001) Amazon history on a Cenozoic timescale. *Paleogeography, Paleoclimatology, Paleoecology* 166, 51–63.

Cordero, E.H. (1943) Oligoquetos sudamericanos de la família Glossoscolecidae, II. Dos nuevas espécies de *Rhinodrilus* del Nordeste del Brasil. *Comunicaciones Zoologicas del Museo de Historia Natural de Montevideo* 1, 1–4.

Cordero, E.H. (1944) Oligoquetos sudamericanos de la família Glossoscolecidae, III. *Rinodrilus francisci* n. sp. de Pernambuco, Brasil. *Comunicaciones Zoologicas del Museo de Historia Natural de Montevideo* 1, 1–4.

Csuzdi, C. (1995) A catalogue of Benhamiinae species (Oligochaeta: Acanthodrilidae). *Annalen des Naturhistorischen Museums in Wien* 97B, 99–123.

Dadalto, G.G. and Costa, L.M. (1990) Relação entre características químicas de solo e excreções de minhocuçu (*Glossoscolex* spp.). *Revista Ceres* 37, 331–336.
Darwin, C. (1881) *Darwin on Earthworms. The Formation of Vegetable Mould through the Action of Worms with Observations on Their Habits.* Bookworm, Ontario, Canada (with a foreword by James P. Martin, PhD, and an introduction by Sir Albert Howard).
Demattê, J.A.M., Mafra, A.L. and Bernardes, F.F. (1998) Comportamento espectral de materiais de solos e de estruturas biogênicas associadas. *Revista Brasileira de Ciência do Solo* 22, 621–630.
Dias, V.S., Brossard, M. and Assad, M.L.L. (1997) Macrofauna edáfica invertebrada em áreas de vegetação nativa da região de cerrados. In: Leite, L.L. and Saito, C.H. (eds) *Contribuição ao conhecimento ecológico do cerrado.* UNB, Brasília, Brazil, pp. 168–173.
Dionísio, J.A., Tanck, B.C.B., Santos, A., Silveira, V.I. and Santos, H.R. (1995) Avaliação da população de oligochaeta (terrestres) em áreas degradadas. *Revista do Setor de Ciências Agrárias (Curitiba)* 13, 35–40.
Eisen, G. (1900) Researches in the American Oligochaeta, with special reference to those of the Pacific Coast and adjacent islands. *Proceedings of the California Academy of Sciences* 2, 85–276.
Fender, W.M. (1995) Native earthworms of the Pacific Northwest: an ecological overview. In: Hendrix, P.F. (ed.) *Earthworm Ecology and Biogeography in North America.* Lewis Publishers, Boca Raton, Florida, pp. 53–66.
Ferraz, J.A.N. and Guerra, R.T. (1983) Estudo preliminar da infuência da umidade do solo sobre a reprodução de *Pontoscolex corethrurus* (Glossoscolecidae, Oligochaeta). *Acta Amazônica* 13, 289–297.
Fragoso, C. (2001) Las lombrices de tierra de México (Oligochaeta; Annelida): Diversidad, ecología y manejo, *Acta Zoológica Mexicana (nueva série) No especial* 1, 131–171.
Fragoso, C. and Lavelle, P. (1992) Earthworm communities of tropical rain forests. *Soil Biology and Biochemistry* 24, 1397–1408.
Fragoso, C., Brown, G.G., Patrón, J.C., Blanchart, E., Lavelle, P., Pashanasi, B., Senapati, B. and Kumar, T. (1997) Agricultural intensification, soil biodiversity and agroecosystem function in the tropics: the role of earthworms. *Applied Soil Ecology* 6, 17–35.
Fragoso, C., Brown, G.G. and Feijoo, A. (2003) The influence of Gilberto Righi on tropical earthworm taxonomy: the value of a full-time taxonomist. *Pedobiologia* 47, 400–404.
Franchini, J.C., Saraiva, O.F., Brown, G.G. and Torres, E. (2004) Soil management and soil carbon contributions in Brazilian soybean production systems. In: Moscardi, F., Hoffman-Campo, C.B., Saraiva, O.F., Galerani, P.R., Krzyzanowski, F.C. and Carrão-Panizzi, M.C. (eds) *Proceedings of the VII World Soybean Research Conference.* Embrapa Soybean, Londrina, Brazil, pp. 531–535.
Gates, G.E. (1954) Exotic earthworms of the United States. *Bulletin of the Museum of Comparative Zoology, Harvard* 111, 220–258.
Gates, G.E. (1972) Burmese earthworms. An introduction to the systematics and biology of megadrile oligochaetes with special reference to South-East Asia. *Transactions of the American Philosophical Society* 62, 1–326.
Gates, G.E. (1973) Contributions to a revision of the earthworm family Ocnerodrilidae, IX. What is *Ocnerodrilus occidentalis*? *Bulletin of the Tall Timbers Research Station* 14, 13–28.
Gavrilov, K. (1981) Oligochaeta. In: Hurlbert, S.H., Rodriguez, G. and Santos, N.D. (eds) *Aquatic Biota of Tropical South America, Part 2: Anarthropoda.* San Diego State University, San Diego, California, pp. 170–190.
Guerra, R.T. (1982) *Influência de* Pontoscolex corethrurus *(Glossoscolecidae, Oligochaeta) na absorção de fósforo pelas plantas utilizando ^{32}P como traçador.* MSc thesis, INPA, Manaus, Brazil.
Guerra, R.T. (1985) Ecologia dos Oligochaeta da Amazônia. I. Estudo da migração horizontal e vertical de *Chibui bari* (Glossoscolecidae, Oligochaeta), através de observações de campo. *Acta Amazônica* 15, 141–146.
Guerra, R.T. (1988a) Ecologia dos Oligochaeta da Amazônia. II. Estudo da estivação e da atividade de *Chibui bari* (Glossoscolecidae, Oligochaeta), através da produção de excrementos. *Acta Amazônica* 18(1–2), 27–34.
Guerra, R.T. (1988b) Densidade e biomassa de oligochaeta em áreas antrópicas de cidade de Rio Branco Acre. *Cadernos da Universidade Federal do Acre, Série B (Ciência e Tecnologia)* 1, 7–16.
Guerra, R.T. (1994a) Sobre a comunidade de minhocas (Annelida, Oligochaaeta) do campus da Universidade Federal do Acre, Rio Branco (AC), Brasil. *Revista Brasileira de Biologia* 54, 593–601.
Guerra, R.T. (1994b) Earthworm activity in forest and savanna soils near Boa Vista, Roraima, Brazil. *Acta Amazônica* 24, 303–308.

Guerra, R.T. and Asakawa, N. (1981) Efeito da presença e do número de indivíduos de *Pontoscolex corethrurus* (Glossoscolecidae, Oligochaeta) sobre a população de microorganismos do solo. *Acta Amazônica* 11, 319–324.

Guerra, R.T. and Bezerra, D.R.B. (1989) Comportamento de *Pontoscolex corethrurus* Mueller, 1857 (Oligochaeta, Glossoscolecidae) em diferentes substratos. *Revista Brasileira de Biologia* 49, 1057–1064.

Guerra, R.T. and Silva, E.G. (1994) Estudo das comunidades de minhocas (Annelida, Oligochaeta) em alguns ambientes terrestres do Estado da Paraíba. *Revista Nordestina de Biologia* 9, 209–223.

Guimarães, A. (2003) Tráfico de minhocuçús. *Jornal da Minhoca (online)* No. 39.

Hamoui, V. (1991) Life-cycle and growth of *Pontoscolex corethrurus* (Müller, 1857) (Oligochaeta, Glossoscolecidae) in the laboratory. *Révue d'Écologie et de Biologie du Sol* 28, 469–478.

Harada, A.Y. and Bandeira, A.G. (1994) Estratificação e densidade de invertebrados em solo arenoso sob floresta primária e plantios arbóreos na Amazônia Central durante a estação seca. *Acta Amazônica* 24, 103–118.

Hausdorf, B. (2002) Units in biogeography. *Systematic Biology* 51, 648–652.

Hauser, J., Boccasius, M.B. and Kessler, R. (1975) Eine neuartige form von bindegewebe bei *Pheretima schmardae* (Horst, 1883) (Oligochaeta: Prosopora, Megascolecidae) (Bieträge zur Anneliden-Histologie). *Berichte des Naturwissenschaftlich-Medizinischer Vereins in Innsbruck* 62, 53–62.

Höfer, H., Hanagarth, W., Garcia, M., Martius, C., Franklin, E., Römbke, J. and Beck, L. (2001) Structure and formation of soil fauna communities in Amazonian anthropogenic and natural ecosystems. *European Journal of Soil Biology* 37, 229–235.

James, S.W. (1995) Systematics, biogeography and ecology of Nearctic earthworms from eastern, central, southern and southwestern United States. In: Hendrix, P.F. (ed.) *Earthworm Ecology and Biogeography in North America*. Lewis Publishers, Boca Raton, Florida, pp. 29–52.

Jiménez, J.J., Brown, G.G., Decäens, T., Feijoo, A. and Lavelle, P. (2000) Differences in the timing of diapause and patterns of aestivation in tropical earthworms. *Pedobiologia* 44, 677–694.

Kinberg, J.G.H. (1867) Annulata nova. *Ofversigt af Kongliga Vetenskaps-Akademiens Forhandlingar*, Stockholm 23, 97–103, 356–357.

Knäpper, C.F.U. (1972a) Dominanzverhältnisse der verschiedenen arten der gattung *Pheretima* in kulturböden von Rio Grande do Sul. *Pedobiologia* 12, 23–25.

Knäpper, C.F.U. (1972b) Oligoquetas terrestres – uma moderna avaliação. *Instituto André Voisin Publicação No. 1*. Instituto André Voisin, Porto Alegre, pp. 11–19.

Knäpper, C.F.U. (1977) Ecological niches of *P. diffringens* (Baird, 1869) and *E. lucens* (Waga, 1857) at São Francisco de Paula. *Estudos Leopoldenses* 42, 194–196.

Knäpper, C.F.U. (1979) Velhos habitats de *P. corethrurus* (Fr. Müller, 1857). *Estudos Leopoldensis* 16, 39–50.

Knäpper, C.F.U. and Hauser, J. (1969) Eine anomalie bei *'Allolobophora caliginosa'* (Savigny, 1826) (Oligochaeta). *Revista Brasileira de Biologia* 29, 411–412.

Knäpper, C.F.U. and Porto, R.P. (1979) Ocorrência de Oligoquetas nos solos do Rio Grande do Sul. *Acta Biológica Leopoldensia* 1, 137–166.

Kobiyama, M. (1994) *Influência da minhoca louca (*Amynthas *spp. Rosa, 1891) sobre o movimento da água no solo, relacionado ao crescimento da bracatinga (*Mimosa scabrella *Benth.)*. PhD thesis, Universidade Federal do Paraná, Curitiba, Brazil.

Kobiyama, M., Barcik, C. and Santos, H.R. (1995) Influência da minhoca (*Amynthas hawayanus*) sobre a produção de matéria seca de Bracatinga (*Mimosa scabrella* Benth). *Revista do Setor de Ciências Agrárias, Curitiba* 13, 199–203.

Krabbe, E.L., Driemeyer, D.J., Antoniolli, Z.I. and Giracca, E.M.N. (1993) Avaliação populacional de oligoquetas e características físicas do solo em diferentes sistemas de cultivo. *Ciência Rural, Santa Maria* 23, 21–26.

Krabbe, E.L., Driemeyer, D.J., Antoniolli, Z.I. and Giracca, E.M.N. (1994) Efeitos de diferentes sistemas de cultivo sobre a população de oligoquetas e características físicas do solo. *Ciência Rural, Santa Maria* 24, 49–53.

Kusdra, J.F. (1998) *Influência do Oligochaeta edáfico* Amynthas *spp. e do* Rhizobium tropici *no feijoeiro (*Phaseolus vulgaris *L.)*. MSc thesis, Universidade Federal do Paraná, Curitiba, Brazil.

Langenbach, T., Inacio, M.V.S., Aquino, A.M. and Brunninger, B. (2002) Influência da minhoca *Pontocolex corethrurus* na distribuição do acaricida dicofol em um Argissolo. *Pesquisa Agropecuária Brasileira* 37, 1663–1668.

Lapied, E. and Lavelle, P. (2003) The peregrine earthworm *Pontoscolex corethrurus* in the east coast of Costa Rica. *Pedobiologia* 47, 471–474.

Lasta, E.F., Giracca, E.M.N., Eltz, F.L.F., Antoniolli, Z.I., Benedetti, E.L. and Weber, M.A. (2002) Meso e macrofauna em solo sob plantio direto com diferentes doses de calcário. In: *Resumos da Fertbio 2002*. SBCS, Rio de Janeiro. CD-ROM.

Lavelle, P. and Lapied, E. (2003) Endangered earthworms of Amazonia: an homage to Gilberto Righi. *Pedobiologia* 47, 419–427.

Lee, K.E. (1985) *Earthworms. Their Ecology and Relationships with Soils and Land Use*. Academic Press, Sydney.

Lenko, K. (1972) Minhocas e sanguessugas (Annelida: Oligochaeta & Hirudinea) em ninhos de *Camponotus rufipes* (Insecta, Hymenoptera: Formicidae). *Revista Brasileira de Entomologia* 16, 7–12.

Lewinsohn, T.M. and Prado, P.I. (2002) *Biodiversidade Brasileira. Síntese do estado atual do conhecimento*. Contexto, São Paulo, Brazil.

Ljungström, P.-O. (1972a) Uma nova espécie de Glossoscolecidae do gênero *Alexidrilus* (Oligoqueta) para o RS-Brasil. *Ciência e Cultura* 24, 357.

Ljungström, P.-O. (1972b) Biology of *Glossoscolex uruguayensis* (Glossoscolecidae, Oligochaeta). A new species for Argentina. *Studies on the Neotropical Fauna* 7, 195–205.

Ljungström, P.-O., Emiliani, F. and Righi, G. (1975). Notas sobre oligoquetos (lombrices de tierra) argentinos). *Revista de la Asociación de Ciencias Naturales del Litoral* 6, 1–42.

Luederwaldt, H. (1927) A coleção de minhocas (Oligochaeta) no Museu Paulista. *Revista do Museu Paulista* 15, 545–556.

Mafra, A.L., Albuquerque, J.A., Medeiros, J.C., Rosa, J.D., Fontoura, S.M.V., Costa, F.S. and Bayer, C. (2002) Manejo do solol e fauna edáfica em experimento de longa duração na região de Guarapuava, PR. In: *Anais da XIV Reunião Brasileira de Manejo e Conservação do Solo e da Água*. SBCS&UFMT, Cuiabá, Brazil, pp. 1–5. CD-ROM.

Mafra, A.L., Miklós, A.A.W., Melfi, A.J., Eschenbrenner, V. and Volkoff, B. (2006) Ação das minhocas na estrutura e composição química de um solo arenoso hidromórfico do Amazonas. In: Brown, G.G. and Fragoso, C. (eds) *Biodiversidade e ecologia das minhocas na América Latina*. Embrapa Soja, Londrina, Brazil (in press).

Martinez, A.A. (1998) *A grande e poderosa minhoca. Manual prático do minhocultor*. Funep, Jaboticabal, Brazil. 148 pp.

Michaelsen, W. (1891) Oligochaeta des Naturhistorischen Museums in Hamburg. IV. *Jahrbuch der Hamburgischen Wissenschaftlichen Anstalten, Hamburg* 8, 3–42.

Michaelsen, W. (1892) Terricolen der Berliner Zoologischen Sammlung, II. *Archiv für Naturgeschiechte, Berlin* 58, 209–261.

Michaelsen, W. (1900a) *Das Tierreich*, Vol. 10: *Oligochaeta*. Friedländer & Sohn, Berlin.

Michaelsen, W. (1900b) Zur kenntnis der Geoscoleciden Südamerikas. *Zoologischer Anziger* 23, 53–56.

Michaelsen, W. (1903) *Die geographische Verbreitung der Oligochaeten*. Friedländer & Sohn, Berlin.

Michaelsen, W. (1910) Oligochäten von verschiedenen gebieten. *Mitteilungen aus dem Naturhistorischen Museum Hamburg* 27, 1–172.

Michaelsen, W. (1918) Die Lumbriciden. *Zoologische Jahrbücher, Abteilung für Systematik* 41, 1–398.

Michaelsen, W. (1925) Zur kenntnis einheimischer und ausländischer Oligochäten. *Zoologischer Jahrbucher Systematik* 51, 255–328.

Michaelsen, W. (1927) Die Oligochatenfauna Brasiliens. *Sonderabdruck Aus den Abhandlungen der Senckenbergischen Gesellschaft* 40, 367–375.

Michaelsen, W. (1928) Miscelanea oligochaetologica. *Archiv för Zoologi* 20, 1–15.

Michaelsen, W. (1931) Zwei neue ausseuropaische Oligochäten des Senckenberg-Museums. *Senckenbergiana* 13, 78–86.

Michaelsen, W. (1934) Opisthopore oligochäten des königlichen naturhistorischen museums von Belgien. *Bulletin du Museum Royal d'Histoire Naturelle de Belgique* 10, 1–29.

Miklós, A.A.W. (1992) biodynamics of the landiscape: biopedological organization and functioning. Part I. Role and contribution of the soil fauna to the organization and dynamics of a pedological cover in Botucatu, state of São Paulo, Brasil. In: Köpke, U. and Schulz, D.G. (eds) *Proceedings of the 9th International Scientific Conference 'Organic Agriculture, a key to a sound development and a sustainable society'*. IFOAM, St Wendel, Germany, pp. 74–86.

Miklós, A.A.W. (1996) Contribuição da fauna do solo na gênese de latossolos e de 'stone lines'. In: *Solo-Suelo 1996, Congresso Latino-Americano de Ciência do Solo*. SBCS, SLCS, Piracicaba, Brazil. CD-ROM.

Minette, S. (2000) *Étude de l'impact des techniques de semis direct sur les caractéristiques physiques et biologiques des sols des cerrados Brésiliens.* Memoire de fin d'études. École Nationalle Superieure Agronomique, Rennes, France.

MMA (2003) Lista de espécies da fauna brasileira ameaçada de extinção. *Anexo à instrução normativa No 3 de 27 de maio de 2003, do Ministério do Meio Ambiente.* MMA, Brasília. Available at: www.biodiversitas.org.br

Moreira, C. (1903) Vermes oligochetos do Brasil. *Archivos do Museu Nacional do Rio de Janeiro* 12, 125–136.

Moreira, M.A.B., Schwengerber, D.R. and Morais, J.W. (1998) Caracterização da macrofauna do solo em diferentes usos da terra em Roraima. In: *Resumos da Fertbio 1998.* CBCS and UFLA, Lavras, Brazil, p. 771.

Moreno, A.G. and Mischis, C.C. (2003) The status of Gilberto Righi's earthworm collection at the Museum of São Paulo. Righi Memorial: tropical ecology. *Pedobiologia* 47, 413–418.

Müller, F. (1857a) *Lumbricus corethrurus,* Bürstenschwanz. Archiv für Naturgeschiechte 23, 113–116 (The original paper could not be obtained. An English translation was consulted and is published in the *Annals and Magazine of Natural History* 2(20), 13–15, 1857.)

Müller, F. (1857b) II–Description of a new species of earthworm (*Lumbricus corethrurus*). *Annals and Magazine of Natural History* 20, 13–15.

Müller, F. (1877) Der Minhocão. *Zoologischer Garten,* pp. 208–302.

Nooren, C.A.M., van Breemen, N., Stoorvogel, J.J. and Jongmans, A.G. (1995) The role of earthworms in the formation of sandy surface soils in a tropical forest in Ivory Coast. *Geoderma* 65, 135–148.

Nunes, D., Pasini, A., Benito, N.P. and Brown, G.G. (2004) Ocorrência de minhocas em agroecossistemas de Jaguapitã – PR. In: *Resumos da Fertbio 2004.* SBCS, Lages, Brazil CD-ROM.

Pacheco, S.M., Junqueira, I.C., Widholzer, R.M.B.F., Esmério, M.E. and Nunes, C.N. (1992) Contribuição ao conhecimento da fauna de Oligochaeta das áreas de alagamento das unisas hidrelétricas de Itá-Machadinho (RS, SC) e Campos Novos (SC). *Communicações do Museu de Ciências da Pontifícia Universidade Católica do Rio Grande do Sul (PUCRS), Série Zoologia (Porto Alegre)* 5, 23–28.

Pasini, A., Fonseca, I.B., Brossard, M. and Guimarães, M.F. (2003) Macrofauna invertebrada do solo em pastagens no cerrado de Uberlândia, MG, Brasil. In: Brown, G.G., Fragoso, C. and Oliveira, L.J. (eds) *Anais do Workshop, O uso da macrofauna edáfica na agricultura do século XXI: a importância dos engenheiros do solo.* Embrapa Soja, Série Documentos, No. 224. pp. 160–166.

Pasini, A., Benito, N.P., Roessing, M. and Brown, G.G. (2004) Macrofauna invertebrada do solo em pastagens do Norte do Estado do Paraná. In: *Resumos da Fertbio 2004.* SBCS, Lages, Brazil CD-ROM.

Peixoto, R.T. and Marochi, A.I. (1996) A influência da minhoca *Pheretima* sp. nas propriedades de um latossolo vermelho escuro álico e no desenvolvimento de culturas em sistema de plantio direto em Arapoti-PR. *Revista Plantio Direto* 35, 23–35.

Peneireiro, F.M. (1999) *Sistemas agroflorestais dirigidos ple sucessão natural: um estudo de caso.* MSc thesis, University of São Paulo, Piracicaba, Brazil.

Perrier, E. (1877) Les vers de terre du Brésil. *Bulletin de la Société Zoologique* 2, 241–246.

Pimentel, M.S., Aquino, A.M., Ricci, M.S., Almeida, D.J. and De-Polli, H. (2002) Estudo preliminar sobre a ocorrência de macrofauna em solos submetidos à cafeicultura orgânica, pastagem e floresta. In: *Resumos da Fertbio 2002.* SBCS, Rio de Janeiro. CD-ROM.

Primavesi, A.M. and Covolo, G. (1968) Comparação entre a atividade dos cupins (Termitidae) e minhocas (*Lumbricus* sp.) com relação a estrutura e nutrientes do solo. In: Primavesi, A. (ed.) *Progressos em biodinâmica e produtividade do solo.* Universidade Federal de Santa Maria, Santa Maria, Brazil, pp. 149–154 (This earthworm is very likely not *Lumbricus,* but rather either another representative of the Lumbricidae family (e.g. *Aporrectodea* sp.), an *Amynthas* sp. or *P. corethrurus* (both the latter spp. known from many locations in RS). Only two species are presently known from Santa Maria: *Amynthas gracilis* and *Glossoscolex wiengreeni* (Knäpper and Porto, 1979).

Quadros, R.M.B., Dionísio, J.A. and Bellote, A.F.J. (1998) Resultados preliminares sobre as características físico-químicas dos coprólitos de minhocas nativas sob *Eucalyptus grandis.* In: *Resumos da Fertbio 1998.* SBCS and UFLA, Lavras, Brazil, p. 422.

Resende, M., Curi, N., Rezende, S.B. and Corrêa, G.F. (2002) *Pedologia. Base para distinção de ambientes.* NEPUT, Viçosa, Brazil.

Ressetti, R.R. (2004) *Determinação da dose de alil isotiocianato em substituição à solução de formol na extração de oligochaeta edáficos.* MSc thesis, Universidade Federal de Curitiba, Curitiba, Brazil.

Reynolds, J.W. (1994) Earthworms of the world. *Global Biodiversity* 4, 11–16.

Ricci, M.S.F. (1996) Manual de vermicompostagem. Série Documentos No. 31, Embrapa Rondônia, Porto Velho, Brazil.

Righi, G. (1967a) Über die Oligochätengattung *Eukerria*. *Beitrage Neotropical Fauna* 5, 178–185.

Righi, G. (1967b) Descrição de *Rhinodrilus priollii*, sp. n. Glossoscolecidae. In: Lent, H. (ed.) *Atas do Simpósio sôbre a Biota Amazônica (Zoologia)*, Vol. 5 Conselho Nacimal de Pesquisas, Rio de Janeiro, pp. 475–479.

Righi, G. (1967c) Sobre algumas Lumbricidae (Oligochaeta) do Estado do Rio Grande do Sol. *Ciência e Cultura* 19, 342.

Righi, G. (1967d) O gênero *Pheretima* Kinberg (Oligochaeta) no Brasil. *Ciência e Cultura* 19, 342–343.

Righi, G. (1967e) *Eudrilus eugeniae* (Kinberg, 1867). Oligochaeta terrícola novo para o Brasil. *Ciência e Cultura* 19, 341–342.

Righi, G. (1968a) Sôbre duas espécies novas de Oligochaeta do Brasil. *Anais da Academia Brasileira de Ciências* 40, 545–549.

Righi, G. (1968b) Sobre alguns Oligochaeta do Brasil. *Revista Brasileira de Zoologia* 28, 369–382.

Righi, G. (1969) Sur une espéce aberrante des Glossoscolecidae, *Thamnodrilus matapi*, sp. n. *Pedobiologia* 9, 42–45.

Righi, G. (1970) Sobre o gênero *Andioscolex* (Oligochaeta, Glossoscolecidae). *Revista Brasileira Biologia* 30, 371–376.

Righi, G. (1971a) Sôbre a Família Glossoscolecidae (Oligochaeta) no Brasil. *Arquivos de Zoologia, São Paulo* 20, 1–96.

Righi, G. (1971b) Sôbre alguns Oligochaeta brasileiros. *Papéis Avulsos de Zoologia* 25, 1–13.

Righi, G. (1971c) A new genus and species of Ocnerodrilinae (Oligochaeta: Acanthodrilidae) from Brazil. *Zoologischer Anzeiger* 186, 388–391.

Righi, G. (1972a) Additions to the genus *Glossoscolex* (Oligochaeta, Glossoscolecidae). *Studies on the Neotropical Fauna* 7, 37–47.

Righi, G. (1972b) Contribuição ao conhecimento dos Oligochaeta brasileiros. *Papéis Avulsos de Zoologia* 25, 148–166.

Righi, G. (1972c) Bionomic considerations upon the Glossoscolecidae (Oligochaeta). *Pedobiologia* 12, 254–260.

Righi, G. (1974) Notas sobre as Oligochaeta Glossoscolecidae do Brasil. *Revista Brasileira de Biologia* 34, 551–564.

Righi, G. (1975) Some Oligochaeta from the Brazilian Amazonia. *Studies on the Neotropical Fauna* 10, 77–95.

Righi, G. (1977) Ecología e modo de vida das minhocas. *O Estado de São Paulo,* 13 February 1977. Suplemento Cultural No. 18, pp. 14–16.

Righi, G. (1979) Introducción al estudio de las lombrices del suelo (Oligoquetos Megadrilos) de la Provincia de Santa Fe (Argentina). *Revista de la Asociación de Ciencias Naturales del Litoral* 10, 89–155.

Righi, G. (1980a) Alguns Oligochaeta, Ocnerodrilidae e Glossoscolecidae do Brasil. *Papéis Avulsos de Zoologia* 33, 239–246.

Righi, G. (1980b) Alguns Megadrile (Oligochaeta, Annelida) brasileiros. *Boletim de Zoologia* 5, 1–8.

Righi, G. (1982a) Oligochaeta, Glossoscolecidae do Parque Nacional da Amazônia, Tapajós. *Revista Brasileira de Biologia* 42, 107–116.

Righi, G. (1982b) Adições ao gênero *Glossodrilus* (Oligochaeta, Glossoscolecidae). *Revista Brasileira de Zoologia,* 1, 55–64.

Righi, G. (1983) Três Ocnerodrilidae (Oligochaeta) da Amazônia brasileira. *Acta Amazônica* 13(5–6), 927–936.

Righi, G. (1984a) On a collection of Neotropical Megadrili Oligochaeta. I. Ocnerodrilidae, Acanthodrilidae, Octochaetidae, Megascolecidae. *Studies on Neotropical Fauna and Environment* 19, 9–31.

Righi, G. (1984b) On a collection of Neotropical Megadrili Oligochaeta. II. *Studies on Neotropical Fauna and Environment* 1, 99–120.

Righi, G. (1984c) *Pontoscolex* (Oligochaeta, Glossoscolecidae), a new evaluation. *Studies on Neotropical Fauna and Environment* 19, 159–177.

Righi, G. (1984d) Oligochaeta Megadrili da região Centro-Oeste de Mato Grosso, Brasil. *Boletim de Zoologia* 8, 189–213.

Righi, G. (1984e) Oligochaeta Megadrili da Chapada do Guimarães, Mato Grosso. *Boletim de Zoologia* 8, 17–23.

Righi, G. (1985) Sobre *Rhinodrilus* e *Urobenus* (Oligochaeta, Glossoscolecidae). *Boletim de Zoologia* 9, 231–257.

Righi, G. (1986a) Sobre o gênero *Andiorrhinus* (Oligochaeta, Glossoscolecidae). *Boletim de Zoologia* 10, 123–151.
Righi, G. (1986b) Alguns Oligochaeta, Glossoscolecidae, de Rondônia, Brasil. *Boletim de Zoologia* 10, 283–303.
Righi, G. (1988a) Adições à drilofauna de Rondônia, Brasil. *Revista Brasileira de Zoologia* 48, 119–125.
Righi, G. (1988b) Uma coleção de Oligochaeta da Amazônia brasileira. *Papéis Avulsos de Zoologia* 36, 337–351.
Righi, G. (1988c) *Pontoscolex* (P). *pydanieli*, spec. nov. (Oligochaeta, Glossoscolecidae) and its parasite *Pessoaella pontoscolecis*, gen. nov., spec. nov. (Eugregarinida, Aikinetocystidae). *Studies on Neotropical Fauna and Environment* 23, 71–76.
Righi, G. (1989a) Três Oligochaeta, Glossoscolecidae da Amazônia. *Amazoniana* 10(4), 393–399.
Righi, G. (1989b) Alguns Oligochaeta da Amazônia. *Boletim do Museu Paraense Emílio Goeldi (Zoologia)* 5, 1–8.
Righi, G. (1990a) *Minhocas de Mato Grosso e de Rondônia*. Programa Polonoroeste, Relatório de Pesquisa no 12. SCT/PR-CNPq, Programa do Trópico Úmido, Brasília, Brazil.
Righi, G. (1990b) Oligochaeta da Estação Ecológica de Maracá, Roraima, Brasil. *Acta Amazônica* 20, 391–398.
Righi, G. (1993) Venezuelan earthworms and consideration on the genus *Andiorrhinus* Cognetti 1908 (Oligochaeta Glossoscolecidae). *Tropical Zoology, Special Issue* 1, 125–139.
Righi, G. (1994) On new and old-known Oligochaeta genera from Paraiba state, Brazil. *Revue Suisse de Zoologie* 101, 89–106.
Righi, G. (1995a) Colombian earthworms. In: van der Hammen, T. and Santos, A.G. (eds) *Studies on Tropical Andean Ecosystems*, Vol. 4. Cramer, Berlin, pp. 485–607.
Righi, G. (1995b) A new earthworm (Ocnerodrilidae, Oligochaeta) from a Brazilian cave and considerations about *Belladrilus*. *Revue Suisse de Zoologie* 102, 361–365.
Righi, G. (1996) Some Venezuelan Oligochaeta Glossoscolecidae and Octochaetidae. *Revue Suisse de Zoologie* 103(3), 677–684.
Righi, G. (1997) Minhocas da América Latina: diversidade, função e valor. In: *Anais do XXVI Congresso Brasileiro de Ciência do Solo*. SBCS, Rio de Janeiro. CD-ROM.
Righi, G. (1998a) Earthworms of the Ilha de Maracá. In: Milliken, W. and Ratter, J. (eds) *Maracá: The Biodiversity and Environment of an Amazonian Rainforest.* John Wiley & Sons, Chichester, UK, pp. 391–397.
Righi, G. (1998b) Oligoquetas. In: Machado, A.B.M., da Fonseca, G.A.B., Machado, R.B., de Aguiar, L.M.S. and Lins, L.V. (eds) *Livro vermelho das espécies ameaçadas de extinção da fauna de Minas Gerais.* Fundação Biodiversitas. Belo Horizonte, pp. 571–583.
Righi, G. (1999) Oligochaeta. In: Brandão, C.R. and Cancello, E.M. (eds) *Biodiversidade do Estado de São Paulo, Brasil: Síntese do conhecimento ao final do século XX. 5. Invertebrados Terrestres.* FAPESP, São Paulo, Brazil, pp. 13–21.
Righi, G. and Ayres, I. (1975) Alguns Oligochaeta sul brasileiros. *Revista Brasileira de Biologia* 35, 309–316.
Righi, G. and Ayres, I. (1976) *Meroscolex marcusi*, sp. n. Oligochaeta, Glossoscolecidae da Amazônia. *Boletim de Zoologia (n.s.)* 1, 257–263.
Righi, G. and Caballero, M.E.S. (1970) Duas espécies brasileiras dos gêneros *Wegeneriona* e *Neogaster* (Oligochaeta, Octochaetidae). *Revista Brasileira de Biologia* 30, 91–96.
Righi, G. and Guerra, A.T. (1985) Alguns Oligochaeta do norte e noroeste do Brasil. *Boletim de Zoologia* 9, 145–157.
Righi, G. and Knäpper, C.U.F. (1965) O gênero *Pheretima* Kinberg, no Estado de Rio Grande do Sol. *Revista Brasileira de Zoologia* 25, 419–427.
Righi, G. and Knäpper, C.U.F. (1966) Ciclo annual de *Pheretima indica* (Horst, 1893). *Revista Brasileira de Zoologia* 26, 341–343.
Righi, G. and Moraes, P.H.F. (1990) *Rhinodrilus pitun*, sp. n. Oligochaeta, Glossoscolecidae de Pernambuco. *Revista Brasileira de Zoologia* 50, 519–522.
Righi, G. and Lobo, D.A. (1979) Nova contribuição ao gênero *Glossoscolex* com sinopse do grupo *giganteus* (Oligochaeta, Glossoscolecidae). *Revista Brasileira de Zoologia* 39, 947–959.
Righi, G., Ayres, I. and Bittencourt, E.C.R. (1976) Glossoscolecidae (Oligochaeta) do Instituto Nacional de Pesquisas da Amazônia. *Acta Amazônica* 6, 335–367.
Righi, G., Ayres, I. and Bittencourt, E.C.R. (1978) Oligochaeta (Annelida) do Instituto Nacional de Pesquisas da Amazônia. *Acta Amazônica* 8(1), 1–49.

Rodrigues, K.M., Correia, M.E.F., Alves, B.J.R. and Aquino, A.M. (2004) Efeitos do manejo do solo e sucessão de culturas na abundância e na dieta da minhoca, *Pontoscolex corethrurus*, por marcação isotópica. In: *Resumos da Fertbio 2004*. SBCS, Lages, Brazil. CD-ROM.

Römbke, J., Meller, M. and Garcia, M. (1999) Earthworm densities in central Amazonian primary and secondary forests and a polyculture forestry plantation. *Pedobiologia* 43, 518–522.

Rosa, D. (1894) Perichetini nuovi o meno noti. *Atti della Reale Accademia di Scienze di Torino* 29, 773.

Rosa, D. (1895) Viaggio dei dottor Alfredo Borelli nella Repuglica Argentina e nel Paraguay. *Bolletino dei Musei di Zoologia ed Anatomia Comparata della Reale Università di Torino* 10 (204), 1–3.

Rosa, D. (1896) Contributo allo studio dei terricoli neotropicali. *Memorie della Reale Accademia delle Scienze di Torino* 45, 89–153.

Sá, J.C.M., Cerri, C.C., Dick, A., Lal, R., Venske-Filho, S.P., Piccolo, M.C and Feigl, B.E. (2001) Organic matter dynamics and carbon sequestration rates of a tillage chronosequence in a Brazilian Oxisol. *Soil Science Society of American Journal* 65, 1486–1499.

Santos, A.F. (1995) *Efeito da atividade do Oligochaeta* Amynthas *spp. (minhoca louca), na produção do feijoeiro (*Phaseolus vulgaris *– FT 120) e da aveia preta (*Avena strigosa *var Canton), com relação a algumas propriedades físicas, químicas e biológicas do solo.* MSc thesis, Universidade Federal do Paraná, Curitiba, Brazil.

Santos, D., Teixeira, W.G., Marques, J.J.G.S.M. and Curi, N. (1996) Parâmetros químicos de excreções de minhocuçu e do solo adjacente. In: *Anais da XXII Reunião Brasileira de Fertilidade do Solo e Nutrição de Plantas*. SBCS, Manaus, Brazil, pp. 608–609.

Sautter, K.D., Tanck, B.C.B., Dionísio, J.A. and Santos, H.R. (1995) Estudo da população de Orbatei (Acari: Cryptostignata), collembola (Insecta) e Oligochaeta, em diferentes ambientes de um solo degradado pela mineração de xisto a céu aberto. *Revista do Setor de Ciências Agrárias (Curitiba)* 13, 171–174.

Sautter, K.D., Brown, G.G., Pasini, A., Benito, N.P., Nunes, D.H. and James, S.W. (2006) Taxonomia e ecologia de minhocas no estado do Paraná, Brasil. In: Brown, G.G. and Fragoso, C. (eds) *Biodiversidade e ecologia das minhocas na América Latina*. Embrapa Soja, Londrina, Brazil (in press).

Savigny, J.C. (1826) Analyse d'un Mémoir sur les Lombrics par Cuvier. La multiplicité des èpeces de ver de terre. *Mémoirs d'Académie Royale des Sciences Institute de France* 5, 176–184.

Silva, J.J.C., Souza, R.M., Fontanela, E., Prates, E.D. and Lima, A.C.R. (2003) Monitoramento da qualidade de solo hidromórfico através de indicadores biológicos. Desenvolvimento de protocolo. In: Brown, G.G., Fragoso, C. and Oliveira, L.J. (eds) *Anais do Workshop, O uso da macrofauna edáfica na agricultura do século XXI: a importância dos engenheiros do solo*. Embrapa Soja, Série Documentos, No. 224, pp. 117–123.

Soares, M.T.S. and Lambais, M.R. (1998) Efeito da minhoca endogeica *Pontoscolex corethrurus* e da matéria orgânica no crescimento de *Brachiaria decumbens*. In: *FERTBIO 1998, Resumos*. SBCS and UFLA, Lavras, Brazil, p. 220.

Soares, M.T.S., Bernardes, F.F., Pereira, J.C., Sparovek, G. and Dias, C.T.S. (1997) Produção de coprólitos da minhoca endogeica *Pontoscolex corethrurus* (Müller, 1857) (Oligochaeta, Glossoscolecidae) em função da textura do solo e do teor de matéria orgânica. In: *Anais do XXVI Congresso Brasileiro de Ciência do Solo*. SBCS, Rio de Janeiro. CD-ROM.

Sparovek, G., Lambais, M., Silva, A.P. and Tormena, C.A. (1999) Earthworm (*Pontoscolex corethrurus*) and organic matter effects on the reclamation of an eroded oxisol. *Pedobiologia* 43, 698–704.

Tanck, B.C.B., Santos, H.R. and Dionísio, J.A. (2000) Influência de diferentes sistemas de uso e manejo do solo sobre a flutuação populacional do oligoqueta edáfico *Amynthas* spp. *Revista Brasileira de Ciência do Solo* 24, 409–415.

Ude, H. (1893) Beitrage zur kenntnis ausländischer Regenwürmer. *Zeitschrift für Wissenschaftliche Zoologie* 57, 57–75.

Uzêda, M.C. and Garcia, M.A. (2006) Análise das relações entre populções de enchytraeidae e minhocas e seu uso como bioindicador da qualidade do solo. In: Brown, G.G. and Fragoso, C. (eds) *Biodiversidade e ecologia das minhocas na América Latina*. Embrapa Soja, Londrina, Brazil (in press).

Vanucci, M. (1953) Biological notes I. On the Glossoscolecid earthworm *Pontoscolex corethrurus*. *Dusenia* 4, 287–301.

Voss, M. (1986) Populações de minhocas em diferentes sistemas de plantio. *Plantio Direto* 4, 6–7.

Zicsi, A. (1995) Revision der gattung *Glossodrilus* Cognetti, 1905 auf grund der arten aus dem andengebiet (Oligochaeta: Glossoscolecidae). Regenwürmer aus Südamerika, 25. *Opuscula Zoologica Budapest* 27–28, 79–116.

Zicsi, A. and Csuzdi, C. (1987) Neue und bekannte Glossoscoleciden-Arten aus Südamerika. 2. Oligochaeta: Glossoscolecidae. *Acta Zoologica Hungarica* 33, 269–275.
Zicsi, A. and Csuzdi, C. (1999) Neue und bekannte regenwürmer aus verschiedenen Teilen Südamerikas. Regenwürmer aus Südamerika 26. *Senckenbergiana Biologica* 78, 123–134.
Zicsi, A., Römbke, J. and Garcia, M. (2001) Regenwürmer (Oligochaeta) aus der Umgebung von Manaus (Amazonien). *Revue Suisse de Zoologie* 108, 153–164.

5 Termite Diversity in Brazil (Insecta: Isoptera)

R. Constantino[1] and A.N.S. Acioli[2]
[1]*Department of Zoology, University of Brasília, 70910-900 Brasília, DF, Brazil, e-mail: constant@unb.br;* [2]*PPG Entomologia – INPA, Cx. Postal: 478, 69011-970 Manaus, AM, Brazil*

Introduction

Termites are social insects of the order Isoptera, which contains about 2800 known species. Best known for their economic importance as pests of wood and other cellulosic materials, termites are also important members of the soil fauna of tropical ecosystems. They are among the most abundant animals in some tropical forests and savannahs, playing an important role in litter decomposition and nutrient cycling (Eggleton *et al.*, 1996). Their uncommon ability to digest cellulose allows them to redirect a considerable proportion of the energy flow in the system, reaching high biomass and, at the same time, being an important food item for many animals (Wood and Sands, 1978). They also improve the soil structure by cementing particles and moving material vertically (Lee and Wood, 1971). Termitaria are important structural elements in tropical forests and savannahs, sheltering an extraordinary diversity of animals (Redford, 1984; Martius, 1994). Due to their ability to modify the structure of their habitat, termites are considered ecosystem engineers (Lavelle *et al.*, 1997), organisms that can affect the availability of resources to other species through physical changes in their habitat. The elimination of termites from an ecosystem may, therefore, cause the loss of other species that rely on them for survival or reproduction.

Brazil has one of the most diverse termite faunas, with nearly 300 known species. Termites are particularly abundant in tropical lowland forests, savannahs and grasslands. Some species are important structural or agricultural pests, and some of them have been introduced from other regions. Despite the limitations on the knowledge of the taxonomy, biology and geographical distribution of Brazilian termites, Brazil is the only country in Latin America with a relatively solid tradition in termite studies.

In this chapter, the current knowledge of the termite faunas in various parts of Brazil is reviewed. More or less distinct termite faunas can be recognized in the major biomes, with significant differences in diversity, abundance and importance.

Termite Biology

General termite biology was reviewed by Krishna and Weesner (1969–1970) and Grassé (1982–1986). A brief summary is presented here. All known termites are eusocial, living in colonies with reproductive division of labour, morphological castes and overlap of generations. Typical colonies are a large family with a reproductive couple (queen and king) who are the parents of all the others and a large number of sterile individuals (workers and soldiers). Workers execute tasks such as nest construction, foraging and feeding reproductives, soldiers and immatures. Soldiers protect the colony against enemies using either their enlarged mandibles or chemical weapons, or both. However, the Kalotermitidae lack a true worker caste and this role is performed by immatures (nymphs or pseudoworkers). Soldiers are totally absent in Neotropical Apicotermitinae and their workers are also responsible for colony defence.

Reproduction within the colony is performed by a few individuals. The primary reproductives are the ones that found the colony, usually a pair that meets after the swarming flight. Winged reproductives are produced seasonally by mature colonies and swarm in large numbers. Very few of them succeed in founding new colonies. Besides the primary ones, secondary reproductives of three different types may arise inside the colony, depending on the species, age and size of the colony and other factors. They can be replacement reproductives, when they arise after the death of a primary one, or supplementary reproductives, when they arise in the presence of the primary reproductives. The most common type of secondary reproductive is the neotenic nymphoid, a nymph that becomes reproductive without completing its normal development. Adultoids are secondary reproductives that are identical to the primary ones, except that they do not swarm and stay in the colony where they were born. More rarely, some workers may become reproductives and they are called ergatoids.

Soldiers are highly modified individuals that defend the colony against enemies (Fig. 5.1). Some have large heads with powerful mandibles while others have sophisticated chemical weapons. The diversity of defensive mechanisms results in high morphological diversity, and for this reason the soldier is the most important caste for taxonomy. The most common types of soldiers are nasutes (head modified as a squirting gun of defensive secretion); phragmotic heads (used to plug the tunnels, present in some Kalotermitidae); various types of biting mandibles; snapping mandibles, either symmetric or asymmetric. Workers also may show defensive behaviour such as biting, exploding the abdomen or releasing defensive secretions.

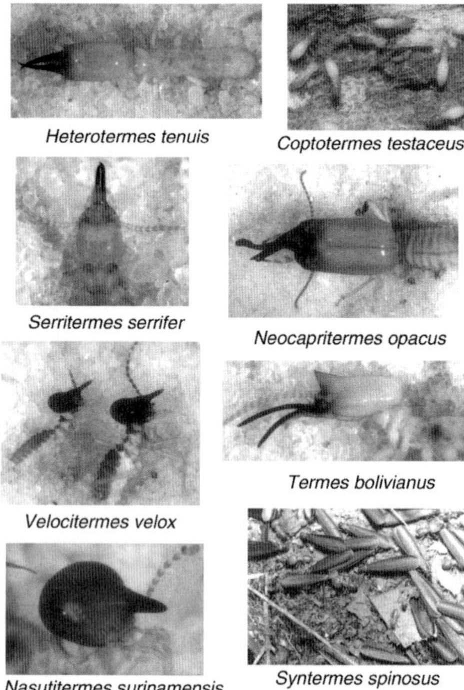

Fig. 5.1. Representatives of common termite genera in Brazil (soldiers). All photos by R. Constantino.

Nesting habits are highly variable and range from diffuse galleries in wood or soil to large and complex nests, either subterranean, epigeic or arboreal. Termite constructions are usually called termitaria, and some can be large and complex. The nests and covered tunnels are made of their own faeces or soil cemented with saliva, or a combination of both. Old termitaria are commonly occupied by many animals, including other termites. True termitophiles are animals that live with the termites inside their galleries, such as several specialized beetles of the family Staphylinidae. Termitariophiles are animals that use the termitarium as a shelter or nest but do not interact with the termites, such as spiders, lizards, rats, bees and birds. The term inquiline is used to refer to termites that occupy the nests of another termite species. Some inquilines may be highly specialized, such as *Inquilinitermes* spp., which live only in nests of *Constrictotermes* spp.

The feeding habits of termites are variable and distributed along a gradient from wood feeding to humus feeding. Some termites have specialized habits, such as leaf or grass litter (Fig. 5.2), lichen (e.g. some *Constrictotermes*) and symbiotic fungi (Macrotermitinae, absent in the Americas). The relative importance of each feeding guild varies among regions and habitats. Termites of all families except Termitidae depend on symbiotic protozoa to digest cellulose. They are flagellates and live in the hind gut of these termites. The Termitidae, the largest family with more than 70% of all species, lack protozoa and can digest cellulose with their own enzymes (Hogan *et al.*, 1988). The role of intestinal bacteria is not clear, but there is evidence that they are important in the digestion of lignin and humus (Bignell, 1994). Among the humus feeders, different species seem to select different particles and may have a distinct gut microflora associated with their feeding habits (Noirot, 1992).

Nitrogen fixation by symbiotic bacteria occurs in the gut of several termites, and seems to be particularly important in the Kalotermitidae. The proportion of fixation relative to other sources of nitrogen has not been determined for most species. Some dry-wood termites seem to acquire more than 50% of their nitrogen from the atmosphere (Tayasu *et al.*, 1994). The ecosystem-level importance of this source of nitrogen is uncertain, but it may be significant because termites comprise an important food item for many animals.

Methods for Sampling Termites

Termite colonies are distributed according to microhabitat preferences, such as abundance of wood and litter, moisture and soil type. Their spatial distribution within the environment is, therefore, not random. Subterranean nests may be 2–3 m below ground and some arboreal nests are high in the canopy. Some species live completely

Fig. 5.2. *Ruptitermes reconditus*, a soldierless litter-feeding species (Termitidae: Apicotermitinae). Photo by R. Constantino.

hidden inside wood, while others forage on the surface and carry pieces of litter to their nest. Due to their highly aggregate distribution, termite sampling is much more difficult than in non-social soil animals such as earthworms, collembolans and mites.

Termite nests can be classified into four groups:

1. Epigeic nests: discrete, above-ground termitaria, with a variable subterranean part.
2. Subterranean nests: either discrete or diffuse, completely subterranean nests.
3. Arboreal nests: discrete nests attached to trees at various heights.
4. Single-piece wood nest: the whole colony is inside a single piece of wood, without external constructions or connection to the soil.

For sampling purposes, termites of groups 1–3 can also be collected in the soil, or foraging outside their nests. Termites of group 4 (Kalotermitidae) are usually not found in the soil and never leave their nests.

Qualitative sampling: Intensive casual sampling is the best method for a rapid estimate of local species richness. The limitation is that training and experience of the collectors may cause discrepancies and results are usually not comparable. Considering that the collectors are the same, comparisons may be made based on the collecting effort measured in person × time, which is very simple. The other limitation is that the product is a simple list of species. This method was used, for example, by Constantino and Schlemmermeyer (2000) for rapid comparison of the faunas of several habitats.

Soil cores and monoliths: Although these methods are commonly used to sample soil fauna, they are not adequate for termites due to their highly aggregate distribution. Their use results in a high proportion of samples without termites and, among the samples with termites, most have only a few workers, making identification very difficult. They can be used in combination with other methods, but are not recommended as a general technique to sample termites.

Plot and transect sampling: Transects and square plots have been used by many authors (Mathews, 1977; Coles, 1980; Constantino, 1992). Termites are sampled intensively in plots or transects, producing both a species list and an estimate of their relative abundance. Abundance and biomass can be estimated for mound-building and arboreal nesting species. Biomass can be estimated based on the relation between size of the nest and of the colony. DeSouza and Brown (1994) used 110 × 3 m transects divided into 22 plots of 3 × 5 m. The collecting effort was standardized in each plot as one person × hour. A modified version of this method was adopted by Eggleton *et al.* (1995), consisting of a 100 × 2 m transect divided into 20 plots of 2 × 5 m, with a monolith of 20 × 20 × 50 cm dug in the middle of each plot. Eggleton's method has been extensively used in forests of South-East Asia and Africa. More recently, Jones and Eggleton (2000) modified their protocol, replacing the monolith with 12 soil scrapes per section. This later protocol is becoming the standard sampling method for forest termites.

Baiting: Various sorts of baits have been used to sample termites (French and Robinson, 1981). Baits are highly selective and do not collect some functional groups. Eggleton and Bignell (1995) argued that because of several limitations, baiting is not appropriate as a general sampling method. This argument, however, is not based on experimental comparison of baiting with other methods, and some studies have shown that baits can be useful for sampling subterranean termites. The major advantages of baiting are that: (i) it is simple and cheap; (ii) it is not affected by training or experience; and (iii) sampling design is simple and results are easily comparable among sites. Several different baiting materials can be used simultaneously to sample species with various preferences (e.g. wood, cardboard, cow dung). One important disadvantage is that effective baiting requires several months and is not adequate for short-term surveys in remote places.

Quantitative sampling and biomass estimates: Estimating total termite abundance is extremely difficult and there are very limited data available. Accurate estimates are nearly impossible. Martius (1994) reviewed the available estimates for Amazonian forests and arrived at an average of 2–2.5 g/m^2. This is based on combined estimates of soil cores, wood and nests.

Identification: Termite classification is more complete than the classification for most other groups of soil arthropods and most common species have been at least described. However, identifying termite species is usually difficult. Termite taxonomy is heavily based on soldier morphology, and soil samples often contain only workers. At the moment there is no identification key for workers of Neotropical termites, such as there is for African termites (Sands, 1998). Many studies have identified termites only to 'morphospecies'. This is very limited, as the results of one study cannot be compared with others.

Current State of the Taxonomy of Brazilian Termites

Taxonomy has been a major impediment to the study of termites in South America due to the large number of species, high proportion of undescribed or poorly described species and lack of proper taxonomic training of most entomologists. Very few specialists are able to identify Neotropical termite species and many important groups lack proper taxonomic revision, resulting in a large number of incorrect, doubtful or incomplete identifications. Without sound taxonomy it is impossible to accurately store and retrieve information on the biology and ecology of each species, to recognize biogeographic patterns or to make generalizations.

Among the termite taxa present in Brazil, the taxonomy of Kalotermitidae, Apicotermitinae and the genus *Nasutitermes* are the most problematic. The Kalotermitidae live only inside wood and their colonies are usually small. They are difficult to find and collect, resulting in poor collections. The Neotropical Apicotermitinae are soldierless termites, which are abundant in the soil. Termite taxonomy is traditionally based on the soldier caste, which is morphologically more variable. For this reason, soldierless termites have been neglected by taxonomists. Also, many species of this group were described from the alates alone, which are not commonly found with workers. In the absence of reproductives, it is nearly impossible to identify most species of this group. Lastly, the dominant genus *Nasutitermes* is the largest in terms of the number of species and its taxonomy is chaotic. Many species of this genus are very abundant and some are even important urban pests. Species identification is difficult due to the large number of very similar species.

Brazil, due to its large extension and diversity of biomes, has a rich termite fauna (Table 5.1). Until the end of the 19th century, information on the termite fauna of South America was limited to fragmentary data reported by European naturalists. During the first half of the 20th century, studies by several European and American entomologists formed the foundation for the scientific study of Neotropical termites: the Italian entomologist Silvestri (1903) on termites from Mato Grosso, Argentina and Paraguay; the Swedish termitologist Holmgren (1906, 1910) on South American termites; the American termitologist Emerson (1925) on the termites from Guyana; the American termitologist Snyder (1926) on termites collected by the Mulford Expedition to Peru, Bolivia and Brazil. During the second half of the 20th century, the work conducted by foreigners was gradually replaced by studies of native termitologists. Araujo (1970, 1977) studied Brazilian termites from 1950 to 1978, and assembled the termite collection of the Museu de Zoologia da Universidade de São Paulo (MZUSP), today the most important of Latin America. Mathews (1977) studied termite taxonomy and ecology of Serra do Roncador, Mato Grosso, a previously unexplored region in a transition zone between the Amazonian region and the cerrado. More recent work conducted by

Table 5.1. Number of termite species recorded in the world, the Neotropical Region, Brazil, Brazilian Amazonia and Brazilian cerrado.

Family/subfamily	World	Neotropical	Brazil	Brazilian Amazonia	Cerrado
Hodotermitidae	19	0	0	0	0
Kalotermitidae	448	122	28	19	3
Mastotermitidae	1	0	0	0	0
Rhinotermitidae	349	27	11	10	6
Serritermitidae	2	2	2	2	1
Termitidae	2021	370	249	207	129
Apicotermitinae	215	41	14	11	8
Macrotermitinae	365	0	0	0	0
Nasutitermitinae	674	248	174	141	85
Termitinae	767	81	61	55	36
Termopsidae	20	1	0	0	0
Total	2860	522	290	238	139

Source: termite database. Available at http://www.unb.br/ib/zoo/docente/constant/catal/

Bandeira and Fontes (1979), Fontes (1982, 1985, 1986), Cancello (1986), Bandeira and Cancello (1992) and Constantino (1995, 1998, 1999, 2001), and a few other authors improved the taxonomic knowledge about several taxa and enhanced collections. Currently, the best collections are at MZUSP (more than 12,000 samples, from all regions), Museu Emílio Goeldi, Belém (about 4000 samples, mainly from Amazonia), Instituto Nacional de Pesquisas da Amazônia, Manaus (about 2000 samples, from Amazonia) and Universidade de Brasília (about 6500 samples, from all regions). Sites with relatively good surveys are indicated in Fig. 5.3 according to the major biomes. The termite fauna of large portions of Brazil remains poorly known and new taxa are frequently discovered.

Major Biomes and Their Termite Fauna

Amazonia

The Amazon region, with the largest extension of tropical forest in the world, is the major biome of South America and Brazil. It has a diverse and abundant termite fauna. The Brazilian 'Legal' Amazonia, which also includes a large portion of savannahs, has 238 known species (Constantino, 2005a), nearly half of the Neotropical fauna. Termites reach high biomass and play a major role in litter decomposition in Amazonian forests (Martius, 1994). Wood feeders are dominant in the forest, especially *Nasutitermes*, and humivores are also important.

Studies in Amazonia have been concentrated near large cities, especially Belém and Manaus, and along the major rivers. Good surveys are available from Manaus (DeSouza and Brown, 1994), Maraã (Constantino, 1991, 1992), Benevides (Bandeira and Torres, 1985), Carajás (Bandeira and Macambira, 1988), Vilhena and Pimenta-Bueno, Rondônia (savannahs, R. Constantino, unpublished data). Local diversity ranges from 10 to 12 species in *várzea* forest (Constantino, 1992) to more than 60 species in *terra firme* forests (Bandeira and Macambira, 1988; Bandeira, 1989; Constantino, 1992; DeSouza and Brown, 1994). These surveys are not comparable, however, due to the different methods employed. Estimates of nest density range from 60 to 123 nests per hectare in *terra firme* and from 37 to 262 nests per hectare in *várzea* forests. These figures refer to visible epigeic and arboreal nests only

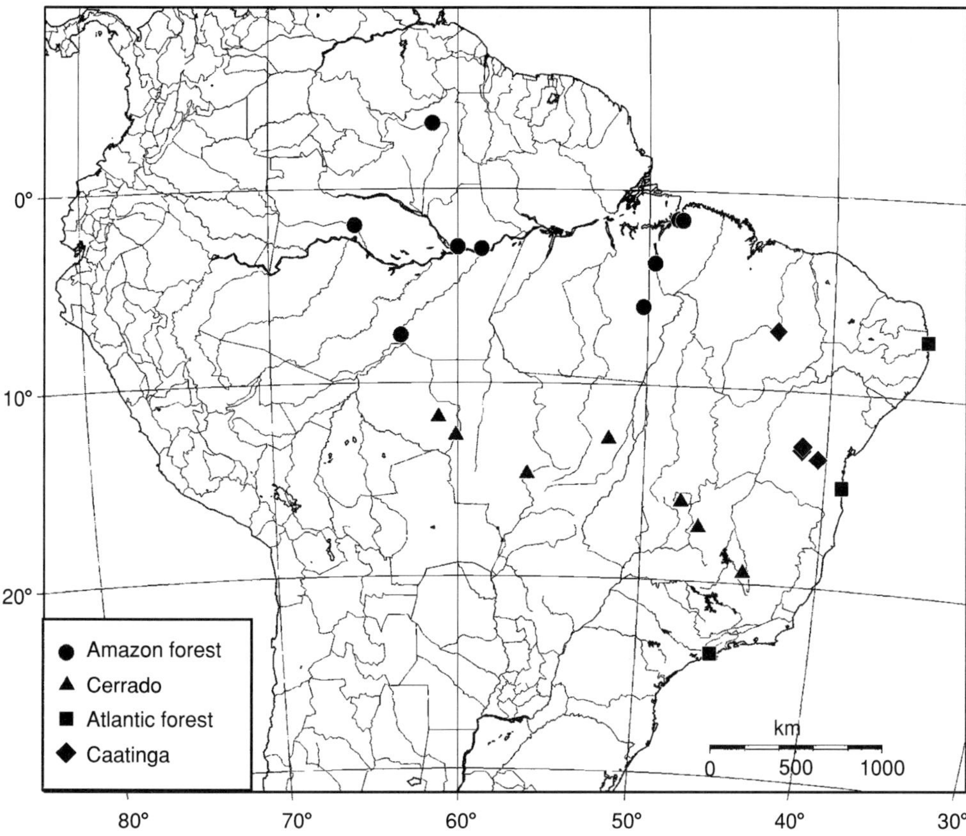

Fig. 5.3. Sites with good termite surveys in Brazil according to major biomes. (Sources: Cancello (1996), Constantino (1998, 2005a,b).)

and do not include subterranean or wood nests. The total number of colonies is, therefore, much higher.

Termite biomass in *terra firme* forests has been estimated at about 2 g/m^2 (Martius, 1994), which corresponds to nearly 20% of the total animal biomass. There are also some estimates of the consumption of cellulosic material by termites, but these are highly variable due to technical difficulties and methodological differences (Martius, 1994). Some estimates are as high as 20% of the total litter carbon input.

Cerrado

The second largest Brazilian biome is the cerrado, a savannah vegetation that covers about 12% of South America. Termites from the cerrado are relatively well known, but studies are concentrated in a few localities. The best surveys are from Brasília (Coles de Negret and Redford, 1982; R. Constantino, unpublished data); Xavantina (Mathews, 1977), Cuiabá (Silvestri, 1903) and Manso (Constantino and Schlemmermeyer, 2000) in Mato Grosso; Sete Lagoas (Gontijo and Domingos, 1991) and Paracatu (R. Constantino, unpublished data) in Minas Gerais. The termite fauna of the cerrado region is very distinct from that of humid forests.

Termites are extremely abundant in the cerrado, with about 140 recorded species (Table 5.1). Termite mounds are a conspicuous part of the landscape, reaching high densities. Mound-building *Cornitermes* can reach impressive densities in some areas. According to Redford (1984), *Cornitermes cumulans* is a keystone species. There is no biomass estimate for termites in the cerrado.

Compared with the Amazon forest, the termite fauna of the cerrado is also dominated by Nasutitermitinae, but the functional groups are different, with a large proportion of species that feed on grass litter on the surface. This group includes all *Syntermes, Velocitermes, Ruptitermes* and *Rhynchotermes*, and some *Cornitermes* and *Nasutitermes*. Humivores are also very abundant and diverse, comprising nearly 30% of the species. Wood feeders are less abundant, and most nests are epigeic or subterranean. *Constrictotermes cyphergaster* is the only common arboreal nesting species.

The cerrado vegetation is subdivided into several types according to the density of trees, and the termite fauna differ among vegetation types. Estimates of colony density in cerrado *sensu stricto* are between 564 per hectare (Coles, 1980) and 972 per hectare (Domingos *et al.*, 1986). Considering that the nests of many species are subterranean or inside wood, the total density of colonies must be higher. Local diversity in cerrado *sensu stricto* ranges from 37 (Coles, 1980) to 47 species (Domingos *et al.*, 1986) in a 50 × 50 m square.

Atlantic forest

The Atlantic forest is the third biome in size, but only about 5% of the original area is preserved. Most of the original forest has been replaced by urban and agricultural landscape. Despite the fact that this is a densely populated region with many important cities, universities and research institutions, the termite fauna of the Atlantic forest, especially the northern portion, is poorly known. The only published local surveys are from Paraíba (Bandeira *et al.*, 1998; Bandeira and Vasconcellos, 2002) and a very limited survey from Espírito Santo (Brandão, 1998). There are also good unpublished surveys represented in collections, especially at MZUSP.

There are several types of forests in this region, but there is also a conspicuous variation in the termite fauna along the latitudinal gradient. The northern part is similar to the Amazon forest, while the southern portion has a much lower diversity. A survey from Viçosa, Minas Gerais, recorded only 16 species (Fadini, 1998). In the northern portion, one site from Paraíba had 43 species (Bandeira *et al.*, 1998). The termite fauna of this region differs from both the Amazonian and cerrado faunas by the virtual absence of epigeic mounds. Kalotermitids seem to be more abundant and diverse in the Atlantic forest than in Amazonia. *Nasutitermes* is the dominant genus, and several species are common in the whole region.

Caatinga

The Caatinga is a semiarid region in northeastern Brazil, with xerophytic open forests and savannahs. The termite fauna of the Caatinga is poorly known. There are only some 20 species recorded for the entire region, but the preliminary report of a survey shows the presence of at least 138 species, nearly 60% of them undescribed (Cancello, 1996). At least in some areas, mounds are rare, just a few per hectare (Martius *et al.*, 1999). The most common genera are *Nasutitermes* and *Heterotermes*.

Pantanal

The Pantanal is a seasonally flooded region in Mato Grosso and Mato Grosso do Sul composed of grasslands with patches of forest and cerrado. The termite fauna of this region is poorly known, but termites do not live in the flooded areas. Apparently, there is no endemic fauna, and the species are the same found in the cerrado and in Amazonia.

Other biomes

The Araucaria forest of Paraná and the grasslands of Santa Catarina and Rio Grande do Sul have a temperate climate and show a much lower termite abundance and diversity. Surveys are very limited, and there is no evidence of a typical or endemic termite fauna. Species recorded in those regions also occur in the Atlantic forest and the cerrado.

Termites and Soils

The relationship between termites and soils has been the subject of numerous studies (Lee and Wood, 1971). Very few of these were conducted in Brazil (Bandeira, 1985; Bandeira and Harada, 1998). In general, these studies deal with the effect of termites on soil, but little is known of the association of particular species to different soil types. Among the surveys indicated in Fig. 5.2, very few indicate the specific type of soil where the termites were collected. Sites surveyed in the Amazon forest are mainly on top of yellow latosol (oxysol) while sites in the cerrado correspond to red latosols and a few to spodosols. There are also a few surveys in seasonally flooded areas with hydromorphic soils. Termite distribution is strongly affected by soil texture and many species do not occur in sandy soil. The fauna of flooded areas is also very distinct from that of dry places (Constantino, 1992). Some species show specific preferences related to soil properties (Brandão, 1991).

Termite Ecology and Management

Studies on termite ecology are still limited in Brazil and most are qualitative. Biomass estimates are available only for a few sites in the Amazon (Martius, 1994) and quantitative studies on their role in decomposition and their impact on the soil structure are missing. However, the data available indicate that termites are very important in the Amazon, in the cerrado and in some parts of the Atlantic forest. In general, there are two conflicting aspects of the impact of termites in agrosystems: their positive services on the soil and the fact that some species can damage living plants.

Termite response to habitat fragmentation and disturbance has been investigated in several places (DeSouza and Brown, 1994; Eggleton *et al.*, 1996; Brandão and Souza, 1998; Bandeira *et al.*, 2003). The common pattern is that termite communities are highly affected by disturbance and fragmentation, and some functional groups are more affected than others. Humus feeders are more affected than average, while some wood-feeding species become more abundant. Moderately disturbed forests may show higher termite abundance due to the increased availability of dead wood. Highly disturbed sites such as annual monoculture, especially under intensive use of pesticides, show very low termite diversity and abundance (Bandeira *et al.*, 2003). Due to their response to habitat modification, termites are considered good bioindicators of land use and management (Barros *et al.*, 2002).

Some termites, especially *Syntermes* spp., are important food items for many Amerindians (Paoletti *et al.*, 2000). Termites are sometimes used as food for domestic animals and Myles (1995) suggested that termites reared with cellulosic wastes could be used to feed poultry. Termite nests are traditionally used as a fertilizer, and the nests of *Nasutitermes* in particular seem to be an excellent substrate for growing orchids.

Termite management in agrosystems, in the broad sense, should maintain positive effects on the soil and at the same time prevent problems with crop damage. About 53 termite species have been reported as agricultural pests in South America, most of them in Brazil (Constantino, 2002). The exact extent of their damage is uncertain, but at least ten species are considered major pests.

Concluding Remarks

Brazil is a large and diverse country, with a rich and abundant termite fauna. Brazilian Amazonia alone contains nearly half of the Neotropical species. Termites comprise important functional groups in most terrestrial ecosystems, and are particularly diverse in the cerrado and Amazonian forests. Several termites are also important urban and agricultural pests (Constantino, 2002).

Brazil is the only country in Latin America with a good tradition in termite studies, a significant number of termite experts and good termite collections. Recent surveys and the effort of several termite research groups are gradually building the foundation for more detailed studies. Nevertheless, our knowledge of taxonomy, geographic distribution, basic biology and ecology of Brazilian termites is still inadequate for termite management and control in urban and agricultural systems.

References

Araujo, R.L. (1970) Termites of the Neotropical region. In: Krishna, K. and Weesner, F.M. (eds) *Biology of Termites*. Academic Press, New York, pp. 527–576.

Araujo, R.L. (1977) *Catálogo dos Isoptera do Novo Mundo*. Academia Brasileira de Ciências, Rio de Janeiro.

Bandeira, A.G. (1985) Cupinzeiros como fonte de nutrientes em solos pobres da Amazônia. *Boletim do Museu Paraense Emílio Goeldi Serie Zoologia* 2(1), 39–48.

Bandeira, A.G. (1989) Análise da termitofauna (Insecta: Isoptera) de uma floresta primária e de uma pastagem na Amazônia Oriental, Brasil. *Boletim do Museu Paraense Emílio Goeldi Série Zoologia* 5(2), 225–241.

Bandeira, A.G. and Cancello, E.M. (1992) Four new species of termites (Isoptera, Termitidae) from the island of Maracá, Roraima, Brazil. *Revista Brasileira de Entomologia* 36(2), 423–435.

Bandeira, A.G. and Fontes, L.R. (1979) *Nasutitermes acangussu*, a new species of termite from Brazil (Isoptera, Termitidae, Nasutitermitinae). *Revista Brasileira de Entomologia* 23(3), 119–122.

Bandeira, A.G. and Harada, A.Y. (1998) Densidade e distribuição vertical de macroinvertebrados em solos argilosos e arenosos na Amazônia Central. *Acta Amazônica* 28, 191–204.

Bandeira, A.G. and Macambira, M.L.J. (1988) Térmitas de Carajás, estado do Pará, Brasil: composição faunística, distribuição e hábito alimentar. *Boletim do Museu Paraense Emílio Goeldi* 4, 175–190.

Bandeira, A.G. and Torres, M.F.P. (1985) Abundância e distribuição de invertebrados do solo em ecossistemas da Amazônia Oriental. O papel ecológico dos cupins. *Boletim do Museu Paraense Emílio Goeldi Série Zoologia* 2(1), 13–38.

Bandeira, A.G. and Vasconcellos, A. (2002) A quantitative survey of termites in a gradient of disturbed forest in northeastern Brazil (Isoptera). *Sociobiology* 39, 429–439.

Bandeira, A.G., Pereira, J.C.D., Miranda, C.S. and Medeiros, L.G.S. (1998) Composição da fauna de cupins (Insecta, Isoptera) em áreas de Mata Atlântica em João Pessoa, Paraíba, Brasil. *Revista Nordestina de Zoologia* 12, 9–17.

Bandeira, A.G., Vasconcellos, A., Silva, M. and Constantino, R. (2003) Effects of habitat disturbance on the termite fauna in a highland tropical forest in the Caatinga domain, Brazil. *Sociobiology* 42, 117–127.

Barros, E., Pashanasi, B., Constantino, R. and Lavelle, P. (2002) Effects of land-use system on the soil macrofauna in western Brazilian Amazonia. *Biology and Fertility of Soils* 35, 338–347.

Bignell, D.E. (1994) Soil-feeding and gut morphology in higher termites. In: Hunt, J.H. and Nalepa, C.A. (eds) *Nourishment and Evolution in Insect Societies*. Westview Press, Boulder, Colorado, pp. 131–158.

Brandão, D. (1991) Relações espaciais de duas espécies de *Syntermes* (Isoptera, Termitidae) nos cerrados da região de Brasília, DF, Brasil. *Revista Brasileira de Entomologia* 35(4), 745–754.

Brandão, D. (1998) Patterns of termite (Isoptera) diversity in the Reserva Florestal de Linhares, state of Espírito Santo, Brazil. *Revista Brasileira de Entomologia* 41, 151–153.

Brandão, D. and Souza, R.F. (1998) Effects of deforestation and implantation of pastures on the termite fauna in the Brazilian 'cerrado' region. *Tropical Ecology* 39, 175–178.

Cancello, E.M. (1986) Revisão de *Procornitermes* Emerson (Isoptera, Termitidae, Nasutitermitinae). *Papéis Avulsos de Zoologia (São Paulo)* 36(19), 189–236.

Cancello, E.M. (1996) Termite diversity and richness in Brazil: an overview. In: Bicudo, C.E.M. and Menezes, N.A. (eds) *Biodiversity in Brazil: A First Approach*. CNPq, São Paulo, Brazil, pp. 173–182.

Coles, H.R. (1980) Defensive strategies in the ecology of Neotropical termites. PhD thesis, University of Southampton, Southampton, UK.

Coles de Negret, H.R. and Redford, K. (1982) The biology of nine termite species (Isoptera: Termitidae) from the cerrado of central Brazil. *Psyche* 89(1–2), 81–106.
Constantino, R. (1991) Termites (Insecta, Isoptera) from the lower Japurá River, Amazonas state, Brazil. *Boletim do Museu Paraense Emílio Goeldi Série Zoologia* 7(2), 189–224.
Constantino, R. (1992) Abundance and diversity of termites (Isoptera) in two sites of primary rain forest in Brazilian Amazonia. *Biotropica* 24, 420–430.
Constantino, R. (1995) Revision of the Neotropical termite genus *Syntermes* Holmgren (Isoptera: Termitidae). *The University of Kansas Science Bulletin* 55, 455–518.
Constantino, R. (1998) Catalog of the living termites of the New World (Insecta: Isoptera). *Arquivos de Zoologia* 35, 135–231.
Constantino, R. (1999) Chave ilustrada para a identificação dos gêneros de cupins (Insecta: Isoptera) que ocorrem no Brasil. *Papéis Avulsos de Zoologia* 40, 387–448.
Constantino, R. (2001) Key to the soldiers of South American *Heterotermes* with a new species from Brazil (Isoptera: Rhinotermitidae). *Insect Systematics and Evolution* 31, 463–472.
Constantino, R. (2002) The pest termites of South America: taxonomy, distribution and status. *Journal of Applied Entomology* 126, 355–365.
Constantino, R. (2005a) Insecta: Isoptera. *Fauna da Amazônia Brasileira* (in press).
Constantino, R. (2005b) Padrões de diversidade e endemismo de térmitas no bioma cerrado. In: Scariot, A., Felfili, J.M. and Souza-Silva, J.C. (eds) *Ecologia e biodiversidade do Cerrado*. Ministério do Meio Ambiente, Brasília, Brazil (in press).
Constantino, R. and Schlemmermeyer, T. (2000) Cupins (Insecta: Isoptera). In: Alho, C.J.R. (ed.) *Fauna silvestre da região do rio Manso – MT*. IBAMA/ELETRONORTE, Brasília, Brazil, pp. 129–151.
DeSouza, O.F.F. and Brown, V.K. (1994) Effects of habitat fragmentation on Amazonian termite communities. *Journal of Tropical Ecology* 10, 197–206.
Domingos, D.J., Cavenaghi, T.M.C.M., Gontijo, T.A., Drumond, M. and Carvalho, R.C.F. (1986) Composição de espécies, densidade e aspectos biológicos da fauna de térmitas de cerrado em Sete Lagoas – MG. *Ciência e Cultura* 38(1), 199–207.
Eggleton, P. and Bignell, D.E. (1995) Monitoring the responses of tropical insects to changes in the environment: troubles with termites. In: Harrington, R. and Stork, N.E. (eds) *Insects in a Changing Environment*. Academic Press, London, pp. 434–497.
Eggleton, P., Bignell, D.E., Sands, W.A., Waite, B., Wood, T.G. and Lawton, J.H. (1995) The species richness of termites (Isoptera) under differing levels of forest disturbance in the Mbalmayo Forest Reserve, southern Cameroon. *Journal of Tropical Ecology* 11, 85–98.
Eggleton, P., Bignell, D.E., Sands, W.A., Mawdsley, N.A., Lawton, J.H., Wood, T.G. and Bignell, N.C. (1996) The diversity, abundance and biomass of termites under differing levels of disturbance in the Mbalmayo Forest Reserve, southern Cameroon. *Philosophical Transactions of the Royal Society of London B* 351, 51–68.
Emerson, A.E. (1925) The termites from Kartabo, Bartica District, Guyana. *Zoologica* 6, 291–459.
Fadini, M.A.M. (1998) *Efeito de fatores locais sobre a diversidade de cupins em florestas neotropicais*. MSc thesis, Universidade Federal de Viçosa, Viçosa, Brazil.
Fontes, L.R. (1982) Novos táxons e novas combinações nos cupins nasutos geófagos da região Neotropical (Isoptera, Termitidae, Nasutitermitinae). *Revista Brasileira de Entomologia* 26(1), 99–108.
Fontes, L.R. (1985) New genera and new species of Nasutitermitinae from the Neotropical region (Isoptera, Termitidae). *Revista Brasileira de Zoologia* 3(1), 7–25.
Fontes, L.R. (1986) Two new genera of soldierless Apicotermitinae from the Neotropical region (Isoptera, Termitidae). *Sociobiology* 12(2), 285–297.
French, J.R.J. and Robinson, P.J. (1981) Baits for aggregating large numbers of subterranean termites. *Journal of the Australian Entomological Society* 20, 75–76.
Gontijo, T.A. and Domingos, D.J. (1991) Guild distribution of some termites from cerrado vegetation in southeast Brazil. *Journal of Tropical Ecology* 7, 523–529.
Grassé, P.P. (1982–1986) *Termitologia. Tome 1–3*. Masson, Paris.
Hogan, M., Veivers, P.C., Slaytor, M. and Czolij, R.T. (1988) The site of cellulose breakdown in higher termites (*Nasutitermes walkeri* and *Nasutitermes exitiosus*). *Journal of Insect Physiology* 34(9), 891–899.
Holmgren, N. (1906) Studien über südamerikanische Termiten. *Zoologische Jahrbücher Abteilung Systematik* 23, 521–676.
Holmgren, N. (1910) Versuch einer Monographie der amerikanische *Eutermes* – Arten. *Jahrbuch der Hamburgischen Wissenschaftlichen Anstalten* 27(2), 171–325.

Jones, D.T. and Eggleton, P. (2000) Sampling termite assemblages in tropical forests: testing a rapid biodiversity assessment protocol. *Journal of Applied Ecology* 37, 191–203.
Krishna, K. and Weesner, F. (1969–1970) *Biology of Termites,* Vols 1–2. Academic Press, New York.
Lavelle, P., Bignell, D., Lepage, M., Wolters, V., Roger, P., Ineson, P., Heal, O.W. and Dhilion, S. (1997) Soil function in a changing world: the role of invertebrate ecosystem engineers. *European Journal of Soil Biology* 33, 159–193.
Lee, K.E. and Wood, T.G. (1971) *Termites and Soils*. Academic Press, London.
Martius, C. (1994) Diversity and ecology of termites in Amazonian forests. *Pedobiologia* 38, 407–428.
Martius, C., Tabosa, W.A.F. and Bandeira, A.G. (1999) Richness of termite genera in a semi-arid region (sertão) in NE Brazil (Isoptera). *Sociobiology* 33, 357–365.
Mathews, A.G.A. (1977) *Studies on Termites from the Mato Grosso state, Brazil.* Academia Brasileira de Ciências, Rio de Janeiro.
Myles, T.G. (1995) The ecological importance of termites and the potential utilization of termites for the decomposition of lignocellulosic wastes. In: Abe, T. (ed.) *The Termite-Symbiont system*. Center for Ecological Research, Kyoto University, Kyoto, Japan, pp. 50–54.
Noirot, C. (1992) From wood- to humus-feeding: an important trend in termite evolution. In: Billen, J. (ed.) *Biology and Evolution of Social Insects*. Leuven University Press, Leuven, Belgium, pp. 107–119.
Paoletti, M.G., Dufour, D.L., Cerda, H., Torres, F., Pizzoferrato, L. and Pimentel, D. (2000) The importance of leaf- and litter-feeding invertebrates as sources of animal protein for the Amazonian Amerindians. *Proceedings of the Royal Society of London B* 267, 2247–2252.
Redford, K. (1984) The termitaria of *Cornitermes cumulans* (Isoptera, Termitidae) and their role in determining a potential keystone species. *Biotropica* 16(2), 112–119.
Sands, W.A. (1998) *The Identification of Worker Castes of Termite Genera from Soils of Africa and the Middle East*. CAB International, Wallingford, UK.
Silvestri, F. (1903) Contribuzione alla conoscenza dei Termiti e Termitofili dell'America Meridionale. *Redia* 1, 1–234.
Snyder, T.E. (1926) Termites collected on the Mulford Biological Exploration to the Amazon Basin 1921–1922. *Proceedings of the US National Museum* 68, 1–76.
Tayasu, I., Sugimoto, A., Wada, E. and Abe, T. (1994) Xylophagous termites depending on atmospheric nitrogen. *Naturwissenschaften* 81, 229–231.
Wood, T.G. and Sands, W.A. (1978) The role of termites in ecosystems. In: Brian, M.V. (ed.) *Production Ecology of Ants and Termites*. Cambridge University Press, Cambridge, UK, pp. 245–292.

6 Patterns of Diversity and Responses to Forest Disturbance by Ground-dwelling Ants in Amazonia

H.L. Vasconcelos
Institute of Biology, Federal University of Uberlândia (UFU), C.P. 593, 38400-902, Uberlândia, MG, Brazil, e-mail: heraldo@umuarama.ufu.br

Introduction

After their evolution from an ancestral vespoid during the Cretaceous, *c.* 80 million years ago, ants radiated and became a major element in most terrestrial ecosystems (Wilson *et al.*, 1967). In the forests of the Amazon Basin, for instance, ants can comprise approximately one-third of the insect biomass, and their biomass is four times greater than that of all vertebrates combined (Fittkau and Klinge, 1973). Species diversity, although not comparable with that of hyperdiverse insect groups such as Coleoptera and Hymenoptera *parasitica*, is high. The remarkable diversification and success of ants is attributed to the fact that they were the first group of predatory social insects that both lived and foraged primarily in the soil and in rotting vegetation on the ground (Hölldobler and Wilson, 1990). Although many species nest in trees, arboreal life appears to represent a secondary, minority adaptation (Hölldobler and Wilson, 1990).

Belonging to the hymenopteran family Formicidae, ants are important in belowground processes through the alteration of the physical and chemical environment and through their effect on plants, microorganisms and other soil organisms (Folgarait, 1998). Ants are major predators of other arthropods, including soil arthropods and herbivorous insects (Yamaguchi and Hasegawa, 1996; Human and Gordon, 1997; Floren *et al.*, 2002). Also, they move considerable amounts of subterranean soil to the surface, and the network of galleries and chambers from ant nests increases soil porosity and drainage, while reducing bulk density (Folgarait, 1998; Fig. 6.1). The soil modified by ants is generally richer in organic matter, N, P and K, and this increase in nutrients can be particularly important for the development of vegetation, especially in poor soils (e.g. Culver and Beattie, 1983). A few ants, such as the leaf-cutter ants, have a direct effect on plants by harvesting leaves and other plant parts (Vasconcelos and Cherrett, 1997). Others eat and move seeds, thus rearranging the seed shadow produced by vertebrate dispersers (Levey and Byrne, 1993). Ants have indirect positive and negative effects on plants as they take nectar produced in extrafloral nectaries, or attend Homoptera for honeydew (Buckley, 1987; Oliveira *et al.*, 1987).

Many factors, operating at regional and local scales, can potentially affect the diversity of ants. In this chapter I examine some of these factors, focusing on Amazonian

Fig. 6.1. Nest of the leaf-cutting *A. sexdens* in the understorey of a secondary forest in Manaus, Brazil. By moving soil from below the ground to the soil surface and by creating a network of chambers and galleries, these ants significantly alter soil chemical and physical properties.

ants. Examples come mainly from studies with ground-dwelling ants, although where necessary studies on canopy ants were included. By ground-dwelling ants I mean ants that forage on the ground, which include not only soil-nesting species but also species that nest in the lower vegetation strata (Fig. 6.2). Nevertheless, as shown above, both groups of species can have major impacts on soil processes and organisms.

I start by briefly presenting the main methods to collect ground-dwelling ants, and then discuss some of the factors affecting ant diversity in Amazonia.

Methods of Collection

A variety of methods have been employed to collect ants for biodiversity studies (reviewed in Bestelmeyer *et al.*, 2000). Currently, the three most commonly employed methods to study ground-dwelling ants are: (i) pitfall traps, (ii) sardine baits, and (iii) the Winkler method (Fig. 6.3). The latter consists of extracting ants from the leaf litter (although it can also be used to extract ants from soil samples as well) by sieving the litter sample through a 0.8 cm mesh and leaving the sifted litter in Winkler

Fig. 6.2. Workers of *Dolichoderus bispinosus*, an arboreal-nesting species, foraging on the forest floor of a central Amazonian forest.

Fig. 6.3. Collection of litter-dwelling species using the Winkler method: (a) sieving of leaf litter in the forest, (b) sifted litter transferred to a mesh bag and (c) Winkler extractors.

bags (Fig. 6.3) for 24 h or more. Pitfall trapping involves the placement of open containers in the ground, and is used to census ants foraging on the soil or litter surface. Traps may be plastic or glass containers. In my own studies I have used plastic cups (6.5 cm diameter; 8 cm depth; 200 ml volume) as containers, as these are readily available and inexpensive. The containers are filled with a killing and preservative agent, such as ethanol (with glycerol to retard evaporation), or simply with water and a few drops of detergent if traps are left in operation for a maximum of 48 h. Sardine canned in vegetable oil is one of the best baits for generalist ants. The ants attracted to the baits are collected, usually 1 h after baiting. These methods, especially the Winkler method and the pitfall traps, tend to complement each other, and their use in combination can thus better characterize the ant fauna of tropical forests. A recently developed ant collecting protocol, known as Ants of the Leaf Litter (ALL) Protocol, indicates that a sample size of 20 litter samples (1 m^2 each) and 20 pitfall traps is sufficient to sample at least 70% of the ground-dwelling ant fauna of a given study site (Agosti and Alonso, 2000).

Ant identification to genus level is relatively easy. Taxonomic revisions exist for a number of genera, but many are still in need of revision. Good collections exist in Brazil, including the largest ant collection in South America, which is located at the Zoology Museum of the University of São Paulo (Museu de Zoologia da USP).

Vertical Partitioning of the Forest Ant Fauna

Species that nest or forage on the ground of tropical forests tend to be quite distinct from those living in the canopy. For instance, of the 524 ant species from a forest in Sabah, Malaysia, 75% were exclusively found on the ground or in the canopy (Brühl *et al.*, 1998). Even when comparisons are made between leaf-litter ants only, i.e. those taken from the leaf litter on the ground, and those from the litter that accumulates under epiphyte mats in the canopy, strong differences are found (Longino and Nadkarni, 1990). In Amazonian forests, a large proportion of the ground-dwelling species belong to the genus *Pheidole*, which accounts for approximately one-quarter of the species recorded. Other species-rich genera include *Pachycondyla*, *Crematogaster*, *Trachymyrmex*, *Paratrechina*, *Solenopsis* and *Hypoponera* (Table 6.1). By contrast, relatively few *Pheidole* species are found in the canopy (Table 6.1), which is characterized by typically arboreal ant genera such as *Camponotus*, *Pseudomyrmex*, *Cephalotes*, *Azteca* and *Dolichoderus* (Harada and Adis, 1997; Tobin, 1997).

When looking at the ground-dwelling fauna only, some differences exist in terms of species composition when comparing the fauna found below and above the ground (Table 6.1). The terms epigaeic and hypogaeic have been employed to distinguish, respectively, those species that forage primarily on the soil surface versus those foraging and nesting underground (e.g. Fowler and Delabie, 1995). Species in the genera

Table 6.1. Proportional representation of different ant genera in different strata of central Amazonian forests. Values represent the per cent of the total number species in each genus.

Genus	Forest strata			
	Below-ground	Litter layer	Soil surface	Canopy
Ponerinae				
Amblyopone	1.89	0.68	0.00	0.00
Anochetus	1.89	2.72	2.82	0.00
Centromyrmex	0.94	0.00	0.00	0.00
Discothyrea	0.94	1.36	0.70	1.01
Ectatomma	1.89	2.04	1.41	1.01
Gnamptogenys	3.77	3.40	4.23	0.00
Heteroponera	0.94	0.00	0.00	0.00
Hypoponera	6.60	4.76	3.52	0.00
Leptogenys	1.89	0.00	2.11	0.00
Odontomachus	1.89	2.04	2.82	0.00
Pachycondyla	4.72	4.08	6.34	4.04
Platythyrea	0.00	0.00	0.70	0.00
Prionopelta	0.94	0.68	0.70	0.00
Typhlomyrmex	2.83	0.68	0.00	0.00
Cerapachyinae				
Acanthostichus	2.83	0.68	0.00	0.00
Cerapachys	0.94	0.68	0.00	0.00
Sphinctomyrmex	0.94	0.00	0.00	0.00
Ecitoninae				
Eciton	0.00	0.68	0.70	0.00
Labidus	0.00	0.00	0.70	0.00
Neivamyrmex	0.94	0.00	2.11	0.00
Leptanilloidinae				
Asphinctanilloides	1.89	0.00	0.00	0.00
Myrmicinae				
Apterostigma	0.94	1.36	2.82	0.00
Atta	0.00	0.00	0.70	0.00
Blepharidatta	0.94	0.68	0.70	0.00
Carebara	0.94	0.00	0.70	0.00
Cephalotes	0.00	0.00	0.70	13.13
Crematogaster	2.83	4.08	4.93	9.09
Cyphomyrmex	0.00	1.36	2.11	2.02
Daceton	0.00	0.68	0.00	0.00
Hylomyrma	0.00	2.04	0.00	0.00
Lachnomyrmex	0.00	1.36	0.00	0.00
Leptothorax	0.00	0.00	0.70	3.03
Monomorium	0.00	0.00	0.00	1.01
Megalomyrmex	0.00	4.08	2.11	0.00
Mycocepurus	0.94	0.68	0.00	0.00
Myrmicocrypta	0.94	1.36	1.41	0.00
Ochetomyrmex	1.89	1.36	1.41	0.00
Octostruma	0.00	0.68	0.70	0.00
Oligomyrmex	0.94	0.68	0.70	0.00
Oxyepoecus	0.00	0.68	0.70	0.00
Pheidole	23.58	25.17	22.54	5.05
Pyramica	0.00	3.40	2.81	0.00
Rogeria	2.83	3.40	0.00	0.00
Sericomyrmex	0.00	1.36	1.41	0.00
Solenopsis	5.66	3.40	3.52	9.09

Continued

Table 6.1. Proportional representation of different ant genera in different strata of central Amazonian forests. Values represent the per cent of the total number species in each genus. – cont'd

	Forest strata			
Genus	Below-ground	Litter layer	Soil surface	Canopy
Strumigenys	2.83	2.04	2.11	0.00
Talaridris	0.00	0.68	0.00	0.00
Trachymyrmex	2.83	4.08	3.52	0.00
Tranopelta	0.94	0.00	0.00	0.00
Wasmannia	0.94	0.68	1.41	0.00
Formicinae				
Acropyga	2.83	1.36	0.70	0.00
Brachymyrmex	0.94	1.36	0.70	0.00
Camponotus	3.77	0.68	4.23	19.19
Myrmelachista	0.00	0.00	0.00	2.02
Paratrechina	3.77	4.08	2.82	0.00
Dolichoderinae				
Azteca	0.00	0.68	0.70	6.06
Dolichoderus	0.00	0.68	2.11	3.03
Dorymyrmex	0.00	0.68	0.00	0.00
Linepithema	0.00	0.00	0.70	1.01
Tapinoma	0.00	0.00	0.00	5.05
Pseudomyrmecinae				
Pseudomyrmex	0.00	0.68	1.41	15.15
Total number of species	106	147	142	99

Source: data on canopy ants are from Harada and Adis (1997). Remaining data are from Vasconcelos and Delabie (2000).

Acropyga, *Amblyopone*, *Centromyrmex*, *Tranopelta* and *Typhlomyrmex*, as well as many Cerapachyinae and Leptanilloidinae, for instance, are classified as hypogaeic. These species are relatively common in soil samples, but are rare in samples using pitfall traps (Table 6.1), which, as indicated before, are used to census ants foraging on the soil surface. The ant fauna associated with the litter layer is intermediate, and contains both epigaeic and hypogaeic species (Table 6.1). Preliminary evidence, based on data collected in central Amazonia (Vasconcelos and Delabie, 2000), indicates that although the diversity of species collected in soil samples is lower than that collected in litter samples or in pitfall traps (106 versus 142–147 species, respectively), within-habitat variations in local diversity are similar. Sites rich in litter-dwelling species or in species foraging above the ground also tend to be rich in soil-dwelling species, the same being observed in relatively species-poor sites.

General Patterns of Diversity

At the time of the publication of the seminal book on ants by Hölldobler and Wilson (1990), approximately 8800 species of ants had been described worldwide (a more recent figure counts 11,826 species[1]). Hölldobler and Wilson (1990) estimated that 20,000 or more species, representing 350 genera, exist in the world. For the Neotropics, Kempf (1972) listed 2233 species. If we assume that the proportion of described to undescribed species is the same for all biogeographic regions, then the estimated number of ant species for the Neotropics should be around 5000. How many of these species are found in the Amazon region is a matter of speculation, since species distribution maps are scarce and in general incomplete. New records with substantial changes in range distribution

[1]http://research.amnh.org/entomology/social_insects (10 March 2005)

are common, especially for cryptic, soil-dwelling species (e.g. Delabie *et al.*, 2001). Based on existing information (e.g. Kempf, 1972; MacKay, 1993), it is likely that no more than 70% of the Neotropical species occur in the Amazon. Regardless of the exact figure, it is clear that a reasonably large proportion of this regional (gamma) diversity can be found in relatively small areas of the Amazon. For instance, Verhaagh (1990) found 520 ant species in 10 km^2 of forest in Peru, whereas Benson and Harada (1988) recorded 307 species in the municipality of Manaus. The fact that so many species are found in one single locality, relative to the number of species expected to be found in the entire basin, suggests that: (i) our estimates of regional diversity of ants in the Amazon are inaccurate, or (ii) many of the species that coexist locally have a wide distribution. Some lines of evidence suggest the latter may be the case. First, some studies have shown that species turnover (beta diversity) among different forest habitats is low, in spite of high within-habitat (alpha) diversity (Wilson, 1987; Majer and Delabie, 1994). Second, preliminary evidence indicates that large overlap exists when comparing the species compositions of distant sites with similar vegetation cover. For instance, 60.7% of the species recorded in a rapid inventory in the Jaú National Park in central Amazonia have also been recorded in Alter do Chão (Santarém, Pará), about 1500 km east of the park (Vasconcelos *et al.*, 2004).

Local inventories of the ground-dwelling fauna in Amazonia typically produce a species-rank abundance curve like the one in Fig. 6.4, with a few common species and a large number of rare species. The number of unicates and duplicates (i.e. species recorded, respectively, in only one and two samples) usually declines as sampling effort increases, but even in very exhaustive inventories (e.g. Longino *et al.*, 2002) many rare species are found. Some species are 'rare' simply because of methodological problems, i.e. because they are not easily collected with the methods employed in the survey. Others are, as defined by Longino *et al.* (2002), 'geographic edge species', i.e. species rare at the particular locality and/or habitat surveyed but common in other regions or habitats. In many cases, however, rarity remains unexplained (Longino *et al.*, 2002).

Species diversity in the Amazon, as in other tropical areas, is much higher than in the temperate zone, with three to ten times more species found in comparable habitats (Benson and Harada, 1988; Verhaagh, 1990). Forty-three species of ants were recorded in the canopy of a single Amazonian tree, which is about the same number of species found in the entire British Isles (Wilson, 1987). A variety of factors have been proposed to explain latitudinal gradients in species diversity (Rohde, 1992). For ants, diversity of nesting sites appears to be of great importance as a prerequisite for maintenance of species diversity. As Benson and Harada (1988) have pointed out, in the tropics, leaf litter and dead wood can serve as nesting sites, whereas in temperate regions, low temperature and destruction of nests during winter make these sites risky. Similarly, arboreal nesting is common in

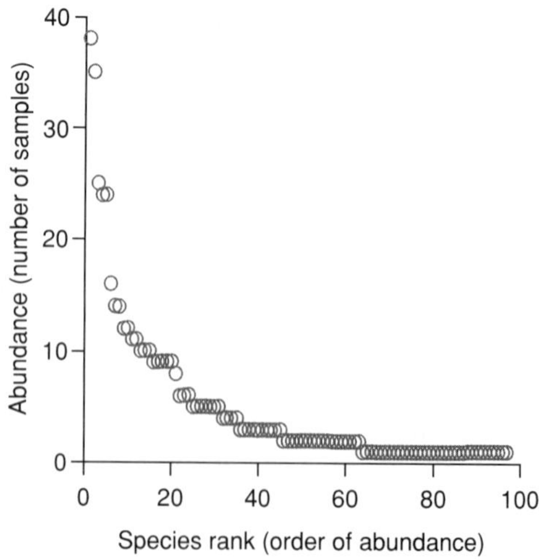

Fig. 6.4. Rank-abundance plot of 97 ant species collected from leaf litter samples in a mature forest near Manaus, in central Amazonia. Abundance is expressed as the number of samples in which the species was recorded. In total 120, 1 m^2 samples of leaf litter were taken. Ants were extracted using the Winkler method (see text for details). (Source: Vasconcelos *et al.*, 2000.)

the tropics, but not in the temperate zone, where soil provides the only secure nesting option (Benson and Harada, 1988). Latitudinal gradients of ant species diversity occur even in relatively small latitudinal spans (Gotelli and Ellison, 2002).

Although no one has yet looked at the relationship between altitude and species richness of Amazonian ants, it is likely that the same pattern found in other tropical regions applies. In these areas, altitude negatively affects species richness, but the relationship is usually not unimodal (Ward, 2000, and included references, but see Brühl *et al.*, 1999, for a counterexample). At elevations under 500 m, species richness increases with altitude, whereas above 500 m it decreases, with very few species being found above 2000 m. Possible explanations for this mid-elevation peak in species richness include greater overlap of faunas at mid-elevations and the coincidence of mid-elevation sites with regions of greater productivity (Ward, 2000).

Species Turnover within and between Habitats

Comparative studies of the composition of the ant fauna in different Amazonian ecosystems are scarce. In one such study, Majer and Delabie (1994) have found that about 50% of the ant species found in the seasonally flooded *várzea* forest are also found in the *terra firme* forest, in spite of the fact that only 5–10% of the tree species are common to both habitats (Majer and Delabie, 1994). Similarly, Wilson (1987), studying the canopy fauna from four forest types in Peruvian Amazonia, has shown that although species diversity at single sites is very high, it is not augmented very much by differences among forest types. Between 57% to 63% of the species found in the forests with lower ant diversity were also present in the forests with higher ant diversity (Wilson, 1987).

In New Guinea, Wilson (1958) recorded that most species show an extensive and unpredictable variation in population densities over short geographic distances, and variation is seen even within the most favoured habitats. A similar pattern was detected in central Amazonia (Vasconcelos and Delabie, 2000), indicating that ant species composition can vary over distances of only a few kilometres in continuous and relatively homogeneous rainforest. The reasons for that are not yet known, but the result is that species tend to occur in small populations (Wilson, 1958).

Some parameters of community structure, such as abundance and richness, are patchy even at smaller scales. In Central America, the densities of litter-dwelling ants vary up to 20-fold at the 1 m^2 scale (Kaspari, 1996), a pattern also seen in the Amazon (Höfer *et al.*, 1996; Vasconcelos, 1999). A small part of this variation in ant abundance is due to variations in litter standing crop, i.e. due to resource tracking (Kaspari, 1996). Other factors accounting for this variation include selective predation by swarm-raiding army-ants (Franks and Bossert, 1983; Kaspari, 1996), spatio-temporal variability in litter moisture content (Levings, 1983; Levings and Windsor, 1984) and topography (Benson and Brandão, 1987; Vasconcelos *et al.*, 2003).

In Amazonia, topographical variations in soil and forest structure are common (Ranzani, 1980; Kahn, 1987; Ribeiro *et al.*, 1994). For instance, near Manaus in central Amazonia, the altitudinal difference between stream valleys and plateaus is of up to 80 m, and this can be reached within a horizontal distance of only a few hundred metres (Ranzani, 1980; Chauvel *et al.*, 1987). Topography has been shown to affect the distribution of Amazonian ants. For instance, Benson and Brandão (1987) reported a higher similarity in ant species composition among 'wetter' sites, presumably in valleys, than between wetter and drier sites of an Amazonian forest. A subsequent study (Vasconcelos *et al.*, 2003) has found similar results. In addition, it was shown that although the number of species per plot (species density) did not differ between different topographic regions, in total more species were recorded in valleys than on plateaus. Species evenness also tended to be greater in valleys than on plateaus. Dominant species were relatively rare in valleys, possibly because litter, an important ant nesting site food resource for ant prey, was present in smaller quantities in valleys than on plateaus (Vasconcelos *et al.*, 2003).

Responses to Natural Disturbances (Flooding, Wildfires and Treefall Gaps)

Some forest types in the Amazon, particularly the *várzea* and the *igapó* forests, are inundated during a few months every year. As expected, species richness in these forests is lower than in *terra firme* forests (Wilson, 1987; Majer and Delabie, 1994), especially when one compares the richness of soil-, litter- and shrub-associated species (Majer and Delabie, 1994). Many ground-dwelling species have adaptations to cope with the inundation phase. The leaf-cutter *Acromyrmex lundi carli* moves its entire nest (including larvae and cultivated fungus) from the soil into the canopy during the high-water season (Adis, 1982).

Terra firme forests, in contrast to the *várzea* and *igapó*, are not affected by these annual floodings. However, they are subject to natural disturbances of both small and large scale, such as treefalls and wildfires (Uhl *et al.*, 1990). The effects of wildfires on forest ants have not been studied, but in the woodland savannahs (cerrados) of Brazil, fire has a negative effect on arboreal species, by killing established colonies and reducing diversity (Morais and Benson, 1988). These effects are probably less intense for ground-nesting species, which establish their nests deep into the soil, and thus are probably immune from the direct effects of fire.

Feener and Schupp (1998) studied the effects of treefall gaps on the ant assemblage of a Panamanian forest. Their results indicate that ground-dwelling ants are not affected by the formation of treefall gaps, in spite of microclimatic differences and differences in plant productivity (and thus of resources for ants) between gaps and the shaded forest understorey.

Anthropogenic Disturbances

As Uhl *et al.* (1990) have pointed out, although natural disturbances in Amazonian forests have been common throughout history, now, because of the dramatic increase in human activities in Amazonia, anthropogenic disturbances have become even more common than natural disturbances in many areas. Each year in the Brazilian Amazon 15,000 to 20,000 km^2 of forest are cleared and converted into agricultural areas or into pastures for cattle ranching (Nepstad *et al.*, 1999). Another 10,000–15,000 km^2 are affected by logging and fire (Nepstad *et al.*, 1999). The remaining forest patches become subject to the diverse, and generally adverse, effects of forest fragmentation (Laurance *et al.*, 2002).

Studies so far indicate that conversion of mature forest into pasture areas causes a dramatic decline in ant species richness (Moutinho, 1998; Vasconcelos, 1999). The number of species found in pasture areas is 50–60% lower than that found in comparable areas in undisturbed forest (Table 6.2). In addition to its effects on species richness, conversion of mature forest into pastures leads to substantial changes in species composition (Moutinho, 1998; Vasconcelos, 1999). Some pest species tend to become abundant in pasture areas, and these species can have a negative effect on the regeneration of the forest in abandoned pasture areas (Nepstad *et al.*, 1990; Vasconcelos and Cherrett, 1997; Moutinho, 1998). Among these species are the leaf-cutters *Atta sexdens* and *Atta laevigata*, as well as the seed predators *Solenopsis saevissima* and *Solenopsis geminata* (Carroll and Risch, 1984).

The effects of logging on Amazonian ants seem to be much less dramatic than those caused by the conversion of the forest into pastures or crops. Although logging may lead to complete clearing of the land, most commonly it does not, as only selected trees of commercial value – such as mahogany – are harvested. Selective logging does not affect the overall abundance of ants on the forest floor nor does it affect ant species richness, as revealed by two inde-

Table 6.2. Effects of changes in land use and land cover on the diversity and abundance of ground-dwelling ants in central Amazonia (modified from Vasconcelos, 1999).

Vegetation type	Total number of ant species recorded	Per cent of species shared with primary forest	Similarity to primary forest (Jaccard index)	Mean number of ants per 0.5 m^2 of leaf litter (range)
Primary forest[a]	81	–	–	7.3 (0–52)
Secondary forest (13 years old)[b]	83	50	0.34	9.0 (0–88)
Secondary forest (10 years old)[c]	62	47.5	0.26	17.4 (0–101)
Pasture	36	26.3	0.085	Not determined

[a]This represents the original vegetation cover.
[b]Established immediately after forest clearing. No burning of the felled vegetation.
[c]Established in a previous pasture area, exploited for 2 years prior to abandonment.

pendent studies, one in central (Vasconcelos *et al.*, 2000) and the other in eastern Amazonia (Kalif *et al.*, 2001). However, changes in species composition are detected, especially in the areas directly impacted by the logging operation (logging gaps and logging tracks), suggesting that the persistence of ant assemblages typical of undisturbed forest in logged plots is likely to depend on the amount of structural change to the forest (Vasconcelos *et al.*, 2000). Species typical of undisturbed forest, such as many soil- and litter-dwelling *Pheidole*, tend to disappear from the most disturbed areas (Kalif *et al.*, 2001), and these can be replaced by invasive, non-forest species (Vasconcelos *et al.*, 2000).

Response of ants to the effects of forest fragmentation seems comparable with those of selective logging. The major effect is on species composition, with no marked changes in species richness or abundance (Carvalho and Vasconcelos, 1999; Vasconcelos and Delabie, 2000; Vasconcelos *et al.*, 2001). Two of the most important mechanisms of change in fragmented forests are area effects and edge effects (Laurance *et al.*, 2002). Area effects are ecological changes that occur as a result of fragment isolation, and are generally proportional in magnitude to fragment area. For example, small fragments tend to have small populations that suffer high rates of extinction due to random genetic and demographic events. Edge effects, in contrast, are caused by gradients in physical and biotic factors near forest edges, and are generally proportional to the distance of the site from the nearest edge (Laurance *et al.*, 2002). Changes in ant species composition in relatively recently isolated (<20 years) forest fragments seem to be explained more by proximity to forest edge than by fragment area *per se* (Carvalho and Vasconcelos, 1999). The observed changes in ant community structure, which can penetrate deep (>100 m) into the forest, are attributed in part to variations in litter depth, which increased markedly near forest edges (Carvalho and Vasconcelos, 1999).

Succession

As indicated earlier, several studies have shown that clearing of mature forest causes a significant decline in species richness and significant changes in species composition. The number of species found in cleared and/or recently planted forest is only one-third to half of that found in nearby mature forest (MacKay *et al.*, 1991; Roth *et al.*, 1994; Moutinho, 1998; Vasconcelos, 1999; Table 6.2). Once these areas are abandoned, forest regeneration proceeds, and with it the original forest fauna gradually recovers.

History of land use appears to be an important determinant of rate of recovery, being slower in areas more intensively used, such as those used for many years as pasture (Moutinho, 1998; Vasconcelos, 1999; Table 6.2). Distance to existing forest patches seems also to have some influence on the re-colonization of the cleared areas by forest ants. In Costa Rica, the rapid recovery (within 25 years) of the ground ant fauna after the abandonment of cocoa plantations has been in part attributed to the proximity of these plantations to mature forest (Roth *et al.*, 1994).

Regeneration of the ant fauna appears to be faster than the regeneration of the woody plant community. Up to 200 years are necessary for abandoned plots in Amazonia to attain mature forest characteristics (Uhl, 1987), while studies in Ghana, Costa Rica and Brazil (Belshaw and Bolton, 1993; Roth *et al.*, 1994; Vasconcelos, 1999) indicate that secondary forests of 25 years or less have an ant fauna very similar to that of undisturbed mature forest.

As forest regeneration proceeds, a temporal replacement of species is usually seen. For instance, the abundance of *A. laevigata* and *A. sexdens* declines sharply as secondary forests grow older (Vasconcelos and Cherrett, 1995). These species are then replaced by *A. cephalotes*, which in Amazonia is typically a mature forest species (Vasconcelos and Cherrett, 1995). Similarly, *Ectatomma brunneum* (= *quadridens*) is replaced by *Ectatomma lugens* and *Ectatomma edentatum* as forest regenerates in abandoned Amazonian plots (Vasconcelos, 1999).

Concluding Remarks

As indicated here, the structure of ground-dwelling ant communities can be affected by a variety of factors. Species richness and/or composition are influenced by altitude, topography, by variations in vegetation cover, depth of the leaf litter, soil moisture, as well as by natural and anthropogenic disturbances. Taxonomic revisions, especially of species-rich genera, are urgently needed to better elucidate these patterns, in particular with regard to broad-scale patterns of diversity and species distribution (e.g. Ward, 2000). The habits and ecological role of many species are still unknown, and this applies especially for most of the cryptic, soil-dwelling species. Therefore, although some information exists on how ant diversity changes according to natural gradients or to gradients of human disturbance, almost nothing is known about what these changes mean in terms of ecosystem functioning and/or the diversity of other organisms.

In terms of biodiversity conservation outside protected areas, what are the best options in terms of land use? Given that the ground-dwelling fauna seems to recover relatively rapidly following low to moderate levels of anthropogenic disturbance, are agroforestry systems able to sustain a diversity of ants comparable with that of mature forest? If so, how structurally and floristically diverse should those systems be?

Finally, an important basic and applied question concerns the dynamics of ground-dwelling ant communities in Amazonia. How variable is species diversity and composition through time? Is, as observed for birds, butterflies and social spiders (Laurance, 2002), species turnover of ants accelerated in disturbed habitats, and if so what are the consequences?

Acknowledgements

I thank Fatima Moreira for inviting me to write this review, and two anonymous referees for reading and commenting on earlier versions of the manuscript.

References

Adis, J. (1982) Eco-entomological observations from the Amazon. III. How do leafcutting ants of inundation forests survive flooding? *Acta Amazonica* 12, 839–840.

Agosti, D. and Alonso, L.E. (2000) The ALL protocol: a standard protocol for the collection of ground-dwelling ants. In: Agosti, D., Majer, J.D., Alonso, L.E. and Schultz, T.R. (eds) *Ants: Standard Methods for Measuring and Monitoring Biodiversity*. Smithsonian Institution Press, Washington, DC, pp. 204–206.

Belshaw, R. and Bolton, B. (1993) The effect of forest disturbance on the leaf litter ant fauna in Ghana. *Biodiversity and Conservation* 2, 656–666.

Benson, W.W. and Brandão, C.R.F. (1987) *Pheidole* diversity in the humid tropics: a survey from Serra dos Carajas, Pará, Brazil. In: Eder, J. and Rembold, H. (eds) *Chemistry and Biology of Social Insects*. Verlag J. Peperny, Munich, Germany, pp. 593–594.

Benson, W.W. and Harada, A.Y. (1988) Local diversity of tropical and temperate ant faunas (Hymenoptera, Formicidae). *Acta Amazonica* 18, 275–289.

Bestelmeyer, B.T., Agosti, D., Alonso, L.E., Brandão, C.R.F., Brown, W.L., Delabie, J.H.C. and Silvestre, R. (2000) Field techniques for the study of ground-dwelling ants: an overview, description, and evaluation. In: Agosti, D., Majer, J.D., Alonso, L.E. and Schultz, T.R. (eds) *Ants: Standard Methods for Measuring and Monitoring Biodiversity*. Smithsonian Institution Press, Washington, DC, pp. 122–144.

Brühl, C.A., Gunsalam, G. and Linsenmair, K.E. (1998) Stratification of ants (Hymenoptera: Formicidae) in a primary rain forest in Sabah, Borneo. *Journal of Tropical Ecology* 14, 295–297.

Brühl, C.A., Mohamed, M. and Linsenmair, K.E. (1999) Altitudinal distribution of leaf litter ants along a transect in primary forest on Mount Kinabalu, Sabah, Malaysia. *Journal of Tropical Ecology* 15, 265–277.

Buckley, R.C. (1987) Interactions involving plants, homoptera and ants. *Annual Review of Ecology and Systematics* 18, 111–135.

Carroll, C.R. and Risch, S.J. (1984) The dynamics of seed harvesting in early successional communities by a tropical ant, *Solenopsis geminata*. *Oecologia* 61, 388–392.

Carvalho, K.S. and Vasconcelos, H.L. (1999) Forest fragmentation in central Amazonia and its effects on litter-dwelling ants. *Biological Conservation* 91, 151–158.

Chauvel, A., Lucas, Y. and Boulet, R. (1987) On the genesis of the soil mantle of the region of Manaus, central Amazonia, Brazil. *Experientia* 43, 234–240.

Culver, D.C. and Beattie, A.J. (1983) Effects of ant mounds on soil chemistry and vegetation patterns in a Colorado montane meadow. *Ecology* 64, 485–492.

Delabie, J.H.C., Vasconcelos, H.L., Vilhena, J.M.S. and Agosti, D. (2001) First record of the ant genus *Probolomyrmex* in Brazil. *Revista de Biologia Tropical* 49, 397–398.

Feener, D.H. and Schupp, E.W. (1998) Effect of treefall gaps on the patchiness and species richness of Neotropical ant assemblages. *Oecologia* 116, 191–201.

Fittkau, E.J. and Klinge, H. (1973) On biomass and trophic structure of the central Amazonian rain forest ecosystem. *Biotropica* 5, 2–14.

Floren, A., Biun, A. and Linsenmair, K.E. (2002) Arboreal ants as key predators in tropical lowland rainforest trees. *Oecologia* 131, 137–144.

Folgarait, P.J. (1998) Ant biodiversity and its relationship to ecosystem functioning: a review. *Biodiversity and Conservation* 7, 1221–1244.

Fowler, H.G. and Delabie, J.H.C. (1995) Resource partitioning among epigaeic and hypogaeic ants (Hymenoptera: Formicidae) of a Brazilian cocoa plantation. *Ecologia Austral* 5, 117–124.

Franks, N.R. and Bossert, W.H. (1983) The influence of swarm raiding army ants on the patchiness and diversity of a tropical leaf litter ant community. In: Sutton, E.L., Whitmore, T.C. and Chadwick, A.C. (eds) *Tropical Rain Forest: Ecology and Management*. Blackwell, Oxford, UK, pp. 151–163.

Gotelli, N.J. and Ellison, A.M. (2002) Biogeography at a regional scale: determinants of ant species density in New England bogs and forests. *Ecology* 83, 1604–1609.

Harada, A.Y. and Adis, J. (1997) The ant fauna of tree canopies in central Amazonia: a first assessment. In: Stork, N.E., Adis, J. and Didham, R.K. (eds) *Canopy Arthropods*. Chapman & Hall, London, pp. 382–400.

Höfer, H., Martius, C. and Beck, L. (1996) Decomposition in an Amazonian rain forest after experimental litter addition in small plots. *Pedobiologia* 40, 570–576.

Hölldobler, B. and Wilson, E.O. (1990) *The Ants*. Belknap Press, Cambridge, Massachusetts.

Human, K.G. and Gordon, D.H. (1997) Effects of argentine ants on invertebrate biodiversity in northern California. *Conservation Biology* 11, 1242–1248.

Kahn, F. (1987) The distribution of palms as a function of local topography in Amazonian terra-firme forests. *Experientia* 43, 251–259.
Kalif, K.A.B., Azevedo-Ramos, C., Moutinho, P. and Malcher, S.A.O. (2001) The effect of logging on the ground-foraging ant community in eastern Amazonia. *Studies on Neotropical Fauna and Environment* 36, 215–219.
Kaspari, M. (1996) Litter ant patchiness at the 1-m^2 scale: disturbance dynamics in three Neotropical forests. *Oecologia* 107, 265–273.
Kempf, W.W. (1972) Catálogo abreviado das formigas da região Neotropical. *Studia Entomologica* 15, 1–344.
Laurance, W.F. (2002) Hyperdynamism in fragmented habitats. *Journal of Vegetation Science* 13, 595–602.
Laurance, W.F., Lovejoy, T.E., Vasconcelos, H.L., Bruna, E.M., Didham, R.K., Stouffer, P.C., Gascon, C., Bierregaard, R.O., Laurance, S.G. and Sampaio E. (2002) Ecosystem decay of Amazonian forest fragments, a 22-year investigation. *Conservation Biology* 16, 605–618.
Levey, D.J. and Byrne, M.M. (1993) Complex ant–plant interactions: rain forest ants as secondary dispersers and post-dispersal seed predators. *Ecology* 74, 1802–1812.
Levings, S.C. (1983) Seasonal, annual, and among-site variation in the ground ant community of a deciduous tropical forest: some causes of patchy species distribution. *Ecological Monographs* 53, 435–455.
Levings, S.C. and Windsor, D.M. (1984) Litter moisture content as a determinant of litter arthropod distribution and abundance during the dry season on Barro Colorado Island, Panama. *Biotropica* 16, 125–131.
Longino, J.T. and Nadkarni, N.M. (1990) A comparison of ground and canopy leaf litter ants (Hymenoptera: Formicidae) in a Neotropical montane forest. *Psyche* 97, 81–93.
Longino, J.T., Coddington, J. and Colwell, R.K. (2002) The ant fauna of a tropical rain forest: estimating richness three different ways. *Ecology* 83, 689–702.
MacKay, W.P. (1993) A review of the New World ants of the genus *Dolichoderus* (Hymenoptera: Formicidae). *Sociobiology* 22, 1–148.
MacKay, W.P., Rebeles, A., Arredondo, H.C., Rodriguez, A.D., Gonzales, D.A. and Vinson, S.B. (1991) Impact of the slashing and burning of tropical rain forest on the native ant fauna (Hymenoptera: Formicidae). *Sociobiology* 18, 257–268.
Majer, J.D. and Delabie, J.H.C. (1994) Comparison of the ant communities of annually inundated and terra firme forests at Trombetas in the Brazilian Amazonia. *Insectes Sociaux* 41, 343–359.
Morais, H.C. and Benson, W.W. (1988) Recolonização de vegetação de cerrado após queimada por formigas arborícolas. *Revista Brasileira de Biologia* 48, 459–466.
Moutinho, P.R.S. (1998) Impactos do uso da terra sobre a fauna de formigas: consequências para a recuperação florestal na Amazônia Oriental. In: Gascon, C. and Moutinho, P. (eds) *Floresta Amazônica: Dinâmica, Regeneração e Manejo.* MCT-INPA, Manaus, Brazil, pp. 155–170.
Nepstad, D., Uhl, C. and Serrão, E. (1990) Surmounting barriers to forest regeneration in abandoned, highly degraded pastures: a case study from Paragominas, Pará, Brazil. In: Anderson, A. (ed.) *Alternatives to Deforestation, Steps toward Sustainable Use of the Amazon Rain Forest.* Columbia University Press, New York, pp. 215–229.
Nepstad, D., Veríssimo, A., Alencar, A., Nobre, C., Lima, E., Lefebvre, P., Schlesinger, P., Potter, C., Moutinho, P., Mendonza, E., Cochrane, M. and Brooks, V. (1999) Large-scale impoverishment of Amazonian forests by logging and fire. *Nature* 398, 505–508.
Oliveira, P.S., Silva, A.F. and Martins, A.B. (1987) Ant foraging on extrafloral nectaries of *Qualea grandiflora* (Vochysiaceae) in cerrado vegetation: ant as potential antiherbivore agents. *Oecologia* 74, 228–230.
Ranzani, G. (1980) Identificação e caracterização de alguns solos da Estação Experimental de Silvicultura Tropical do INPA. *Acta Amazonica* 10, 7–41.
Ribeiro, J.E.L.S., Nelson, B.W., Silva, M.F., Martins, L.S. and Hopkins, M. (1994) Reserva Floresta Ducke: diversidade e composição da flora vascular. *Acta Amazonica* 24, 19–30.
Rohde, K. (1992) Latitudinal gradients in species richness: the search for the primary cause. *Oikos* 65, 514–527.
Roth, D.S., Perfecto, I. and Rathcke, B. (1994) The effects of management systems on ground-foraging ant diversity in Costa Rica. *Ecological Applications* 4, 423–436.
Tobin, J.E. (1997) Competition and coexistence of ants in a small patch of rainforest canopy in Peruvian Amazonia. *Journal of the New York Entomological Society* 105, 105–112.
Uhl, C. (1987) Factors controlling succession following slash-and-burn agriculture in Amazonia. *Journal of Ecology* 75, 377–407.
Uhl, C., Nepstad, D., Buschbacher, R., Clark, K., Kauffman, B. and Subler, S. (1990) Studies of ecosystem response to natural and anthropogenic disturbances provide guidelines for designing sustainable land-

use systems in Amazonia. In: Anderson, A. (ed.) *Alternatives to Deforestation, Steps toward Sustainable Use of the Amazon Rain Forest*. Columbia University Press, New York, pp. 24–42.

Vasconcelos, H.L. (1999) Effects of forest disturbance on the structure of ground-foraging ant communities in central Amazonia. *Biodiversity and Conservation* 8, 409–420.

Vasconcelos, H.L. and Cherrett, J.M. (1995) Changes in leaf-cutting ant populations (Formicidae: Attini) after the clearing of mature forest in Brazilian Amazonia. *Studies on Neotropical Fauna and Environment* 30, 107–113.

Vasconcelos, H.L. and Cherrett, J.M. (1997) Leaf-cutting ants and early forest regeneration in central Amazonia: effects of herbivory on tree seedling establishment. *Journal of Tropical Ecology* 13, 357–370.

Vasconcelos, H.L. and Delabie, J.H.C. (2000) Ground ant communities from central Amazonia forest fragments. In: Agosti, D., Majer, J.D., Alonso, L. and Schultz, T. (eds) *Sampling Ground-Dwelling Ants: Case Studies from the World's Rain Forests*. Curtin School of Environmental Biology Bulletin No. 18, Perth, Australia, pp. 59–69.

Vasconcelos, H.L., Vilhena, J.M.S. and Caliri, G.J.A. (2000) Responses of ants to selective logging of a central Amazonian forest. *Journal of Applied Ecology* 37, 508–515.

Vasconcelos, H.L, Carvalho, K.S. and Delabie, J.H.C. (2001) Landscape modifications and ant communities. In: Bierregaard, R.O. Jr, Gascon, C., Lovejoy, T.E. and Mesquita, R. (eds) *Lessons from Amazonia: the Ecology and Conservation of a Fragmented Forest*. Yale University Press, New Haven, Connecticut, pp. 199–207.

Vasconcelos, H.L., Macedo, A.C.C. and Vilhena, J.M.S. (2003) Influence of topography on the distribution of ground-dwelling ants in an Amazonian forest. *Studies on Neotropical Fauna and Environment* 38, 115–124.

Vasconcelos, H.L., Fraga, N.J. and Vilhena, J.M.S. (2004) Formigas do Parque Nacional do Jaú: uma primeira análise. In: Borges, S.H., Iwanaga, S., Durigan, C.C. and Pinheiro, M.R. (eds) *Janelas para a Biodiversidade no Parque Nacional do Jaú*. Fundação Vitória Amazônica, Manaus, Brazil, pp. 153–160.

Verhaagh, M. (1990) The formicidae of the rain forest in Panguana, Peru: the most diverse local ant fauna ever recorded. In: Veeresh, G.K., Mallik, B. and Viraktamath, C.A. (eds) *Social Insects and the Environment*. Oxford & IBH, Delhi, India, pp. 217–218.

Ward, P.S. (2000) Broad-scale patterns of diversity in leaf litter ant communities. In: Agosti, D., Majer, J.D., Alonso, L.E. and Schultz, T.R. (eds) *Ants: Standard Methods for Measuring and Monitoring Biodiversity*. Smithsonian Institution Press, Washington, DC, pp. 99–121.

Wilson, E.O. (1958) Patchy distribution of ant species in New Guinea rain forests. *Psyche* 65, 26–38.

Wilson, E.O. (1987) The arboreal ant fauna of Peruvian Amazon forests: a first assessment. *Biotropica* 19, 245–251.

Wilson, E.O., Carpenter, F.M. and Brown, W.L. (1967) The first Mesozoic ants. *Science* 157, 1038–1040.

Yamaguchi, T. and Hasegawa, M. (1996) An experiment on ant predation in soil using a new bait trap method. *Ecological Research* 11, 11–16.

7 Soil Mesofauna in Central Amazon

E. Franklin and J.W. de Morais
Instituto Nacional de Pesquisas da Amazônia (INPA), Coordenação de Pesquisas em Entomologia, (CPEn), CP 478, 69011-970 Manaus, AM, Brazil, e-mail: morais@inpa.gov.br; beth@inpa.gov.br

Introduction

Soil invertebrates are an important component of native ecosystems and are sensitive to changes in the habitat (Bromham *et al.*, 1999). The intricate relationship of these animals with their ecological niches in the soil, the fact that many of them live a rather sedentary life and the stability of the community composition at a specific site provide good starting points for bioindication of changes in soil properties and impact of human activities. For example, the density of the oribatid mite *Platynothrus peltifer* as a percentage of total oribatids was used as an indicator of temperate forest vitality (Straalen, 1998). Soil invertebrates are divided into microfauna (Protozoa, Nematoda, Turbellaria, Rotifera, Tardigrada, Crustacea, part of Oligochatea), mesofauna (generally smaller than 2 mm; part of Oligochaeta, Collembola, Protura, Diplura, Pauropoda, Symphyla, Acari, Pseudoscorpionida and Palpigradi) and macrofauna (invertebrates larger than 2 mm). In many habitats in central Amazon, such as primary and secondary forests and flooded forests, the mesofauna, principally their lowest elements Acari (mites) and Collembola (springtails), are the most abundant and frequent group (Franklin *et al.*, 1997a, 2001a).

According to Valdecasas and Camacho (2003), the data for biodiversity and conservation are mainly taxonomic. Therefore, one of the numerous impediments to the study of biodiversity is that some groups include so many species that most of them are still undescribed and require highly qualified and skilled experts, and individuals are often identified to a high level of systematics or as morphospecies (Noti *et al.*, 2003). Principally in the tropics, both the proportion of morphospecies that cannot be assigned to named species and the number of scientist-hours required to process samples increase dramatically for smaller-bodied taxa (Lawton *et al.*, 1998). This obviously applies to the mesofauna. Because of the lack of taxonomists and time, most studies involving soil invertebrates present results as high taxonomic ranks of classifications (class, order or family). Otherwise, mites (identified only to suborder Oribatida or non-oribatid) proved of limited use because these two groups were present in all samples in extremely high numbers that overwhelmed the contribution of other arthropod groups (Nakamura *et al.*, 2003).

A considerable inventory of soil mesofauna has been done in the central Amazon region. Natural (primary forest, secondary forest, flooded forest and campinarana) and anthropogenic environments (plantation and polyculture systems) have been surveyed, using several sampling procedures and extraction methods. The purpose of this chapter is to review the soil mesofauna of central Amazon. The main taxonomic groups, their importance and function, environments studied, survey procedures used and the principal results are reported. Additionally, some recommendations are made, taking into consideration the lack of research on specific subjects.

Diversity of Soil and Litter Mesofauna in Central Amazon

Acari, Palpigradi and Pseudoscorpiones

Thirty-five families, 50 genera and 150–300 described species of Acari were found in Amazonia (Adis, 2002). Studies on the biodiversity of Arachnida were also done by Morais (1985), Ribeiro (1986), Hayek (2000), Santos (2001) and Adis (2002).

The suborder Oribatida (Fig. 7.1), one of the numerically dominant arthropod groups in most soils around the world, has been studied at the species level in Amazonia since 1967 (Beck, 1967, 1968, 1971). A considerable amount of research has been done since, also classifying these mites at the species level (Franklin, 1994; Franklin *et al.*, 1997a,b, 1998; Hayek, 2000; Franklin *et al.*, 2004). In a review of the literature, Woas (2002) listed 260 species of oribatid mites in the Amazon region. The list encompasses 260 species, 43% being morphotypes, pointing to the taxonomic problems of this mega-diverse group. Summarizing the current state of knowledge of the diversity and distribution of oribatid mites in 26 environments in the north in Brazil and in a rainforest in Peru, we realize that the published studies were mostly concentrated in central Amazon (Franklin *et al.*, in press). Most of the registers are from forest environments. Only one report gives results from agricultural polyculture (Hayek, 2000; Franklin *et al.*, 2004). At present, 146 species have been definitively identified from a total of 444 taxa, totalling 188 known genera, reinforcing the notion of a

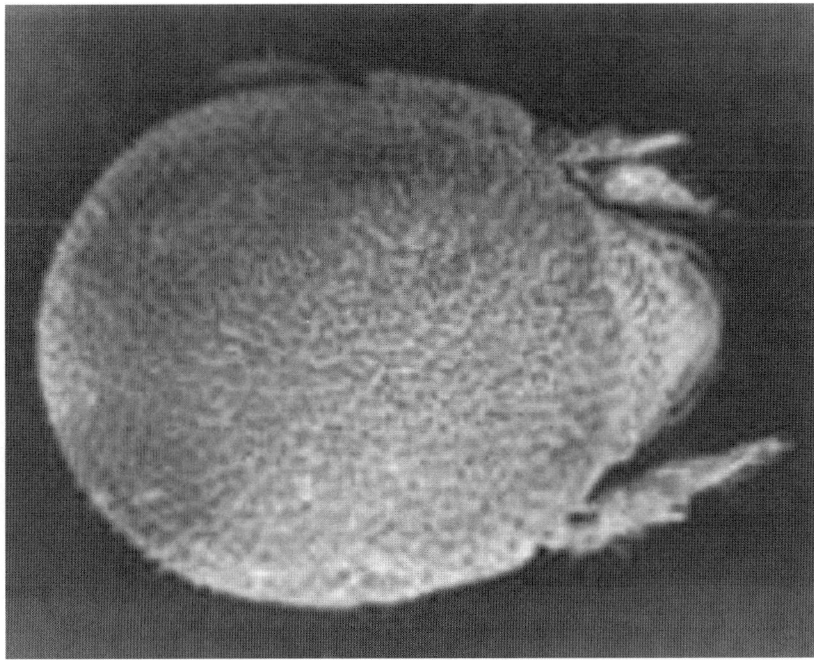

Fig. 7.1. *R. foveolatus* (Acari: Oribatida), one of the most abundant and frequent species in central Amazon. Dorsum SEM × 123. Source: Franklin *et al.* (1997b).

richly biodiverse area. The high number of 298 (67%) non-described species (morphospecies) clearly shows the inadequacy of the actual taxonomic knowledge. The highest diversity (54–155 species/morphospecies) was registered in the soil of primary forests. Eighty-nine species were unique to primary forests, followed by 34 to savannahs, 32 to trees, 10 to '*igapó*', four to caatinga, three to secondary forest, two in '*várzea*' and only one to polyculture. Twenty genera were the most speciose. The species with the largest home ranges were *Rostrozetes foveolatus*, *Scheloribates* sp. A, and *Galumna* sp. A.

Palpigradi presently comprise 2 families, 6 genera and about 80 species worldwide. Two families, two genera and three described species occur in Amazonia. At present only *Eukoenenia janetscheki* Condé, 1997, is known from central Amazon (Adis, 2002).

Twelve families, 31 genera and 75 described species of Pseudoscorpionida occur in Amazonia. From Reserva Florestal Adolpho Ducke (Ducke Reserve), a fragment of forest of 10,000 ha near Manaus, 6 families, 11 genera and 15 described species are known (Morais, 1985; Adis *et al.*, 1988; Morais *et al.*, 1997; Adis, 2002). Searching in a huge extension, 64 km^2, that represented all of the variations of soil, topography and vegetation types of the Ducke Reserve, Gualberto (2003) (Table 7.1) found four families, six genera and seven species in the litter. Table 7.1 provides more details of the study. Encompassing the litter and mineral soil fauna in several habitats in a primary forest on the banks of Urucu River (Coari, Amazonas, Brazil), Aguiar (2000) found 14 species.

Collembola

In central Amazon, several research papers have dealt with taxonomic aspects (Arlé 1959; Arlé and Rufino, 1976; Arlé and Oliveira, 1977) and geographical distribution (Arlé, 1960). Oliveira (1982) identified 28 species of epigeic Collembola, 71% being morphotypes. In 2002, the number of known species increased to 83 (Câmara, 2002); therefore the taxonomic situation is not better than 20 years ago, as 64% of the species listed are composed of morphotypes.

Myriapoda (Pauropoda and Symphyla)

Two families, 8 genera and 55 described species of Pauropoda occur in Amazonia (Adis, 2002). From Reserva Ducke, 2 families, 7 genera and 31 described species are known (Adis, 2002). Symphylans presently comprise 2 families, 13 genera and 200 species worldwide. Both families, 4 genera and 5 described species occur in Amazonia (Adis *et al.*, 1996a; Morais, 1996; Scheller and Adis, 1996, 1997; Adis, 2002).

Methods of Sampling and Extraction of the Soil Mesofauna Employed in Central Amazon

A squared or rounded split corer, which is driven into the soil, has been utilized to take soil and litter samples from the field. The size of the corer varies with the objective of the study, reaching up to 14 cm deep into the soil profile. Until now, two extraction methods have been used to extract the soil mesofauna from the soil and litter samples using the Kempson (Fig. 7.2) unit and the Berlese–Tullgren apparatus. A detailed description of the Kempson unit is given by Adis (1987). The Berlese–Tullgren apparatus, which is the most commonly used, consists of a wooden box of 160 × 50 cm that is divided into two compartments by a support in which rows of plastic funnels are placed. The sieves (8 cm in diameter and 5 cm high) containing the soil and litter samples are placed inside the funnels. The mesh size of the sieves is 1.5 mm, and four holes of 4 mm are arranged to allow the escape of larger animals. Funnels are heated by electric light bulbs (25 W) suspended 14 cm above the sieve. Collecting

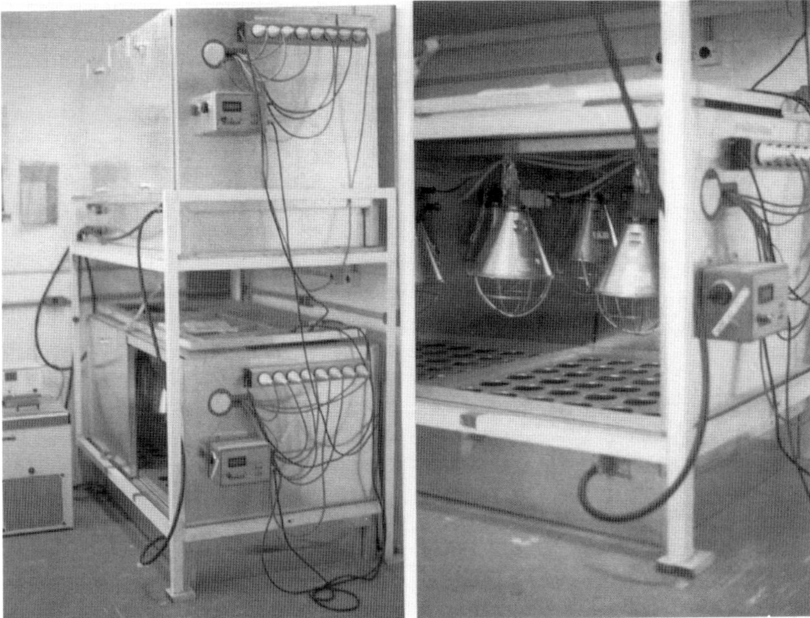

Fig. 7.2. Kempson apparatus: with two cabinets and refrigeration machine (left) and infrared lamps (right).

vials containing the killing/preserving agent (formalin at 5%) placed in the lower compartment receive the animals that are forced to escape from the heat. The temperature is gradually increased from 27°C until approximately 40–45°C. The samples remain in the apparatus until they are completely dry (around 10–14 days).

The efficiency of both methods varies according to the type of soil, substrate, taxonomic group, density and activity of the soil invertebrate (Southwood, 1980; Adis, 2002). The Kempson method is extremely efficient for meso- and microfauna, except copepods, holometabolic insects (Kempson *et al.*, 1963; Southwood, 1980), isopods and thrips (Phillipson, 1971). Therefore, the advantages of the Kempson method are: (i) simultaneous extraction of a large amount of material; (ii) maintenance of cool temperatures in the lower part of the equipment, which does not allow the decomposition of animals; and (iii) the gradual increase of temperature. The method is inconvenient because of the following reasons: (i) the varying results according to the behaviour of the animal; (ii) the influence of weather change; and (iii) eggs and pupae are not extracted, but adults can emerge during the extraction period. The Berlese–Tullgren has the following disadvantages: (i) quick rise in temperature, which can cause the death of animals; (ii) the use of stronger lamps can cause fast drying and the resulting humidity of the soil samples can become lethal to the soil invertebrates; (iii) the lack of cooling in the lower compartment, which can cause premature death of the animals; and (iv) as the Kempson method, eggs and pupae are not extracted, but they can emerge during the extraction period. Therefore, if someone with vast experience is handling the equipment, the problems of heat and humidity are greatly reduced. It is cheaper and simpler to build the Kempson apparatus. The Berlese–Tullgren can also be easily moved to the place of sampling and adjusted for a large quantity of samples.

Bark-brushing, hand aspirators and arboreal photoeclectors were also utilized in a few studies (Franklin, 1994; Table 7.1).

Table 7.1. Studies on soil mesofauna in the central Amazon region.

Reference	System	Objective and duration of the sampling	Sampled area/ number of samples	Method of sampling	Group studied	Taxonomic level
Beck (1967, 1968, 1971), Franklin (1994), Franklin *et al.* (1997a, 2001a)	Flooded forests of *várzea* and *igapó*	Faunal inventories	1 plot at each environment	Berlese–Tullgren		
		5/6 months	Number variable of samples	Arboreal photoeclectors, bark-brushing	Acari: Oribatida	Species
Morais (1985), Rodrigues (1986)	Primary and secondary forest	Faunal inventories	1 plot		Soil invertebrates (Acari and Collembola excluded from the analysis)	Higher categories
		Monthly samples, 12 months	A grid composed of 12 transects separated 5 m from each other; 12 random samples	Kempson		
	Secondary forest	Faunal inventories	1 plot		Pseudoscorpionida	Species
					Soil invertebrates (Acari and Collembola included in the analysis)	
Adis *et al.* (1987a,b)		Two samplings: wet and dry season	6 random samples	Kempson		Higher categories
	Campinarana forest	Faunal inventories	1 plot		Soil invertebrates (Acari and Collembola included in the analysis)	
Adis *et al.* (1989a,b)		2 samples: wet and dry season	6 random samples	Kempson		Higher categories
Oliveira (1982)	Primary forest, secondary forest and pasture	Impact of deforestation	1 plot		Collembola	

		Monthly samples, 22 months	50 random samples	Hand aspirator at each environment		Species
	Forest fragment	Impact of deforestation	1 plot of 1 ha	Berlese–Tullgren	Collembola	
Câmara (2002)		Monthly samples, 12 months	2 plots of 1600 m^2 for each sampling method; 10 random samples	Pitfall traps		Species
	Floodplain forest compared with plantation	Impact of deforestation	1 plot at each environment		Soil invertebrates (Acari and Collembola included in the analysis)	
Adis and Ribeiro (1989)		2 samplings: wet and dry season	1 transect of 300 m at each environment; 6 soil samples	Kempson		Higher categories
Höfer *et al.* (2001), Franklin *et al.* (2001a)	Primary forest, secondary forest and polyculture system (plots A and C)	Impact of deforestation	1 plot (30 × 40 m) at each environment		Soil invertebrates (Acari and Collembola included in the analysis)	Higher categories or functional groups
		Trimestral samplings during 21 months	10 random samples in a primary and 10 random samples in a primary and secondary forest; 5 samples at each polyculture system	Kempson		
Franklin *et al.* (2001a)					Acari: Oribatida	Species
Oliveira and Franklin (1993)	Pasture	Effect of fire	1 plot (4 ha)		Soil invertebrates (Acari and Collembola included in the analysis)	Higher categories or functional groups
		1, 15, 30, 40, 60, 125, 145, 200, 270, 320 and 370 days after the fire	15 random samples (burned and unburned substrate)	Berlese–Tullgren		

Continued

Table 7.1. Studies on soil mesofauna in the central Amazon region. – cont'd

Reference	System	Objective and duration of the sampling	Sampled area/ number of samples	Method of sampling	Group studied	Taxonomic level
Ribeiro (1986), Ribeiro and Schubart (1989)	Primary forests on yellow latosol, in hydromorphic soil and secondary forest	Succession during decomposition of leaves (*Clitoria racemosa*)	1 plot at each environment			
		15, 30, 60, 90, 120 and 150 days from the beginning of the experiment	10 random samples at each environment at each period	Berlese–Tullgren	Oribatid mites	Species
Höfer *et al.* (2001)	Primary forest, secondary forest and polyculture system (plots A and C)	Succession during decomposition of leaves (*Vismia guianensis*)	1 plot (30 × 40 m) at each environment		Soil invertebrates (Acari and Collembola included in the analysis)	Higher categories or functional groups
		Different intervals for more or less 350 days; one series started in October and another in April	10 random samples in a primary and secondary forest; 5 samples at each polyculture system	Berlese–Tullgren	Macro- and mesofauna	
Hayek (2000)	Primary forest, secondary forest and polyculture system (plots A and C)	Succession during decomposition of leaves (*Vismia guianensis*)	1 plot (30 × 40 m) at each environment			
Franklin *et al.* (2004)		26, 58, 111, 174, 278 and 350 days from the beginning of the experiment; one series started in April	10 random samples in a primary and secondary forest; 5 samples at each polyculture system	Berlese–Tullgren	Oribatid mites	Species
Luizão (1995)	Campina (low caatinga)	Succession during decomposition of leaves (*Clitoria*,	1 plot 50 × 50 m at each environment		Soil invertebrates (Acari and Collembola included in the analysis)	Higher categories or functional groups

		Pradosia and *Aldina latifolia*)				
	Campinarana (tall-caatinga) and primary forest	30, 60, 120, 180, 270 and 360 days from the beginning of the experiment	4 samples/each type of leaf/period	Berlese–Tullgren	Macro- and mesofauna	
Santos (2001), Franklin *et al.* (2001a)	Secondary forest with predominance of *Vismia* sp. in the litter layer	Litter manipulation (4 substrates: 1. *Hevea brasiliensis*, 2. *C. guianensis*, 3. mixture of *C. guianensis*, *H. brasiliensis* and *V. guianensis*, 4. original litter)	5 plots (30 × 40 m)		Soil invertebrates (Acari and Collembola included in the analysis)	Higher categories or functional groups
		45, 100, 180, 240 and 300 days from the beginning of the experiment	2 samples at each plot/treatment/period; two layers: litter and mineral soil	Kempson	Macro- and mesofauna	
Franklin and Morais (in development)	Forest, fragment of forest and savannah	Large inventories (3 years)	74 plots composed of 4 transects of 250 m, separated 50 m from each other (26 in forest, 8 in fragment of forest and 40 in savannah		Soil invertebrates (Acari and Collembola included in the analysis)	Higher categories or functional groups
Santos and Franklin (in development)		Independent variables: topography, granulometry and soil chemistry, vegetation structure and composition		Berlese–Tullgren	Acari: Oribatida	Species

Continued

Table 7.1. Studies on soil mesofauna in the central Amazon region. – cont'd

Reference	System	Objective and duration of the sampling	Sampled area/ number of samples	Method of sampling	Group studied	Taxonomic level
Guimarães (2003), Fagundes (2003)	Gradient of vegetation covering 10,000 ha composed of primary forest and Campinarana (tall-caatinga)	Large inventories (6 months)	18 trails 8 km long, 1 km from each other, forming a network of 64 km^2	Berlese–Tullgren	Soil invertebrates (Acari and Collembola included in the analysis)	Higher categories or functional groups
		Independent variables: topography, granulometry and soil chemistry, vegetation structure and composition	72 plots of 250 m separated 1 km from each other; 5 samples at each plot		Macro- and mesofauna	
Gualberto (2003)					Pseudoscorpionida	Species
Fagundes (2003)					Formicidae	Genera/morphospecies

Spatial Distribution Models of the Mesofauna in Natural and Anthropogenic Environments

Soil fauna inventories

Flooded forests of várzea and igapó

Beck (1967, 1968, 1971) began the first inventories (Table 7.1). A considerable amount of research has been done since, identifying mites to the species level (Franklin, 1994; Franklin *et al.*, 1997a,b, 1998). In a *várzea* forest (Ilha de Marchantaria), 73% of the total invertebrates in the soil were oribatid mites. The mean density was around 6300 ind/m^2 both in the dry and in the rainy season. The mean number of total oribatid mite species recorded per month was 7.1±2.1. *Galumna* sp. A and *Paralamellobates* sp. A were the most dominant species. In an *igapó* forest (Rio Tarumã Mirim), about 89% of the total catch was oribatid mites. The mean density was 10,700 (dry season) and 11,500 (wet season) ind/m^2. The mean number of total oribatid mite species recorded per month was 14.6±6.3. *R. foveolatus* and *Eremobelba* sp. A were the most dominant species. Contrary to many terricolous invertebrates that pass the aquatic phase in the trunk/canopy region, vertical migrations in response to the flood pulse are of minor importance to terricolous oribatid mite specialists.

Primary and secondary forest

Disregarding Acari and Collembola, the greatest invertebrate abundance was registered for Pseudoscorpionida in a primary forest at Ducke Reserve, near Manaus (Morais, 1985; Table 7.1). The most abundant species were *Microblothrus tridens* Mahnert (Siaridae, 53.5% of the total of individuals sampled), *Tyrannochthonius minor* Mahnert (Chthoniidae, 17.5%) and *Brazilatemnus browni* Muchmore (Miratemnidae, 11.6%). These species were more abundant during the dry season (59% of the total captured). Other dominant mesofauna groups were Pauropoda (3.4%; 397 ind/m^2), Protura (2.7%; 258 ind/m^2) and Symphyla (3.9%; 455 ind/m^2). Approximately the same pattern of species dominance was registered in secondary forest, in the neighbouring *igapó* forest (Rodrigues, 1986) (Table 7.1). The most abundant species were *M. tridens* (36%), *B. browni* (23%) and *T. minor* (22%). The abundance of arthropods in the soil was 18,992 ind/m^2, which was almost twice as high as that of a primary forest in Reserva Ducke (10,163 ind/m^2). Other abundant mesofauna groups were Protura (9.3%; 1758 ind/m^2), Symphyla (3.2%; 606 ind/m^2) and Pauropoda (6.7%, 1273 ind/m^2).

In a cut and unburned secondary forest (Capoeira) on yellow latosol (Adis *et al.*, 1987a,b) a total of 50,000 and 64,000 ind/m^2 were found during the dry and wet season, respectively (Table 7.1). The dominance of Acari and Collembola over the total invertebrate community oscillated between 72% and 75%. Therefore, during neither the rainy nor the dry season was the abundance of invertebrates in mineral subsoils higher in response to the changing humidity in the organic layer.

Campinarana forest

In the white sand soil of Neotropical campinarana (Adis *et al.*, 1989a,b; Table 7.1), a total of 58,000 and 74,000 ind/m^2 were found during the dry and wet season, respectively. The dominance of Acari and Collembola over the total invertebrate community oscillated between 75% and 80%. Disregarding Acari and Collembola, Formicidae, Pauropoda, Diplura and Protura accounted for more than 50% of the total catch in both seasons. No indication was found that the arthropods migrate to the mineral subsoil as a response to changing abiotic factors in the organic strata, as occurred in the secondary forest studied by Adis *et al.* (1987a,b; see above).

Impact of deforestation

Oliveira (1982) investigated the structure of communities of epigeic Collembola in six habitats (Table 7.1). The diversity of the Collembola community was reduced according to the major degree of human disturbance. The highest densities of all species were found during the dry season. Densities of Collembola depended on the microclimate near the soil. Numbers of Collembola were significantly correlated with temperature and rate of litter fall in the secondary forest. The Shannon–Weaver (bits/ind) and equitability ($e = S'/S$) indices were highly correlated with the scale of disturbance in pasture, secondary forests and primary forests (Shannon–Wiener/equitability indices: pasture = 1.7/0.6; secondary forests = 2.4/0.5–2.6/0.6; primary forest on clay soil = 3.9/0.7; primary forest on sand soil = 3.4/0.8). Analysis of similarity (Mountford) also identified four groups: (i) pasture; (ii) secondary forests; (iii) forest on clay soil; and (iv) forest on sandy soil. The author concluded that the analysis of community structure of leaf litter Collembola offered a useful tool for the objective evaluation of the effect of management practices in forest environments.

Câmara (2002) studied some aspects of the Collembola community in the soil of a forest fragment situated at the Campus of Amazon University in Manaus (Table 7.1). The data obtained were on population densities, species richness and gut contents in relation to precipitation, soil temperature and humidity of the soil samples. There was no definitive pattern of fluctuations in abundance in relation to the dry and wet periods. The guts of epigeic species contained mostly fungal hyphae, in contrast to the hemiedaphic ones that seem to feed mostly on algae and amorphous detritus.

The soil invertebrates of a floodplain forest were evaluated by comparing a forested area with one that had been converted to a plantation (Adis and Ribeiro, 1989, Table 7.1). Both sites are located on Careiro Island, the first island in the Amazon River below the confluence with the Negro River, which is seasonally flooded for 5–6 months each year. Approximately 42,300 ind/m^2 were obtained from the soil of a flood forest, and 13,550 per square metre from a nearby area that had been deforested 15 years ago and since used for agriculture (maize and cassava). Around 71% of all the arthropods were found in the top 3.5 cm of the soil. Mites and Collembola represented 83–89% of the total number of arthropods, but only 1.6–8.5% of the total dry biomass. Abundance, biomass and dominance of arthropod groups varied with the dry and wet seasons.

Recuperation of degraded land for sustainable use in the future was the focus of several projects in the German–Brazilian Studies of Human Impact on Forest and Floodplains in the Tropics (SHIFT) programme in central Amazon. The mesofauna communities were assessed from June 1997 to March 1999, in two plots of a polyculture system (plots A and C) with four tree species (*Hevea* spp., *Schizolobium amazonicu*, *Swietenia macrophylla* and *Carapa guianensis*) and from plots of a primary and a secondary forest (Franklin *et al.*, 2001a; Höfer *et al.*, 2001; Table 7.1). In the two plots of the polyculture systems, spontaneous secondary vegetation (mainly *Vismia* spp.) was introduced between the rows of trees. Samples were separated into the litter layer and the top 5 cm soil layer. The macrofauna mean abundance (mg/m^2) and the standard deviation as percent of the mean were: in the polyculture systems (A and C) = 3745–4266 (37–41%); secondary forest = 3769 (31%); primary forest = 4866 (31%). For the mesofauna the following values were obtained: polyculture systems = 25,033–32,890 (28–62%); secondary forest = 24,703 (40%); primary forest = 24.450 (21%). The dry biomass of the mesofauna (mg/m^2) oscillated around 655–937 in the polyculture systems, 679 in the secondary forest and 609 in the primary forest. In all sites, the mesofauna was strongly dominated by oribatid mites (42–59% of dominance), predatory mites (7–22%) and Collembola (4–13%). Martius (2004a) found that canopy closure strongly determined the litter temperature in the sites and that soil macrofauna biomass was strongly correlated with canopy closure.

From the SHIFT project cited above, the density and biomass results of Acari and Collembola are also published (Franklin *et al.*, 2001a). The values obtained for the wet and dry individual weight of Acari Oribatida, Acari non-Oribatida and Collembola were summarized and compared by the authors with the results obtained by other authors for temperate regions. In the litter layer, they expected the highest values for Acari Oribatida, Acari non-Oribatida and Collembola in the primary forest, but during the 21 month sampling period, the total densities of mesofauna decreased in the following order: polyculture system (plot A) > secondary forest > polyculture system (plot C) > primary forest. The highest densities were obtained in plot A of the polyculture system, due to the dominance of Oribatida. The oribatid mite's density was highly influenced by *Archegozetes longisetosus*. This plot was situated at the north-western extremity of the experimental area and did not receive much shading from the neighbouring plantations, or from the adjacent secondary growth. All these factors resulted in extreme abiotic conditions and contributed to a different annual cycle such as that recorded in plot C of the polyculture system, which is situated near a primary forest and received strong shading from it. The difference in soil mesofauna density between the plots of the polyculture system (A and C) might also be explained by the more extreme microclimatic conditions in polyculture system A (Höfer *et al.*, 2001). Franklin *et al.* (2001a) noticed that densities of Acari Oribatida and Collembola were notably lower in the mineral soil. For non-Oribatid Acari, the same tendency was not so clearly detected. Contrary to the other groups, the highest densities of Collembola were found in the primary forest. In general, densities in the litter layer were higher. The mesofauna population was lowest in 1997 and only in this year was the density in the primary forest and in the polyculture system higher in the soil fraction. The pattern in the secondary forest was not the same because of the higher amount of litter. The differences between 1997 and 1998 were a result of: (i) a possible reaction of the mesofauna that migrated to the mineral soil during the extremely dry period of 1997; and (ii) the litter layer reduction that occurred in 1997, causing lower mesofauna densities. Depending on the physical factors, there are years of higher and lower populations. Extremely wet years could also exert an influence on the soil mesofauna and the authors recommend studies of long-term periods. Later, Martius (2004a), analysing the same data of the SHIFT project cited above, came to the same conclusion, assuming that the soil fauna reacts with short-term vertical migrations to changes in the moisture and temperature conditions of the litter. Adis *et al.* (1987a,b) (see the section 'Primary and Secondary Forest') and Adis *et al.* (1989a,b) (see the section 'Campinarana Forest') did not detect this short-term vertical migration in a secondary forest and in a campinarana forest probably because they made only one series of samples in the dry and in the wet season, covering a period of only 1 year. There was a tendency for the Acari non-Oribatida biomass estimated in the study of Franklin *et al.* (2001a) to be lower than in temperate forest, the values are however higher than values estimated for many tropical forests. On the contrary, Oribatida and Collembola biomass were characterized by lower values as compared with temperate forests. However, since the average individual weights have been measured on only one occasion, which ignored changes in relative population with time (Petersen and Luxton, 1982), the results of the biomass study need to be interpreted with caution. The most recommended procedure would be the estimation for each sampling period of the year in a medium- or long-term study.

Impact of fire

Very little investigation has been done with respect to the effects of fire on soil mesofauna in central Amazon. Oliveira and Franklin (1993) worked in a 4 ha plot, where the forest was felled and burned to form a pasture. The burning was not homogeneous, giving rise to a patchwork of well-burned and unburned areas. The soil fauna of these two types of areas were compared (Table 7.1). The perturbation of the environment resulted in the disappearance of several taxonomic groups, principally during the first periods, with higher diversity in the non-burned areas (Fig. 7.3). The appearance of taxa not registered before stabilized 200 days after the fire.

The dominant (%) groups in the burned and non-burned areas, respectively, were Acari Oribatida adults (10.4/26.0) and immatures (9.8/29.0), Acari non-Oribatida (64.3/23.9), Collembola (3.7/12.1) and Coleoptera adults (4.8/3.9) and immatures (3.9/2.2). In the burned areas, the number of Oribatida was less than the total number of non-oribatid mites. Acari Oribatida and Collembola were more abundant in the non-burned areas. These results mean that the low-intensity fire created islands of non-burned vegetation and stems of trees that constituted refuges for the soil animals. Representatives of Isoptera, Diptera (immatures), Formicidae, Psocoptera, Hemiptera (adults and immatures), Diplura, Protura, Homoptera (immatures), Lepidoptera (immatures), Thysanoptera, Orthoptera and Neuroptera (immatures), listed here in order of numerical dominance, also occurred, but their dominance reached only 2.8% and 3.6% of the total invertebrates sampled in the burned and non-burned areas, respectively.

Soil fauna and litter decomposition

Using the nylon mesh bag technique, Ribeiro (1986) and Ribeiro and Schubart (1989) studied the succession of the Oribatid fauna during the decomposition of *Clitoria racemosa* Benth (Leguminosae) in primary and secondary forests (Table 7.1). The suborder Oribatida amounted to more than half of all the mites collected in the samples of the three sites studied (77% in a primary forest on yellow latosol, 71% in a primary forest on hydromorphic soil and 54% in a secondary forest). The adult oribatids represented 103 species, 79 genera and 41 families. Thirty-six of the species were common to the three sites. Quite marked changes occurred in the most abundant species during the study period both within and between each of the sites. The species diversity was highest at the middle stage of succession, corresponding to the increase in the number of individuals and species and the evenness. The highest degree of similarity (50%) was obtained for the sites in primary forest areas, followed by primary forest in hydromorphic soil and secondary forest (37%) and lastly primary forest in yellow latosol and secondary forest (24%).

The biological conditions of two 5-year-old polyculture tree plantations in Amazonia were compared with a 13-year-old secondary forest and with nearly

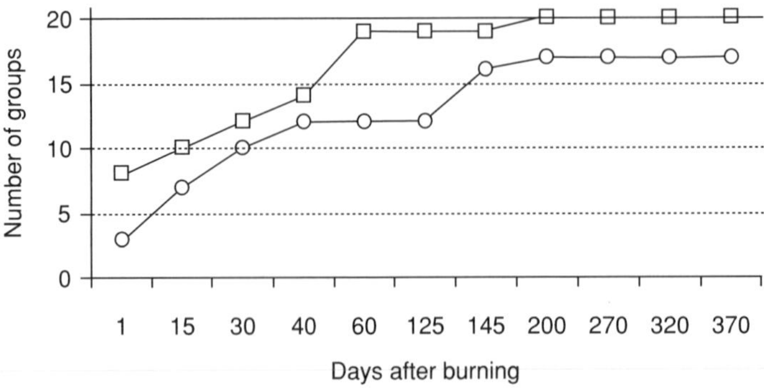

Fig. 7.3. Invertebrates in the soil of a pasture after burning. Circle = burned areas; square = non-burned areas. (Source: Oliveira and Franklin, 1993.)

undisturbed primary forest (Höfer *et al.*, 2001; Table 7.1), as part of the SHIFT programme. Abundance and biomass of functional groups of soil meso- and macrofauna were measured at 3 month intervals over 2 years and litterbag experiments with fauna exclusion were also carried out. The importance of the different size classes of soil fauna in litter decomposition was studied for the first time by using litterbags of three different mesh sizes: coarse (1 cm; faunal activity not restricted), medium (250 μm; to exclude the entrance of macrofauna) and fine (20 μm; to exclude the entrance of meso- and macrofauna). The bags were filled with leaves of *Vismia guianensis*. One litterbag series was started at the end of the dry season (October) and the second at the end of the rainy season (April) and the samplings were done at different intervals for approximately 350 days. The litterbag assays showed that the macrofauna determines the decomposition process in all studied plots. The decomposition rates (per year) were in the range 0.6–1.4 for the anthropogenic sites (secondary forest and culture system) and 2.3 and 3.1 in the primary forest when fauna components were present. When this biological component was excluded from the litterbags, decomposition rates (per year) were in the range 0.3–0.6, regardless of the land use.

This result indicates a strong involvement of the macrofauna, whose activity could not be compensated by an abundant mesofauna. The effect of excluding the macrofauna was strongest in the primary forest where arthropod macrofauna and earthworms were more abundant than in the polyculture and the secondary forest. Significant contributions of soil macrofauna to the decomposition of leaf litter in tropical forests in litterbag experiments have already been recorded for other conditions (Tian *et al.*, 1998; Zicsi *et al.*, 2001).

In the same experiment mentioned above, Hayek (2000) and Franklin *et al.* (2004) evaluated the shift of oribatid mite communities (Table 7.1) in the coarse, medium and fine mesh size litterbags. In all sites and mesh sizes, less than 30% of the original leaf litter had disappeared after the first 26 days of exposure. The greatest weight loss occurred with the leaves enclosed in the coarse mesh size litterbags in the forest (FLO) (Kruskal-Wallis; $H = 52.6$; $P < 0.001$), secondary forest (SEC) (Kruskal-Wallis; $H = 15.4$; $P < 0.001$) and polyculture system (POA) (Kruskal-Wallis; $H = 6.1$; $P = 0.046$). Less than 25% of the leaf material remained after 1 year in the primary forest in the coarse mesh bags. No difference was detected between the decomposition rates of the leaves enclosed in the fine and medium mesh litterbags, where macrofauna was excluded (data of F. Luizão, INPA, Manaus, SHIFT project ENV 052). Therefore, contrary to what was expected, the highest oribatid mite density and diversity were registered in the medium size mesh and not in the coarse size mesh. The fine mesh bags were not completely successful in excluding arthropods. The reasons were:

1. The oribatid mites were capable of penetrating even into the fine mesh litterbags because of the holes originating because of biological factors, like root penetration and invertebrate action.
2. The fine and medium mesh bags caused a protective effect that did not allow the entrance of predator groups.
3. The entrance of immature forms of mites and Collembola that could not escape the fine and medium litterbags after reaching the adult phase.

It was clear that the greatest dominance of the oribatid mites in the medium size mesh bags did not exert influence on the rate of decomposition. A total of 95 species of oribatid mites was organized in distinct communities in relation to the identity of the dominant species in each site and each litterbag mesh. There was a high degree of similarity between the species colonizing the litterbags of the three mesh sizes and a high degree of dominance of the common species. The authors concluded that the greatest abundance of the oribatid mites did not exert an influence on the decomposition process of the leaves. There was not a succession of species during the course of the decomposition, and the results were different from those obtained in temperate forest (Wallwork, 1983), because in the central Amazon forest neither early colonizers nor species that prefer more advanced decomposition stages were found. There were differences in the colonization of species in relation to the mesh size of the litterbag that were closer to the specific habits and habitat of the dominant species. For example, abundant genera in the litterbags, like *Afronothrus*, *Allonothrus* and *Archegozetes*, are restricted to tropical regions. In such artificial conditions as litterbags, remarkable increases in population numbers can be observed. Species of *Archegozetes* and *Allonothrus* seemed to reach higher concentrations in areas with more adverse climatic conditions (Hayek, 2000; Woas, 2002; Franklin *et al.*, 2004).

The influence of the diversity of leaves on the colonization rate of litter animals

Luizão (1995) studied the litter animals colonizing decomposing leaves in campinarana (tall-campina) and lowland evergreen rain forest. Litterbags containing leaves of *Clitoria*, *Pradosia*, and *Aldina* were placed in the field in 1991 (Table 7.1). A total of 41 functional groups of litter animals were found colonizing the decomposing leaves in the litterbags. The five groups most frequently recorded were Acari, Collembola, Pseudoscorpionida, Formicidae and Diptera larvae. Both the number of taxonomic groups and the total numbers of litter animals differed significantly among the three leaf species, being higher in *Clitoria* and lower in *Pradosia*. There was evidence pointing to some specific groups of litter animals as key decomposers. Collembola, possibly including many fungivorous species, was appointed as important in forest and in the tall-caatinga. Acari, Diplopoda, and Symphyla also appear to be key decomposers. In the primary forest, earthworms as important decomposer groups replace Diplopoda, possibly because of the pH, (which is higher than in the tall-caatinga) together with the higher water retention potential of the mineral soil, which contains more clay. Comparing the three leaf species, Acari appear to be decomposers in all of them, while Collembola appear to be important for *Clitoria* and *Aldina*, the Isopoda more important for *Pradosia*, and the Protura especially important for *Aldina*. Diplura are probably important predators in all three of the leaf species.

Litter manipulation

It is probable that not only the amount of litter but also the diversity of leaves available on the soil surface could exert some effect on the soil invertebrate communities. Santos (2001) and Franklin *et al.* (2001a) studied the decomposition processes in a secondary forest (Table 7.1). The effect of different qualities of litter was tested in the development of the invertebrate community. Four substrates were tested: (i) *Hevea brasiliensis* ('seringueira'); (ii) *C. guianensis* ('andiroba'); (iii) a 'mixture' of *Hevea*, *Carapa* and *Vismia* spp. ('lacre', the predominant leaf species on the soil of the secondary forest plots); and (iv) the original and more diversified litter layer. The effects of litter quality on the development of soil invertebrate communities were studied while keeping the other factors stable. The animals were classified in zoological groups (mainly orders or families) and allocated to guilds of decomposers, herbivores, predators, social groups (ants and termites) and others. The density of the groups of soil mesofauna in the litter layer did not differ significantly between the four substrates. Analysis of the multidimensional scaling, using the statistial and pattern analysis software (PATN) programme showed that there was no difference in the fauna composition in the litter and mineral soil between the four treatments. The authors cited above concluded that: (i) the reduction in the diversity of the leaves did not adversely affect the soil community, and this applies principally to Acari and Collembola, which possess a large alimentary spectrum; and (ii) the classification of the animals in higher taxonomic categories were obscuring much more complex habitat partitioning occurring at the species level. Martius (2004b) investigated the biological conditions of two 5-year-old polyculture tree plantations (plots A and C) compared with a secondary forest and with a primary forest (see also Höfer *et al.*, 2001; Table 7.1), as part of the SHIFT programme. They concluded that although the decomposer organisms are most active in the primary forest, from the decomposition coefficients it seems that the polyculture systems (plots A and C) are also inhabited by soil communities that are potentially functional with respect to the task of litter decomposition. Taking the results of Santos (2001), Franklin *et al.* (2001a) and Martius (2004b) as a basis we can suppose that in central Amazon the reduction in the diversity of leaves does not follow the expected pattern of great impact on the soil community, as for example, reduction of their abundance and diversity. Therefore, more detailed studies to classify the invertebrates with better taxonomic resolution are necessary to assess the magnitude of the effect of biotic and abiotic factors on biodiversity.

Medium-scale pattern community structure in central Amazon

The conservation and enhancement of biodiversity is a key objective of forest management. To understand how managed habitats function, one must first understand how the natural environment works on a large scale. The soil invertebrates are affected by numerous factors like topography, soil condition and litter layer characteristics. However, the contribution of these factors has not been well studied in the tropics. Since 1996, the focus of some mesofauna studies has been large inventories in fragmented and forested habitats in an Amazonian savannah, on the right bank of Tapajós River, state of Pará, Brazil (Table 7.1, Franklin and Morais, in development, Santos and Franklin, in development). Another large sampling area (about 10,000 ha) is located at Ducke Reserve, about 30 km from Manaus. The basis of the project is a system of trails that allows access to the entire area (Table 7.1). A team of researchers of the National Institute for Amazonian Research (INPA) carries out both projects. Dr W. Magnusson (Ecology Department/INPA) idealized the projects and experimental designs. Three MSc dissertations are available, related to soil mesofauna (Guimarães, 2003), soil macrofauna with emphasis on Formicidae (Fagundes, 2003) and species of Pseudoscorpionida (Gualberto, 2003), at Ducke Reserve. Studies include the distribution of the invertebrates and their relation to abiotic conditions (topography, clay percentage, the carbon percentage, iron and aluminium content, pH of the soil and quantity of litter from the samples and parcels) to investigate the influence of these predictor variables over the soil animals (Table 7.1).

Guimarães (2003) investigated the differences in the distribution of soil mesofauna from Ducke Reserve in the function of their classification in higher categories (class, order or family), functional groups (immatures and adults) and guilds of decomposers, predators, herbivores and others, followed by ordination analysis (non-metric hybrid multidimensional scaling (SSH-MDS)). The relationships of Acari Oribatida, Acari non-Oribatida and Collembola, which reached 71% of the mesofauna abundance, were also investigated by multiple regression analysis (Pearson). The ordination of the invertebrates in higher taxonomic categories did not show the differences in the way the animals were distributed in the soil of the reserve. The higher taxonomic ranks encompass many species, and it is difficult to measure their relationship with the predictor variables. The positive relation of the Acari Oribatida with the clay percentage, the negative relation of Acari non-Oribatida with the slope of the land and the positive relation of Collembola with the amount of litter in the parcels were significant but the regression coefficients were very weak (lower than 0.3).

In Ducke Reserve, Fagundes (2003), using ordination analysis (SSH-MDS), did not find significant relationships in the soil macrofauna, either for the taxonomic classification in higher taxonomic levels, as occurred with the mesofauna (Guimarães, 2003), or for functional groups (immatures and adults) or guild (decomposers, predators and herbivores) of the soil invertebrates. The higher taxonomic level employed by the author did not show the differences in the way the animals were distributed in the soil of the reserve. However, when the ants were evaluated in lower ranks of genera or morphospecies, the relationships were significant and showed that three guilds of genera were distributed in a different manner in the soil of the reserve. In spite of the huge extension of the sampling area (64 km^2) that represented all of the variations of soil, topography and vegetation types in the reserve, the classification of the macroinvertebrates in high ranks was not enough to show their relationship to environmental factors.

Gualberto (2003) found seven species in the litter of Ducke Reserve. *Ideobisium shusteri* was related to the clay percentage of the soil. *T. minor* and the species richness showed significant relation with litter quantity, but the regression coefficients were very weak (lower than 0.3). The reserve lies on the division of two major watersheds. Each watershed has several drainage basins that are isolated from each other within the reserve by a central plateau. Therefore, they support similar numbers and densities of species of Pseudoscorpionida (Gualberto, 2003). The lack of correlations with the predictor variables and the large home range detected in the medium-spatial-scale pattern study do not point to them as good biological indicator organisms to be used to monitor changes in the reserve. To select faunal indicator taxa criteria for monitoring ecosystem health, criteria such as density, presence or absence, special food/habitat and correlation to ecosystem changes are taken into consideration. In this aspect, the pseudoscorpions of the litter are not a good biological indicator.

Concluding Remarks and Recommendations

In central Amazon, the equipment utilized to sample the mesofauna was the common Berlese–Tullgren and the more sophisticated Kempson apparatus. Sampling procedures have been improved during the last three decades, principally in respect of the number of plots sampled, giving better statistical support for the results, as can be seen by the medium-scale pattern studies (see the section 'Litter Manipulation'). The most recent surveys were aimed at many aspects of biodiversity, and permitted the integration of data collected by different teams.

A considerable inventory of soil mesofauna has been recorded in the central Amazon. Natural (primary forest, secondary forest, flooded forest and campinarana) and anthropogenic environments (plantation and polyculture system) were surveyed. The earliest studies principally listed an inventory of the abundance and diversity of the mesofauna and the influence of annual flooding of the rivers in the central Amazonian region. The studies that made the association of biomass estimations, abundance and the relationship between dry and wet seasons in natural and anthropomorphic environment began in 1989. Very little work has been done in respect of the effects of fire on soil mesofauna. The influence of the diversity of leaves on the colonization rate of litter animals started to be investigated in 1995. In 1996, studies on the structure of the soil fauna in correlation with the abiotic conditions of the sites began. The importance of the different size classes of the fauna in litter decomposition was studied in experiments using litterbags. More recently (2001), the effect of different qualities of litter was tested in the development of the invertebrate community.

In a review of the literature, Hilty and Merenlender (2000) found that many invertebrates are often studied in higher taxonomic categories, encompassing many species, making it difficult to measure specific attributes as biological indicators. Adis and Schubart (1984) indicated that trained specialists capable of classifying and describing the arthropod material collected in the Amazon region were lacking. The great number of new species registered for the region aggravates the problem. The results of Guimarães (2003) and Fagundes (2003) showed that the classification of the invertebrates in high ranks (order, family or functional groups based on their alimentary habits) was not enough to show their relation with the environmental factors in central Amazon.

To complement the above studies, it is necessary to make comparisons between natural and anthropomorphic systems involving:

1. Large-scale inventories, considering the mosaic pattern of the Amazon region, to guide us in determining the extent of soil biodiversity and its relationship with biotic and abiotic factors.
2. Studies on the role of soil mesofauna communities in incorporation of organic inputs to deeper layers and its effect on organic matter transformations.
3. Investigations on the effect of different soil types and also different soil substrates (richness of the leaves) on the soil mesofauna communities and the process of litter decomposition.
4. Training of new taxonomists to identify invertebrates at lower taxonomic levels (genera/species), to make it possible to select the species or groups of species that better satisfy the criteria for identifying useful environmental indicators.
5. The benefits of such information to the farmers and land users in relation to maintaining the soil as a favourable environment for the soil invertebrates.

Acknowledgements

We thank the technical staff (M.I.C. Albuquerque and M.A.A. Pereira), E.M.R. Santos (PhD student) and E.D.L. Eliane (biology student) of INPA for their indispensable support.

References

Adis, J. (1987) Extraction of arthropods from Neotropical soils with a modified Kempson apparatus. *Journal of Tropical Ecology* 3, 131–138.

Adis, J. (2002) *Amazonian Arachnida and Myriapoda*. Pensoft, Sofia-Moscow.

Adis, J. and Ribeiro, M.O.A. (1989) Impacto de desmatamento em invertebrados de solo de florestas inundáveis na Amazônia Central e suas estratégias de sobrevivência às inundações de longo prazo. *Boletim Museu Paraense Emilio Goeldi* 5, 101–125.

Adis, J. and Schubart, H.O.R. (1984) Ecological research on arthropods in central Amazonian forest ecosystems with recommendations for study procedures. In: Cooley, J.H. and Golley, F.B. (eds) *Trends in Ecological Research for the 1980 NATO Conferences*. Plenun Press, New York/London, pp. 111–144.

Adis, J., Morais, J.W. de and Ribeiro, E.F. (1987a) Vertical distribution and abundance of arthropods in the soil of a Neotropical secondary forest during the dry season. *Tropical Ecology* 28, 174–181.

Adis, J., Morais, J.W. de and Mesquita, H.G. (1987b) Vertical distribution and abundance of arthropods in the soil of a Neotropical secondary forest during the rainy season. *Studies on Neotropical Fauna and Environment* 22, 189–197.

Adis, J., Mahnert, V., Morais, J.W. de and Rodrigues, J.M.G. (1988) Adaptation of an Amazonian pseudoscorpion (Arachnida) from dryland forests to inundation forests. *Ecology* 69, 289–291.

Adis, J., Morais, J.W. de, Ribeiro, E.F. and Ribeiro, J.C. (1989a) Vertical distribution and abundance of arthropods from white sand soil of a Neotropical campinarana forest during the rainy season. *Studies on Neotropical Fauna and Environment* 24, 193–200.

Adis, J., Ribeiro, E.F., Morais, J.W. de and Cavalcante, E.T.S. (1989b) Vertical distribution and abundance of arthropods from white sand soil of a Neotropical campinarana forest during the dry season. *Studies on Neotropical Fauna and Environment* 24, 201–211.

Adis, J., Morais, J.W. de and Scheller, U. (1996a) On abundance, phenology and natural history of symphyla from a mixedwater inundation forest in central Amazonia, Brazil. *Acta Myriapodologica* 169, 607–616.

Adis, J., Minelli, A., Morais, J.W. de, Pereira, L.A., Barbieri, F. and Rodrigues, J.M.G. (1996b) On abundance and phenology of Geophilomorpha (Chilopoda) from central Amazonian upland forests. *Ecotropica* 2, 165–175.

Adis, J., Scheller, U., Morais, J.W. de, Rochus, C. and Rodrigues, J.M.G. (1997) Symphyla from Amazonia non-flooded upland forests and their adaptations to inundation forests. *Entomologica Scandinavica Supplement* 51, 307–317.

Aguiar, N.O. (2000) Diversidade e história natural de Pseudoscorpiões (Arachnida), em floresta primária de terra firme, no Alto Rio Urucu, Coari, Amazonas. PhD thesis, Instituto Nacional de Pesquisas da Amazônia/Universidade Federal do Amazonas, Manaus, Amazonas, Brazil.

Arlé, R. (1959) Generalidades e importância ecológica da ordem Collembola (Apterygota). *Atas da Sociedade de Biologia do Rio de Janeiro* 3, 1–7.

Arlé, R. (1960) Notas sobre a família Oncopoduridae, com descrição de duas espécies novas do Brasil. *Arquivos do Museu Nacional* 50, 9–24.

Arlé, R. and Oliveira, M.M. (1977) O gênero *Temeritas* Delamare & Massoud, 1963 na Amazônia (Collembola: Symphypleona). *Boletim do Museu Paraense Emílio Goeldi Zoologia, Nova Série* 87, 1–23.

Arlé, R. and Rufino, E.O. (1976) Contribuição ao conhecimento dos Pseudachorutinae da Amazônia (Collembola). *Acta Amazonica* 6, 99–107.

Beck, L. (1967) Die Bodenfauna des Neotropischen Regenwaldes. *Atlas do Simpósio sobre a Biota Amazônica* 5, 97–101.

Beck, L. (1968) Sobre a Biologia de alguns aracnídeos na Floresta Tropical da Reserva Ducke (INPA, Manaus/Brasil). *Amazoniana* 1, 2247–2250.

Beck, L. (1971) Bodenzoologische Gliederung und Charakterisierung des Amazonischen Regenwaldes. *Amazoniana* 3, 69–123.

Bromham, S., Cardillo, M. and Bannet, A. (1999) Effects of stock grazing on the ground invertebrate fauna of woodland remnants. *Australian Journal of Ecology* 24, 199–207.

Câmara, V.A. (2002) Flutuação populacional, diversidade específica e alguns aspectos ecológicos da comunidade de Collembola (Hexapoda) em um fragmento florestal urbano – Manaus-AM, Brasil. MSc thesis, Instituto Nacional de Pesquisas da Amazônia/Universidade Federal do Amazonas, Manaus, Amazonas, Brazil.

Fagundes, E.P. (2003) Efeitos de fatores do solo, altitude e inclinação do terreno sobre os invertebrados da serapilheira, com ênfase em Formicida (Insecta, Hymenoptera) da Reserva Ducke, Manaus, Amazonas, Brasil. MSc thesis, Instituto Nacional de Pesquisas da Amazônia/Universidade Federal do Amazonas, Manaus, Brazil.

Franklin, E.N. (1994) Ecologia de oribatídeos (Acari: Oribatida) em florestas inundáveis da Amazônia Central. PhD thesis, Instituto Nacional de Pesquisas da Amazônia/Universidade Federal do Amazonas, Manaus, Amazonas, Brazil.

Franklin, E.N., Schubart, H.O.R. and Adis, J.U. (1997a) Ácaros (Acari: Oribatida) Edáficos de duas florestas inundáveis da Amazônia Central: Distribuição vertical, abundância e recolonização do solo após a inundação. *Revista Brasileira de Biologia* 57, 501–520.

Franklin, E.N., Adis, J. and Woas, S. (1997b) The Oribatid mites. In: Junk, W.J. (ed.) *Central Amazonian River Floodplains: Ecology of a Pulsing System*. Springer-Verlag, Berlin/Heidelberg/Alemanha, pp. 331–349.

Franklin, E.F., Woas, S., Schubart, H.O.R. and Adis, J. (1998) Ácaros oribatídeos (Acari: Oribatida) arborícolas de duas florestas inundáveis da Amazônia Central. *Revista Brasileira de Biologia* 58, 317–335.

Franklin, E.F., Morais, J.W. and Santos, E.M.R. (2001a) Density and biomass of Acari and Collembola in primary forest, secondary forest and polycultures in central Amazonia. *Andrias* 15, 141–153.

Franklin, E.F., Guimarães, R.L., Adis, J. and Schubart, H.O.R. (2001b) Resistência a submersão de ácaros (Acari: Oribatida) terrestres de florestas inundáveis e de terra firme na Amazônia Central em condições experimentais de laboratório. *Acta Amazonica* 31, 285–298.

Franklin, E., Hayek, T., Fagundes, E.P. and Silva, L.L. (2004) Oribatid mites (Acari: Oribatida) contribution to decomposition dynamic of litter in primary forest, second growth and polyculture in the central Amazon *Brazilian Journel of Biology,* 64(1), 59–72.

Franklin, E., Santos, E.M.R. and Albuquerque, M.I.C. (in press) Diversity and distribution of orbital mites (Acari: Ribatida) in a lowland rainforest of Peru and in several environments of the Brazilian States of Amazones, Rondönia, Roraina and Pará. *Brazilian Journal of Biology* 67(3) (in press).

Gualberto, T.L. (2003) Pseudoscorpiões (Arachnida) da serapilheira e suas relações com fatores do solo, da Reserva Florestal Adolpho Ducke, Manaus, Amazonas, Brasil. MSc thesis, Instituto Nacional de Pesquisas da Amazônia/Universidade Federal do Amazonas, Manaus, Brazil.

Guimarães, R.L. (2003) Topografia, serapilheira e nutrientes do solo: análise dos seus efeitos sobe a mesofauna do solo na Reserva Florestal Adolpho Ducke, Manaus, Am, Brasil. MSc thesis, Instituto Nacional de Pesquisas da Amazônia/Universidade Federal do Amazonas, Manaus, Brazil.

Hayek, T.F. (2000) Soil mites: (Acari: Oribatida): diversity, abundance and biomass in the litter decomposition in primary forest, secondary forest and a mixed culture system in the central Amazonian region. MSc thesis, Instituto Nacional de Pesquisas da Amazônia/Universidade Federal do Amazonas, Manaus, Brazil.

Hilty, J. and Merenlender, A. (2000) Faunal indicator taxa selection for monitoring ecosystem health. *Biological Conservation* 92, 185–197.

Höfer, H., Hanagarth, W., Garcia, M., Martius, C., Franklin, E., Römbke, J. and Beck, L. (2001) Structure and function of soil fauna communities in Amazonian anthropogenic and natural ecosystems. *European Journal of Soil Biology* 37, 229–235.

Kempson, D., Lloyd, M. and Chelardi, R. (1963) A new extractor for woodland litter. *Pedobiologia* 3, 1–21.

Lawton, J.H., Bignell, D.E., Bolton, B., Bloerners, G.F., Eggleton, P., Hammond, P.M., Hodda, M., Holt, R.D., Larsen, T.B., Mawdsley, N.A., Stork, N.E., Srivastava, D.S. and Watt, A.D. (1998) Biodiversity inventories, indicator taxa and effects of habitat modification in tropical forest. *Nature* 391, 72–75.

Luizão, F.J. (1995) Ecological studies in contrasting forest types in central Amazonia. PhD thesis, University of Stirling, UK.

Martius, C. (2004a) Microclimate in agroforestry systems in central Amazonia: does canopy closure matter to soil organisms? *Agroforestry Systems* 60, 291–304.

Martius, C. (2004b) Litter fall, litter stocks and decomposition rates in rainforest and agroforestry sites in central Amazonia. *Nutrient Cycling in Agroecosystems* 68, 137–154.

Morais, J.W. de (1985) Abundância e distribuição vertical de Arthropoda do solo numa floresta primária não inundada. MSc thesis, Instituto Nacional de Pesquisas da Amazônia/Universidade Federal do Amazonas, Manaus, Brazil.

Morais, J.W. de (1996) Abundância, distribuição vertical e fenologia da fauna de Arthropoda de uma região de água mista, próxima de Manaus, AM. PhD thesis, Escola Superior de Agricultura 'Luiz de Queiroz'/Universidade de São Paulo, Piracicaba, SP, Brazil.

Morais, J.W., Adis, J., Mahnert, V. and Berti-Filho (1997) Abundance and phenology of pseudoscorpiones (Arachnida) from a mixedwater inundation forest in central Amazonia, Brazil. *Revue Suisse de Zoologie* 104, 475–483.

Nakamura, A., Proctor, H. and Catterall, C. (2003) Using soil and litter arthropods to assess the state of rainforest restoration. *Ecological Management & Restoration* 4, 20–28.

Noti, M.I., André, H.W., Ducarme, X. and Lebrun, P. (2003) Diversity of soil oribatid mites (Acari: Oribatida) from high katanga (Democratic Republic of Congo): a multiscale and multifactor approach. *Biodiversity and Conservation* 12, 767–785.

Oliveira, E.P. (1982) Colêmbolos (Insecta: Collembola) epigéicos como indicadores ecológicos em ambientes florestais. MSc thesis, Instituto Nacional de Pesquisas da Amazônia/Universidade Federal do Amazonas, Manaus, Brazil.

Oliveira, E.P. and Franklin, E.N. (1993) Efeito do fogo sobre a mesofauna do solo: recomendações em áreas queimadas. *Pesquisa Agropecuária Brasileira* 28, 357–369.

Petersen, H. and Luxton, M. (1982) A survey of the main animal taxa of detritus food web. *Oikos* 39, 293–294.

Phillipson, J. (1971) *Methods of Study in Quantitative Soil Ecology: Population, Production and Energy Flow.* Blackwell, London.

Ribeiro, E.F. (1986) Oribatídeos (Acari: Oribatida) colonizadores de folhas em decomposição sobre o solo de três sítios florestais da Amazônia Central. Manaus. MSc thesis, Instituto Nacional de Pesquisas da Amazônia/Universidade Federal do Amazonas, Manaus, Brazil.

Ribeiro, E.F. and Schubart, H.O.R. (1989) Oribatídeos (Acari; Oribatida) colonizadores de folhas em decomposição de três sítios florestais da Amazônia Central. *Boletim do Museu Paraense Emílio Goeldi* 5, 243–276.

Rodrigues, J.M.G. (1986) Abundância e distribuição vertical de Arthropoda do solo em Capoeira de terra firme. MSc thesis, Instituto Nacional de Pesquisas da Amazônia/Universidade Federal do Amazonas, Manaus, Brazil.

Santos, E.M.R. (2001) Densidade, diversidade e biomassa da fauna do solo em serapilheira manipulada numa floresta secundária na Amazônia Central. MSc thesis, Instituto Nacional de Pesquisas da Amazônia/Universidade Federal do Amazonas, Manaus, Brazil.

Scheller, U. and Adis, J. (1996) A pictorial key for the Symphylan families and genera of the Neotropical Region south of Central Mexico (Myriapoda, Symphyla). *Studies on Neotropical Fauna and Environment* 31, 57–61.

Southwood, T.R.E. (1980) *Ecological Methods: with Particular Reference to the Study of Insect Populations*. Chapman & Hall, London/New York.
Straalen, N.M. (1998) Evaluation of bioindicator systems derived from soil arthropod communities. *Applied Soil Ecology* 9, 429–437.
Tian, G., Adejuyigbe, C.O., Adeoye, G.O. and Kang, B.T. (1998) Role of soil microarthropods in leaf decomposition and N release under various land-use practices in the humid tropics. *Pedobiologia* 42, 33–42.
Valdecasas, A.G. and Camacho, A.I. (2003) Conservation to the rescue of taxonomy. *Biodiversity and Conservation* 12, 1113–1117.
Wallwork, J.A. (1983) Oribatids in forest ecosystems. *Annual Review of Entomology* 28, 109–130.
Woas, S. (2002) Acari. In: Adis, J. (ed.) *Amazonian Arachnida and Myriapoda*. Pensoft, Sofia-Moscow, pp. 21–291.
Zicsi, A., Römbke, J. and Garcia, M. (2001) Regenwürner (Oligochaeta) aus der Umgebung von Manaus (Amazonien). Regenwürmer aus Südamerika 32, *Verhandlungen Suisse Zoologie* 108, 1–12.

8 Nematode Communities in Soils under Different Land Use Systems in Brazilian Amazon and Savannah Vegetation

[†]S.P. Huang and J.E. Cares
Universidade de Brasília, Instituto de Ciências Biológicas, Departamento de Fitopatologia, Caixa Postal 4457, CEP 70.904-970, Brasília, DF, Brazil, e-mail: cares@unb.br[†]in memoriam

Introduction

Nematodes are small and ubiquitous invertebrate animals. The nematode community is characterized by five major functional groups: plant parasites, bacterial feeders, fungal feeders, predators and omnivores (Freckman and Caswell, 1985). The plant-parasitic nematodes may cause enormous annual yield losses in crops. Bacterial feeder nematodes can regulate the available quantity of nitrogen and phosphorus for plants, influence *Rhizobium* nodulation and consume and disseminate beneficial and plant-pathogenic bacteria. Some bacterial feeders, which are typical r-strategists, being capable of reproducing and exploding in number within a short time in an enriched soil, are considered an indicator of soil fertility (Ferris *et al.*, 2001). Some fungal feeders, besides feeding on saprophytic, pathogenic, beneficial and mycorrhizal fungi, are also considered as plant facultative parasites. On the other hand, most nematode predators are polyphagous, feeding on other nematodes, protozoa, rotifers, tardigrades, bacteria and fungal spores. Omnivores may feed on all food resources, including fungi, bacteria, plant roots, algae and other nematodes. One of the major roles of soil nematodes in ecosystems is the release of nutrients in soil for absorption by plant roots (Coleman *et al.*, 1984). Due to their short life span (mostly about 1 month per life cycle) and different feeding functions, soil nematodes reflect environmental changes in their community structure and composition.

Some nematodes occur in a wide range of habitats, whereas others are more restricted (Schmitt and Norton, 1972). Degree and diversity of plant ground cover was considered as the most important in determining nematode community structure (Niblack and Bernard, 1985). Soil nematode community composition reflects agricultural management practices (Ferris and Ferris, 1974; Wasilewska, 1989; Freckman and Ettema, 1993). Although nematode population density has not been related to its diversity (Yeates, 1979), Ferris and Ferris (1974) found less nematode diversity and greater abundance in intensively managed agroecosystems than in native ecosystems.

Due to complicated taxonomy at species level, the analysis of community structure has focused on trophic composition (Bernard, 1992). Many research workers have detected differences in trophic structures in

 Soil Biodiversity in Amazonian and Other Brazilian Ecosystems (eds F.M.S. Moreira *et al.*)

different ecosystems or treatments (Ferris, *et al.*, 1996; Todd, 1996; Freckman and Huang, 1998) and confirmed functional groups as good indicators of agroecosystem management (Todd, 1996). The functional groups, plant parasites, bacterial and fungal feeders, constitute the majority of nematodes in most ecosystems (Freckman and Caswell, 1985).

Bongers (1990) classified soil nematodes from colonizers (c) to persisters (p) (similar to r- to K-strategists) on a cp-scale of 1 to 5. The cp 1 colonizers were characterized by short generation time, production of many small eggs, presence of dauerlarvae and growth under food-rich conditions. In contrast, the cp 5 persisters were distinguished by long generation time, production of few but large eggs, low motility, absence of dauerlarvae and sensitivity to pollutants and other disturbance factors (Bongers and Bongers, 1998). Two indices were then created, the maturity index (MI) (including only free-living nematodes) and the plant parasitic index (PPI) (including only plant parasites), which sums up relative frequency × cp value of each nematode species in a community (Bongers, 1990). The higher the values of both indices, the lesser the soil disturbance level. Yeates (1994) made the modified maturity index (mMI) to include both free-living nematodes and plant parasites. The three indices along with other diversity indices have been widely used to assess soil disturbance level (Freckman and Ettema, 1993; Neher and Campbell, 1996; McSorley, 1997). Also, these indices have been used to assess changes in nematode communities under different land management practices (Hyvönen and Persson, 1990; Ettema and Bongers, 1993; Korthals *et al.*, 1996). Further, Bongers and Bongers (1998) combined the cp scaling with five trophic groups to analyse community structure, and used MI_{2-5} (same as MI, but cp 1 nematodes excluded) to evaluate soil stress and PPI/MI to assess soil fertility.

Nematode communities have been studied in two distinct regions in Brazil: the Amazonian tropical and humid region with sampling sites in two western states, Rondônia and Acre, and the tropical savannah (or so-called cerrado) region in three central states, Goiás, Minas Gerais and Distrito Federal. This chapter reviews research results on seven aspects: periodically flooded versus non-flooded ecosystems, native versus cultivated systems, soybean–maize rotation sequence, tillage versus no-tillage practices combined with two soybean cultivars, temporal versus spatial samplings, edaphic factors and fire. The communities were analysed for abundance (population density and relative abundance), diversity (generic richness, Shannon–Weaver's and Simpson's diversity indices and evenness of both indices, trophic diversity and dominance), soil disturbance (MI, mMI and PPI) and soil decomposition pathway (ratios of fungivores/bacteriovores and of (fungivores + bacteriovores)/plant parasites)). The most relevant results are discussed in the following sections.

Basic Taxonomy of Nematodes

Nematodes (in Greek, *Nema* = thread, *eides* = similar) are multicellular triploblastic unsegmented invertebrate animals (kingdom Animalia) with five distinct systems: digestive, nervous, muscular, excretory and reproductive (subkingdom Eumetazoa). Their body shows bilateral symmetry (division Bilaterata). During embryogenesis the blastopore develops into the mouth (subdivision Protostomia). Nematodes possess a pseudocoelom, a body cavity filled with liquids, which together with the cuticle and muscles maintains their body shape in a so-called hydrostatic skeleton (superphylum Pseudocoelomata). After the embryo develops to the first-stage juvenile, it has to pass through four moultings to become an adult. The adult male is characterized by the presence of a cloaca with gonopore and anus in the same opening, but both are separated in females (phylum Nemata Cobb, 1919). So far the phylum Nemata is classified into two classes, Secernentea (oesophagus in three parts and sense organs, phasmids present) and Adenophorea (oesophagus in one, two or rarely three parts and phasmids absent) with a total of 19 orders, but only 13 orders associated with the soil environment, as shown in Table 8.1.

Currently, the nematode classification system is based on morphological and morphometrical characters. Other characters have also been considered for taxonomic identification or phylogeny studies, including nematode host range, host reaction, cytogenetics, immunology and molecular biology (Cares and Huang, 2000). Morphology and morphometry performed with light microscopy, and both scanning and transmission electron microscopy, are of capital importance for nematode identification, from higher taxonomic ranks to the species level. Approaches based on the host range, histopathology, cytogenetics, patterns of isoenzymes, ELISA, restriction fragment length polymorphisms (RFLPs), PCR and random amplified polymorphic DNA-PCR (RAPD-PCR), combined with morphological and morphometrical features, have been considered for nematode identification to the species level, and have been used also for identification of infraspecific variations among nematodes, such as races and biotypes.

Generally, nematode trophic groups can be differentiated by their stomatal structures (Fig. 8.1). Bacterial feeders possess a tube-like or small cavity-like stoma, plant-parasitic nematodes a stoma modified into a strong stylet with knobs, fungal feeders a stoma modified into a needle-like stylet, omnivores an odontostylet with aperture equal to or smaller than half of stylet length, and predators an odontostylet with aperture bigger than half of stylet length or a big stomatal cavity equipped with one or more teeth. But one taxon of nematodes often shows more than one type of trophic function, and one feeding apparatus may be found in different functional groups. Yeates *et al.* (1993) catalogued all known soil nematodes into their possible feeding habits (Table 8.1). Nevertheless, in most cases, the morphology of the stomatal apparatus is still the easiest and most important character to identify nematode feeding habits.

Resources for Nematode Identification in Brazil

In Brazil, nematode identification activities started with the work of Goeldi published in 1887 on the root-knot nematode

Table 8.1. Soil nematodes and their trophic groups.

Order	Trophic group
Class Secernentea	
Rhabditida	Bacteriovores, predators, insect parasites, substrate ingestion
Diplogasterida	Bacteriovores, predators, substrate ingestion
Tylenchida	Plant parasites, alga parasites, fungivores, predators, insect parasites
Class Adenophorea	
Enoplida	Bacteriovores, predators, alga parasites, omnivores
Isolaimida	Bacteriovores
Mononchida	Predators, bacteriovores
Dorylaimida	Fungivores, plant parasites, predators, omnivores
Triplonchida	Fungivores, plant parasites
Stichosomida	Parasites of arthropods and other invertebrates
Chromadorida	Alga parasites, bacteriovores, predators
Desmoscolecida	Bacteriovores
Monhysterida	Bacteriovores, predators, algae feeders
Araeolaimida	Bacteriovores

Source: modified from Yeates *et al.* (1993).

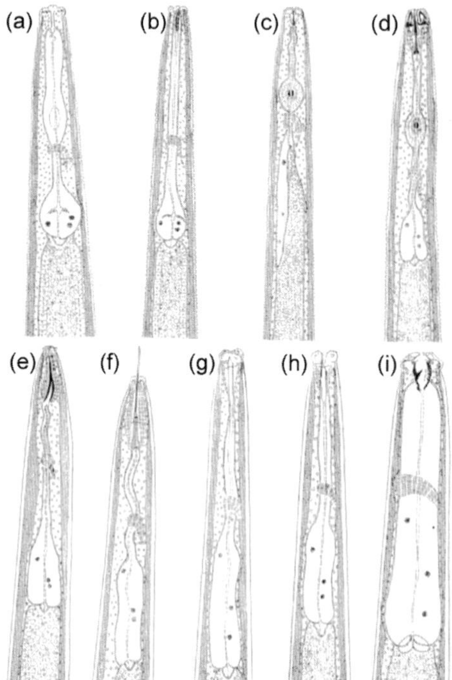

Fig. 8.1. Five trophic groups: Bacterial feeders: (a) rhabditid and (b) cephalobid; fungal feeders: (c) aphelenchid; plant parasites: (d) tylenchid, (e) trichodorid, and (f) longidorid; omnivores: (g) quadsianematid; predators: (h) aporcelaimid and (i) mononchid.

(*Meloidogyne*) in coffee roots in Rio de Janeiro. In 1974 the activities of the Brazilian Society of Nematology started, and it has been responsible for the yearly Brazilian Congress of Nematology and for the publication of the scientific journal *Nematologia Brasileira*.

Currently, more than 20 teaching and research centres are involved with nematode identification to different degrees, as well as with training for identification of soil nematodes, particularly plant parasites. Most of the activities related to nematode identification are directed towards important agricultural plant parasites. Expertise on nematode identification is mainly available in universities with graduate programmes in plant pathology or crop production, including Universidade Federal do Ceará, Universidade Federal Rural de Pernambuco, Universidade Estadual do Maranhão, Universidade Estadual do Norte Fluminense, Universidade Federal Rural do Rio de Janeiro, Escola Superior de Agricultura Luiz de Queiroz (Universidade de São Paulo), Universidade Estadual de São Paulo (Jaboticabal and Botucatu), Universidade Federal de Viçosa, Universidade Federal de Lavras, Universidade Federal de Uberlândia, Universidade Estadual do Paraná (Londrina and Maringá), Universidade Federal de Pelotas, Universidade Federal de Goiás and Universidade de Brasília. Training activities on nematode identification also occur in other research centres, including the ones of the Empresa Brasileira de Pesquisa Agropecuária (EMBRAPA), Embrapa Recursos Genéticos e Biotecnologia (CENARGEN), Embrapa Hortaliças, Embrapa Cerrados, Embrapa Mandioca e Fruticultura, Embrapa Soja, Embrapa – Dourados, as well as at Instituto Agronômico de Campinas and Instituto Agronômico do Paraná.

There is a paucity of information regarding nematode collections in Brazil. Most of the universities listed above have their own nematode collections, mainly for teaching purposes. The nematode collection at Universidade de Brasília holds more than 6000 entries of nematode taxa of all trophic groups of soil nematodes, including type specimens of Brazilian species.

Assessment of Nematode Communities

Once total nematode numbers have been counted, 100 individuals are picked randomly from each sample for identification to the generic level with the aid of taxonomic information (Goseco *et al.*, 1974a,b; Bongers, 1987; Fortuner *et al.*, 1988; Jairajpuri and Ahmad, 1992). Ecological parameters and indices are calculated in the following ways:

1. Abundance: Total abundance is assessed by counting the total number of nematodes in a certain amount of sampled soil. Absolute and relative abundance of each nematode genus in the community are also assessed.

2. Diversity: There are five diversity indices. Generic richness (d) is calculated by $d = (S - 1)/\log N$ (where S = number of genera and N = total number of nematodes in each sample) (Magurran, 1988). Shannon–Weaver's diversity index (H') is obtained by $H' = -\Sigma$ ($Pi \times \log_e (Pi)$) (where Pi = relative abundance of genus '*i*') (Pielou, 1977) and its evenness index (J') by $J' = H'/H'_{max}$ (where $H'_{max} = \log_2 S$) (Elliot, 1990). Both give higher weights to rare nematodes such as predators and omnivores. Simpson's diversity index (Ds) is formed with $Ds = 1-\Sigma(Pi)^2$, and its evenness index (Es) by $Es = Ds/Ds_{max}$ (where $Ds_{max} = 1 - 1/S$) (Elliot, 1990). Both indices give bigger weights to highly abundant nematodes, such as plant parasites and bacterial feeders (Pielou, 1977).

3. Trophic habits: Based on trophic habits, soil nematodes can be allocated into five major groups (plant parasites, bacterial feeders, fungal feeders, omnivores and predators). Algal feeders with total abundance frequently lower than 0.1% have been neglected in most cases. Many nematodes possess two or three feeding habits, for example, mononchid nematodes are predators but commonly reported as bacteriovores, and many dorylaimids are predators, but frequently reported as omnivores (Yeates *et al.*, 1993).

4. Soil disturbance: The three indices, MI, PPI and mMI, are calculated by the same formula, $\Sigma\ v(i) \times f(i)$ (where $v(i)$ = cp value from 1 to 5 for genus '*i*' or family '*i*,' and $f(i)$ = relative frequency of genus '*i*' or family '*i*'). The index mMI is applied to all soil nematodes (Wasilewska, 1994; Yeates, 1994), whereas the PPI index only to the plant-parasitic nematodes in Tylenchina, Trichodoridae and Longidoridae, and the MI index is applied to all soil nematodes except the plant-parasitic nematodes. According to Bongers (1990), a high value of the indices indicates less soil stress, and a low value denotes high soil disturbance.

5. Decomposition pathway: Two ratios, fungal feeders/bacterial feeders (FF/BF) and (fungal feeders + bacterial feeders)/ plant parasites ((FF + BF)/PP), are used to indicate the decomposition pathway (Wasilewska, 1994).

6. Index of similarity: Two indices are commonly used. One is Jaccard's index ($Isj = c/(a + b + c)$), where c equals the number of genera common for two areas, and a and b equal the number of genera found only in area a or area b, relatively (Norton, 1978). Another is Bray and Curtis' index (C), being formulated as $C = 2w/(a + b)$, where w = the sum of the least quantitative values in a common taxon (genus or family), and a and b are the total densities in habitats a and b.

7. Correspondence analysis: This type of factorial analysis is applied to frequency data in order to establish a relationship between the elements in a row (nematode taxa) with a given property in a column (sampling sites, depths of sample collected, etc.). After mathematical transformations on a matrix with n rows and p columns, a symmetrical matrix is obtained and the data are plotted as a two-axes graphic (x, y). Each axis is the resultant of the contribution by the frequency of the elements (nematode taxa) and represents a contributing factor with a percentage of explanation for total variations in the experiment (Lebardt *et al.*, 1982).

Land Use Systems and Edaphic Conditions of the Studied Areas

In this section we review the studies that have been conducted in two Brazilian regions: the Amazonian region comprising three states, Amazonas, Rondônia and Acre, and the savannah biome ('cerrado') of central Brazil, located in Distrito Federal, and states of Minas Gerais and Goiás.

In Amazonas state, two distinct ecosystems are considered: 'várzea' (the plain periodically flooded by raising rivers) and 'terra firme' forest (the non-flooded upper lands). The 'várzea' lands are located at Xiborena, an island at the joining point of the two rivers, Solimões (clear water river) and Rio Negro (dark water), to form the Amazon River. The white-water 'várzeas' are areas under the influence of the river Solimões and some of its tributaries coming from the Andes mountains and carrying large amounts of nutrients in sediments. During the flooding season, from November to April, large amounts of sediments are deposited on the floodplains that are suitable for cultivation of annual and perennial crops. The dark-water 'várzeas' are the lands alongside the Rio Negro and its tributaries that carry less sediment and are not suitable for agricultural purposes. The dark colour results from high concentrations of leaching compounds from decomposed plant materials. The 'terra firme' forest is characterized by low pH and poor-nutrient Oxisols, covered with a highly diverse flora. It is located in the upland and is thus not influenced by flooding, but markedly affected by slash-and-burn agriculture.

In the states of Rondônia and Acre, five common land use systems were studied: forest, annual crops, pasture, fallow and agroforestry. In this tropical humid region, after forests are slashed and burned, annual crops (rice, beans, maize, cassava, etc.) or pasture (mostly Brachiaria grass) is planted. But, due to poor soil fertility and physical structure, which are easily damaged by tropical rainfall, the plantations have to be converted to fallow for 3–6 years before being returned for cropping. Also, the agroforestry system, which combines forestry species with agricultural annual and perennial cultivated species, and/or livestock in the same plantation, is widely considered as an alternative way not only for maintaining high agricultural productivity for a longer time but also for reducing ecological damage. In this region during the period of 1982–1994, mean annual precipitation was 1970.9 mm (maximum 2165 mm and minimum 1705 mm), with 80% of annual precipitation occurring from November to April. The highest medium temperature was 25.6°C, measured in October, and the lowest, 22°C, observed in July. Average relative air humidity was about 79% (Scerne *et al.*, 1996).

Another area studied is in the 'cerrado' (savannah) ecosystem that occupies about 20% of Brazilian territory with 80% of total area localized in three states (Minas Gerais, Goiás and Mato Grosso) and Distrito Federal. Generally, the savannah soil is characterized by its high acidity, high exchangeable and saturated aluminium and poor availability of essential nutrients. There are five major types of savannah soils: dark-red latosols (containing more than 9% of iron), yellowish red latosols (less than 9% of iron) (both latosols occupying 46% of total savannah), arenosols (15%), acrisols (6%) and gleysols (2%) (Adámoli *et al.*, 1986; Azevedo and Adámoli, 1988). The two common latosols are characterized by low amounts of organic matter and diverse textures. The arenosols contain more than 80% sand and less than 15% clay. The acrisols are composed of sandy clay in the upper layer and clay in the lower layer, possessing good permeability and low capacity of water retention. The gleysols are flat organic soils resulting from alluvial sediments of rivers, and characterized by a high silt content and poor water drainage.

There are two distinct seasons: dry (May to September) and rainy (October to April), with soil temperatures fluctuating between 20°C and 26°C, and annual precipitation around 1000–2000 mm (Garrido *et al.*, 1982; Azevedo and Adámoli, 1988). Six families of plants (Fabaceae, Poaceae, Asteraceae, Rubiaceae, Arecaceae and Cyperaceae) dominate the native savannah vegetation (Goodland, 1970), with grassland native vegetation in some parts.

There are four major types of native vegetation: 'cerrado *sensu stricto*' (trees of height lower than 7 m, covering 10–70% of the surface) occupying 68.8% of the total savannah area, 'cerradão' (8 to 15 m high tree canopy, covering 70% of the surface) occupying 10.3% of savannah, 'campo cerrado' (with trees covering less than 10% of the surface, subdivided into 'campo sujo' (with some scattered trees) and 'campo limpo' (with just a few trees)) occupying 12%, and gallery forest in valley bottoms following a brook, (adjacent to wet grassy land) occupying only 5% of the savannah area (Eiten, 1978, 1979, 1984; Adámoli *et al.*, 1986). The plant diversity is higher in 'cerrado *sensu stricto*', followed by 'cerradão', then by 'campos' and lower in gallery forest. After agricultural conversion of native savannah, perennial plants (such as *Pinus* spp., *Eucalyptus* spp., coffee), annual crops (such as maize, rice, soybean, vegetable plants) and pasture species such as *Brachiaria* spp. have replaced the native and highly diverse vegetation.

Nematode Community Structure

In Amazon

Periodically flooded and non-flooded ecosystems

Cares (1984) studied communities of plant-parasitic nematodes, and fungal feeding aphelenchids in four sites near Manaus. Species belonging to 30 genera were identified (Table 8.2). *Aphelenchoides* spp. (in 50.8% of samples), *Discocriconemella* spp. (45.0%), *Xiphinema ensiculiferum* (21.6%) and *Tylenchorhynchus* spp. (19.7%) were frequent in native vegetation, while *Criconemella* spp. (45.0%), *Aphelenchus* spp. (37.0%), *Aphelenchoides* spp. (32.0%), *Meloidogyne* spp. (30.9%), *Helicotylenchus pseudorobustus* (22.2%), *Coslenchus costatus* (21.7%), *Meloidogyne javanica* (19.9%) and *Xiphidorus amazonensis* (17.6%) were common in agroecosystems. Only *Aphelenchoides* species adapted well in both native and cultivated areas.

There were more species of plant-parasitic nematodes found in non-flooded land ('terra firme') than in periodically flooded plains ('várzea') (Cares, 1984). Similarly, in 'terra firme', there were more species associated with native forest and tropical fruit trees (34 species) than with annual and perennial crops (23 species). But in the 'várzea', only 16 species were identified from native vegetation on the side of the island with dark water, as compared with 33 species from samples mostly collected in cropping fields at the island side with white water. There were no differences in nematode vertical distribution between the two soil profiles (0–15 cm and 15–30 cm). Although most of the nematode species (80.07% of total species) were recovered from both soil depths, there was a tendency for most of them to concentrate more in the lower layer. More species (10.62%) were restricted to this layer compared with those restricted to the upper layer (6.31%).

Five land use systems studied in western Brazilian Amazon

Huang *et al.* (1998) investigated nematode communities in five land use systems (forest, annual crops, pasture, fallow and agroforestry) in three sites/systems located in Rondônia and Acre. Fifteen thousand nematodes were observed and assigned to 159 genera and 59 families. There were 113 genera and 45 families identified in fallow, 108 genera and 44 families in agroforestry, 102 and 42 in disturbed forest, 97 and 41 in annual crops and 79 and 37 in *Brachiaria* pasture. As compared with nematode communities in forest systems, there were more genera in the communities under fallow and agroforestry, and the reverse was found in pasture and annual crop systems. Total nematode abundance was high in pasture, followed by disturbed forest and fallow and low in agroforestry and annual crops (Huang and Cares, 2000a). The accumulated abundance curves showed that 50% of abundance was accounted for by only four genera in forest and in pasture, by five in annual crops, by seven in agroforestry and by nine genera in fallow. The rare genera (relative abundance <1%) occupied 20% of total abundance under agroforestry, 19.2% under fallow, 14.8% under crop, 14.1% under pasture and 11.5% under disturbed forest.

Discocriconemella (33.2% of total abundance) and *Helicotylenchus* (10%) dominated in disturbed forest, *Discocriconemella* (25.7%) in annual crops, *Dorylaimellus* (20.5%), *Belondirella* (16.2%) and *Helicotylenchus* (11.5%) in pasture and *Helicotylenchus* (16.2%) in agroforestry.

Table 8.2. Plant and fungal-feeding nematodes from periodically flooded plains and from non-flooded forest in the Amazon.

Family	Species	Family	Species
Tylenchidae	*Tylenchus* spp.	Criconematidae	*Criconema* sp.
	Cucullitylenchus sp.		*Criconema* sp.
	Ecphyadophora spp.		*Criconema* sp.
	Echphiadophoroides spp.		*Criconemella* spp.
	Coslenchus costatus		*Discocriconemella* spp.
	Chitinotylenchus sp.		*Hemicriconemoides* spp.
Anguinidae	*Ditylenchus* spp.		*Hemicycliophora* spp.
Belonolaimidae	*Tylenchorhynchus* spp.	Tylenchulidae	*Tylenchulus* spp.
	Trophurus spp.		*Trophotylechulus* spp.
Pratylenchidae	*Pratylenchus brachyurus*		*Trophonema* spp.
	P. loosi		*Paratylenchus leptus*
	P. zeae		*P. salubris*
	Pratylenchus spp.		*Gracilacus punctata*
Hoplolaimidae	*Helicotylenchus dihystera*	Aphelenchidae	*Aphelenchus* spp.
	H. multicinctus	Aphelenchoididae	*Aphelenchoides* spp.
	H. pseudorobustus	Longidoridae	*Xiphinema brasiliense*
	Helicotylenchus spp.		*X. clavicaudatum*
	Rotylenchus spp.		*X. ensiculiferum*
	Aorolaimus spp.		*Xiphinema* spp.
	Hoplolaimus spp.		*Xiphidorus amazonensis*
	Rotylenchulus reniformis	Trichodoridae	*Paratrichodorus minor*
Heteroderidae	*Meloidogyne incognita*		
	M. javanica		
	Meloidogyne spp.		

Source: modified from Cares (1984).

The abundance of *Aporcelaimellus* was the highest (8.3%) in free fallow in which nematode communities were shared by 20 genera in gradually decreasing abundance, but with none having less than 1% abundance.

Nematode diversity as evaluated by generic richness, Simpson's and Shannon's indices, and both indices of evenness, was in a decreasing sequence: fallow > agroforestry system > disturbed forest > annual crops > pasture (Huang and Cares, 2000a,b). MI indicated more soil disturbance in pasture and crop systems, followed by fallow and agroforestry systems, and less in forest systems. The values of mMI and PPI were high in pasture, low in forest and intermediate in the other three systems.

Plant parasites were the biggest trophic group comprising about 30–75% of total abundance, followed by bacterial feeders (about 10–25%), while fungivores, predators and omnivores were below 20% in abundance. In the fallow plots, there was a high abundance of bacterial feeders and predators, and a low abundance of plant parasites, whereas the contrary was found in the plots of forestry and pasture systems. The following sequence demonstrates the tendency of decreasing plant-parasitic abundance and trophic dominance, but increasing trophic diversity: forest, pasture, annual crops, agroforestry and fallow. There was a high ratio of FF/BF in forest and a low ratio in fallow; the other three land use systems were intermediate.

In the savannah region

Native and cultivated systems

Cares and Huang (1991) investigated communities of plant-parasitic nematodes in native and cultivated savannah soils. Of the

42 genera of plant-parasitic nematodes and fungal feeding aphelenchids found in the central region of Brazil, 37 were present in native savannah vegetation, 24 in native gallery forest, 23 in cultivated perennial plants and only 13 in annual crops, indicating the greater diversity of nematode fauna in less disturbed soils. *Tylenchus*, *Helicotylenchus*, *Discocriconemella*, *Ecphyadophora*, *Hemicriconemoides*, *Trophotylenchulus*, *Coslenchus*, *Meloidogyne* and *Xiphinema* occurred with high frequency in native savannah, whereas *Ecphyadophora*, *Helicotylenchus*, *Meloidogyne*, *Tylenchus* and *Xiphinema* occurred in gallery forest, and *Ditylenchus*, *Helicotylenchus*, *Meloidoygne* and *Tylenchus* occurred in cultivated lands. In native savannah, the nematode abundance was generally high in the surface layer (0–20 cm), moderate at 20 to 40 cm depth and low at 40–60 cm. Of the 22 genera found in a place with three adjacent native vegetation forms, humid grassland, gallery forest and 'campo sujo' (grassland), three adapted well to all forms, seven to only two vegetation types and the others to only one (Fig. 8.2). Many plant-parasitic nematodes disappeared or became rare when the native vegetation was replaced by pine, eucalyptus, soybean or rice (Table 8.3). However, some nematode populations were higher in these cultivated plantations than in their native vegetation, indicating that some nematodes, originally in the native vegetation, can adapt to cultivated lands. On the other hand, three common species of *Meloidogyne* (*M. javanica*, *M. incognita*, and *M. arenaria*) identified by the bands of esterase phenotypes in electrophoresis were found in native cerrado vegetation (Huang *et al.*, 1991; Souza *et al.*, 1994). Possibly, they can adapt from native to agroecosystems and be the first inoculum source for introduced crops.

Nematode communities associated with a savannah wood tree species, 'sucupira branca' (*Pterodon pubescens* Benth) in Distrito Federal, were composed of 40% plant parasites, 30% omnivores, 20% bacterial feeders, 7% fungal feeders and 3% predators (Huang *et al.*, 1996). *Trophotylenchulus*, *Coslenchus* and *Meloidogyne* in the 32 genera of plant parasites comprised 50% of total abundance, whereas *Tylenchus* and *Ditylenchus* in six fungal

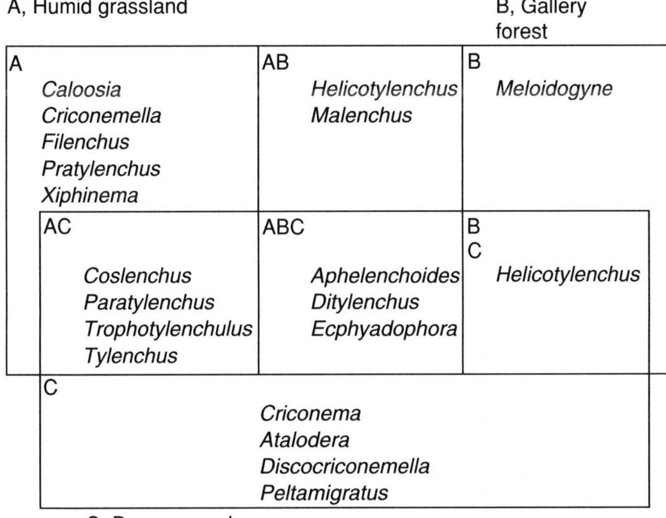

Fig. 8.2. Adaptation of plant-parasitic nematodes in three adjacent systems. (Source: modified from Cares and Huang, 1991.)

Table 8.3. Density of plant-parasitic nematodes in native vegetation (NV) and different crops in the Brazilian cerrado.[a]

Nematode	Number of nematodes/1 litre of soil							
	NV	Pine	NV	*Eucalyptus*	NV	Soybean	NV	Rice
Superfamily Criconematoidea								
Criconema sp. 1	16	0	157	0	175	16	61	16[b]
Criconema sp. 2	62	0	–	–	16	0	–	–
Criconemella sp. 1	69	0	83	75	206	0[b]	245	72[b]
Criconemella sp. 2	23	0	51	16	9	0	9	0
Discocriconemella sp.	105	0[b]	–	–	277	0[b]	523	65[b]
Gracilacus sp.	95	0[b]	23	0	–	–	16	0
Hemicriconemoides sp.	183	0[b]	194	55[b]	331	0[b]	175	0[b]
Hemicycliophora sp.	–	–	23	0	213	0[b]	215	0[b]
Nothocriconemoides sp.	113	0[b]	113	0[b]	30	0	–	–
Ogma sp.	33	0	45	13	79	0	49	0
Paratylenchus sp.	95	0[b]	101	10	300	0[b]	606	0[b]
Trophotylenchulus sp.	760	137[b]	677	106[b]	883	0[b]	1611	0[b]
Tylenchulus sp.	–	–	65	0	–	–	–	–
Others								
Aphelenchoides sp.	–	–	–	–	95	521[b]	224	128
Aorolaimus sp.	–	–	46	0	–	–	–	–
Coslenchus sp.	358	94[b]	334	121[b]	491	143[b]	604	49[b]
Ditylenchus sp. 1	127	123	153	342	568	707	697	593
Ditylenchus sp. 2	–	–	–	–	–	–	0	348[b]
Ecphyadophora sp.	391	161[b]	605	417	930	16[b]	999	16[b]
Helicotylenchus sp.	23	0	216	241	60	454[b]	16	0
Meloidogyne sp.	39	379	223	262	281	39[b]	137	0
Merlinius sp.	–	–	–	–	182	0	–	–
Pratylenchus sp.	0	16	23	0	0	282[b]	16	0
Pseudhalenchus sp.	–	–	0	46	–	–	16	0
Tylenchus sp.	115	175	112	23	494	871[b]	542	303[b]
UGH[c]	–	–	–	–	56	0	–	–
Xiphinema sp.	–	–	–	–	119	16[b]	–	–
Total number of genera	17	7	19	13	21	10	20	9

[a]Means of 15 replications were transformed by $x-1$ for statistical analyses.
[b]Means at the same horizontal line are significantly different according to the Tukey test ($P \leq 0.05$).
[c]UGH = unidentified genus belonging to subfamily Heteroderinae.
Source: after Cares and Huang (1991).

feeders comprised 75% of abundance. In this study, the high populations of dorylaimids in these native soils confirm the quality of this persistent group as an indicator for undisturbed soils.

Plant-parasitic nematodes associated with five types of vegetation ('cerradão', 'cerrado *sensu stricto*', gallery forest, cultivated annual and perennial plants) in the savannah region were investigated. The genera *Helicotylenchus*, *Meloidogyne*, *Ecphyadophora*, *Discocriconemella*, *Trophotylenchulus* and *Tylenchus* dominated the communities (Cares, 1990; Cares and Huang, 1991). The abundance of nematodes was higher in soil cultivated with perennial plants than in the others, which might relate to high root mass. There were 28, 22, 22, 15 and 13 genera of plant parasites observed in 'cerrado *sensu stricto*', 'cerradão', gallery forest, cultivated perennial and annual plants, respectively. It appears that the nematode generic diversity is related to plant diversity. Based on nematode density and frequency in cluster analyses, 'cerrado' and 'cerradão' were a closed group, which was near to gallery forest and far from the cultivated group with perennial and annual crops (Huang and Cares, 1995). The absence of nematodes belonging to the superfamily Criconematoidea in the cultivated soil might be one reason for this cluster pattern.

Nematode community structure could be divided into three distinct groups: group 1 having only 16.8% of total generic number and 62.2% of total prominence value (PV = density × frequency$^{1/2}$) (Norton, 1978), group 2 having 26.7% of generic number and 27.1% of PV, and group 3 with several rare genera having 56.5% of total generic number but only 10.7% of total PV (Table 8.4) (Huang and Cares, 1995). The three groups can be related to cp 2 (colonizer) to cp 5 (persisters) nematodes (Bongers, 1990). In this study, *Meloidogyne* species, such as *M. javanica* and *M. arenaria*, could be allocated to group 1 in cultivated annual crops, to group 2 in gallery forest and to group 3 in native 'cerrado' and 'cerradão', indicating that these nematodes can be colonizers under favourable conditions, or persisters in a different environment.

In Brazilian central savannah, more studies on nematode communities were done in four native systems ('cerrado', 'cerradão', gallery forest and native grassland) and four cropping systems (eucalyptus, coffee, maize and tomato) by sampling on five different locations in each type of vegetation during a rainy season (Mattos, 1999; Mattos, *et al.*, 2000a,b). The studies identified a total of 115 genera and 39 families, including 80 genera in 'cerrado', 74 in 'cerradão', 77 in gallery forest, 77 in native grasslands, 72 in *Eucalyptus*, 75 in coffee, 70 in maize and only 46 in tomato fields (Tables 8.5 and 8.6). The nematode populations were higher in tomato and maize cropping systems, and lower in *Eucalyptus*. *Trophotylenchulus*, *Discocriconemella*, *Helicotylenchus*, *Ditylenchus*, *Tylenchus*, *Dorylaimellus* and *Tylencholaimus* were more abundant in the four native systems, whereas *Acrobeles*, *Panagrolaimus*, *Rhabditis*, *Xiphinema*, *Helicotylenchus*, *Meloidogyne* and *Cephalobus* were common in the four cultivated systems (Huang and Mattos, 2000). Plant parasites dominated in the four native systems, whereas bacterial feeders did in the cropping systems (except in maize). The abundance of some nematodes could be well differentiated among the eight vegetation systems, for example, *Trophotylenchulus* being abundant in all four native systems, *Discocriconemella* dominant in 'cerrado' and 'cerradão', *Xiphinema* in *Eucalyptus* and *Helicotylenchus* in native grassland and in both annual cropping fields (Table 8.7). Nematode communities were similar among the four native systems (Bray and Curtis' similarity index C = 0.56–0.76) and less similar among the four cultivated systems (C = 0.10–0.46). These diversity indices and the relative abundance of criconematids (belonging to the superfamily Criconematoidea) were high in the native systems, but low in the cultivated systems (the lowest one in tomato fields). The MI and the mMI indices classified the four native systems as slightly disturbed, the two cultivated perennial systems as intermediate and the two annual systems as disturbed, but the PPI

Table 8.4. Categories[a] of plant-parasitic nematodes found in five different kinds of vegetation and soils in central Brazil (after Huang and Cares, 1995).

	Vegetation[b]					Soil[b]				
Genus	CD	CE	GF	CA	CP	DRL	YRL	ARS	GLS	ACS
Aphelenchoides	3	3	3	2	2	3	3	0	3	3
Aphelenchus	0	3	0	3	0	3	3	0	0	0
Atalodera	3	3	3	0	3	3	3	3	3	3
Basiria	0	0	2	0	0	0	0	0	2	0
Caloosia	3	3	0	0	0	3	3	1	0	3
Coslenchus	2	2	3	3	3	2	2	2	3	3
Criconema	3	3	3	0	3	2	3	3	3	3
Criconemella	3	3	3	3	3	3	3	3	3	3
Discocriconemella	1	1	3	3	3	1	2	3	3	2
Discotylenchus	0	0	2	0	0	0	0	0	2	0
Ditylenchus	2	3	3	2	3	2	2	3	3	3
Ecphyadophora	2	1	2	2	1	1	2	2	2	2
Filenchus	0	3	3	0	0	3	0	3	0	3
Gracilacus	1	3	0	0	0	3	3	0	0	0
Helicotylenchus	1	1	1	1	1	2	1	1	1	1
Hemicriconemoides	3	2	3	0	0	2	2	3	3	3
Hemicycliophora	1	2	2	0	0	3	1	3	2	0
Hoplolaimus	0	3	0	0	0	3	0	0	0	0
Malenchus	0	2	0	0	3	3	3	0	0	0
Meloidogyne	3	1	2	1	1	1	1	2	2	3
Paratrichodorus	3	3	0	0	0	3	3	0	0	0
Paratylenchus	3	3	3	0	0	3	3	0	0	3
Peltamigratus	0	3	2	0	3	3	3	0	2	2
Pratylenchus	3	3	0	3	0	3	3	3	0	0
Pseudhalenchus	3	3	0	0	0	3	3	0	0	0
Scutellonema	3	3	1	0	0	3	3	0	0	1
Trichodorus	0	0	3	0	0	0	0	0	3	0
Trophotylenchulus	1	1	2	2	2	1	1	1	2	2
Tylenchorhynchus	0	3	0	0	0	3	3	0	0	0
Tylenchus	2	1	2	1	2	1	2	1	2	2
Xiphinema	3	2	2	3	3	2	2	3	2	3
Total genera	22	28	22	13	15	28	26	17	19	19
Total samples	24	224	20	12	16	110	144	12	18	12

[a]Prominence value percentages grouped into three categories, one (PV% > 50), two (PV% = 17.1–50) and three (PV% < 17). (0) indicates the absence of the nematode.

[b]Vegetation: CD, cerradão; CE, cerrado; GF, gallery forest; CA, cultivated annual plants; CP, cultivated perennial plants. Soil: DRL, dark-red latosols; YRL, yellowish red latosols; ARS, arenosols; GLS, gleysols; ACS, acrisols.

did not differentiate the eight systems well. Also, on the basis of trophic structure, the three disturbance indices and the relative abundance of 20 more abundant genera and families, cluster analysis grouped the four native vegetation systems closely together, *Eucalyptus*, coffee and maize systems as less associated and tomato system as isolated.

Managed soybean plantations

Gomes *et al.* (2003) investigated the composition of nematode communities by surveying 23 soybean plantations in a soybean-growing region in the Brazilian Federal District and identified 41 genera from 115 soil samples. In the nematode community, plant-parasitic nema-

Table 8.5. Nematode genera found in eight land use systems in central Brazil (modified from Mattos, 1999).[a]

Achromadora[1,5,6]	*Discocriconemella*[1,2,3,4,6]	*Mylodiscus*[3]
Acrobeles[1,2,3,4,5,6,7,8]	*Discolaimium*[1,3,4,5,6,7,8]	*Mylonchulus*[1,2,3,4,5,6,7,8]
Acrobeloides[1,2,3,4,5,6,7,8]	*Discolaimoides*[1,3,4,5,6,8]	*Nygolaimoides*[6]
Akrotonus[1,2,3,4,5,8]	*Discolaimus*[2,3,4,7]	*Opisthodorylaimus*[5]
Alaimus[1,2,3,4,5,6,7,8]	*Discomyctus*[1,2,3,4,5,6,7,8]	*Oriverutus*[2,5,6,7,8]
Allodorylaimus[3]	*Ditylenchus*[1,2,3,4,5,6,7,8]	*Oxydirus*[1]
Aphanolaimus[2,7]	*Dorylaimellus*[1,2,3,4,5,6,7,8]	*Panagrolaimus*[1,2,3,4,7,8]
Aphelenchoides[1,2,3,4,5,6,7,8]	*Dorylaimoides*[1,2,3,4,5,6,7,8]	*Paractinolaimus*[2,5,6]
Aphelenchus[1,2,3,4,5,6,7,8]	*Ecphyadophora*[1,2,3,4,5,6,7]	*Paratylenchus*[1,2,3,4,5,6,7,8]
Aporcedorus[2,3,5]	*Enchodelus*[1,2,3,5,6,7]	*Paraxonchium*[1,3,4,5,6,7,8]
Aporcelaimellus[1,2,3,4,5,6,7,8]	*Eucephalobus*[1,2,3,4,6,7,8]	*Paroriverutus*[2,3]
Aporcelaimium[3]	*Eudorylaimus*[1,2,3,4,5,6,7]	*Pelodera*[5,7]
Aporcelaimoides[4]	*Ficulenchus*[1]	*Plectus*[1,2,3,4,5,6,7]
Atalodera[1,3,4]	*Filenchus*[1,2,3,4,5,7,8]	*Pratylenchus*[1,2,3,4,5,6,7]
Aulolaimus[1,2,4,6,7]	*Gopalus*[3,4,7]	*Prionchulus*[1,3,4,5,6,7]
Axonchium[1,2,4,6,7]	*Gracilacus*[1,2,3,4,5]	*Prismatolaimus*[1,2,3,4,5,6,7,8]
Basiria[1,5,7]	*Granonchulus*[1,2,3,4,5,6]	*Prodorylaimium*[1,2,4]
Basirotyleptus[1,2,3]	*Helicotylenchus*[1,2,3,4,5,6,7,8]	*Protorhabditis*[5,8]
Belondira[1,2,3,4,5,7]	*Hemicriconemoides*[1,2,3,4,5,6]	*Punctodora*[3,5]
Belondirella[3,4,7]	*Hemicycliopora*[3]	*Pungentus*[3,4,5,7,8]
Brachonchulus[2,6]	*Heterocephalobus*[1]	*Rhabditis*[1,2,3,4,5,6,7,8]
Bunonema[1,2,3,4,6]	*Iotonchus*[1,2,4,6,7]	*Seinura*[2,3,4,5,6,7,8]
Calcaridorylaimus[4]	*Labronema*[1,2,3,4,5,6,7,8]	*Sicaguttur*[1,2,3,4,5,6,7,8]
Caloosia[1,3,4]	*Labronemella*[2,5,6]	*Sicorinema*[2]
Carcharoides[3]	*Lelenchus*[1,2,3,4,5,6,7]	*Teratocephalus*[2,3,4,5,6,7,8]
Carcharolaimus[1,3,4,5,6,7]	*Leptolaimus*[1,2,3,4,5,6,7,8]	*Thonus*[2,3,4]
Cephalenchus[1,3,4,5,6,7,8]	*Leptonchus*[1,2,5,6]	*Thorneella*[5]
Cephalobus[1,2,3,4,5,6,7,8]	*Longidorella*[5]	*Thornia*[1]
Chrysonema[4]	*Lordellonema*[7]	*Timmus*[1,2,7]
Clarkus[3,4,5,6]	*Makatinus*[1,3,4,5,6,8]	*Trichodorus*[1,2,3,4,5,6,7,8]
Cobbonchus[1,2,3,4,6]	*Malenchus*[1,2,4,6,7]	*Trophotylenchulus*[1,2,3,4,5,6,7]
Coomansinema[1]	*Meloidogyne*[1,2,3,4,5,6,7,8]	*Tylencholaimus*[1,2,3,4,5,6,7,8]
Coslenchus[1,2,3,4,5,6,7,8]	*Mesodorylaimus*[1,2,3,4,5,6,7,8]	*Tylenchorhynchus*[1,2,3,4,5,6,7]
Crassolabium[1,2,3,4,5,6,7,8]	*Metateratocephalus*[5,6,7]	*Tylenchus*[1,2,3,4,5,6,7,8]
Criconemella[1,2,3,4,5,6,7,8]	*Microdorylaimus*[2,5,6,7]	*Tyleptus*[1,6,7]
Cryptonchus[1,2,6,7]	*Monhystera*[1,3,4,5,6,8]	*Wilsonema*[1,2,3,4,5,6,7,8]
Cylindrolaimus[1]	*Mononchus*[1,2,3,4,5,6,7]	*Xiphinema*[1,2,3,4,5,6,7,8]
Diphtherophora[1,2,3,4,5,6,7,8]	*Moshajia*[1,2,4,5,6,7,8]	
Diploscapter[8]	*Mydonomus*[1,2,3,5]	

[a]Numbers following the genera indicate that the nematode is associated with cerrado (1), cerradão (2), gallery forest (3), native grassland (4), coffee (5), eucalyptus (6), maize (7) and tomato (8).

todes accounted for 52% of total abundance, followed by bacterial feeders (35%), while predators, fungal feeders and omnivores each only accounted for less than 6%. *Helicotylenchus* (40% of total abundance), *Acrobeles* (15%), *Cephalobus* (7.6%), *Meloidogyne* (5.6%) and *Pratylenchus* (4.9%) dominated the nematode community.

Jorge (1999) studied the changes of nematode communities in two sequences of rotation (maize–fallow–soybean, and soybean–fallow–maize) in a savannah soil. Nematode abundance increased during crop cultivation and decreased during the fallow period. In both rotation sequences, there were higher numbers of genera found in the soybean planting period than in the maize

Table 8.6. Nematode families found in the eight land use systems studied in central Brazil.

Achromadoridae	Bunonematidae	Heteroderidae	Paratylenchidae
Actinolaimidae	Carcharolaimidae	Hoplolaimidae	Plectidae
Alaimidae	Cephalobidae	Leptolaimidae	Pratylenchidae
Anguinidae	Chromadoridae	Leptonchidae	Prismatolaimidae
Aphelenchidae	Criconematidae	Longidoridae	Qudsianematidae
Aphelenchoididae	Diphthorophoridae	Monhysteridae	Rhabditidae
Aporcelaimidae	Diplopeltidae	Mononchidae	Teratocephalidae
Aulolaimidae	Diploscapteridae	Mydonomidae	Trichodoridae
Bathyodontidae	Dorylaimidae	Nordiidae	Tylenchidae
Belondiridae	Halaphanolimidae	Nygolaimellidae	Tylencholaimidae
Belonolaimidae	Hemicycliophoridae	Panagrolaimidae	Tylenchulidae

Source: modified from Mattos (1999).

Table 8.7. Dominant genera in the nematode community in eight land use systems in central Brazil.

	Land use system							
Genera	Cerrado	Cerradão	Gallery forest	Grassland	Coffee	Eucalyptus	Maize	Tomato
Acrobeles					x		x	
Alaimus						x		
Aphelenchus								x
Aporcelaimellus	x					x	x	
Basirotyleptus	x			x				
Caloosia				x				
Coslenchus		x						
Discocriconemella	x			x				
Ditylenchus			x					
Dorylaimellus		x	x	x				
Dorylaimoides	x	x	x			x	x	
Enchodelus					x			
Helicotylenchus			x	x	x		x	x
Meloidogyne								x
Panagrolaimus					x	x		x
Rhabditis					x		x	
Rotylenchus								x
Trophotylenchulus		x	x					
Xiphinema	x	x				x		

x: present in samples.
Source: after Mattos (1999).

planting period. The populations of plant parasites were relatively stable and did not show differences in most of the sampling periods, whereas those of bacterial feeders were higher in the rainy season (cultivation period), but fungal feeders were higher in the dry season (fallow period). Soil water contents influenced the abundance of bacterial feeders positively ($r = 0.94$, $P \leq 0.05$) and those of fungal feeders negatively ($r = -0.95$, $P \leq 0.05$). There were few changes in nematode diversity and soil disturbance indices in the two rotation sequences. Also, some nematodes occupied functional roles in different steps of rotation, for example, the density of *Aphelenchoides* increased only in the fallow periods, that of *Acrobeles* increased only in soybean-cultivated cycles, but those of *Criconemella*, *Helicotylenchus* and *Meloidogyne* only in the maize-growing seasons.

Jorge (1999) studied the composition of nematode communities in combinations of two planting systems (tillage and no-tillage) and two soybean cultivars (cv. Cristalina and cv. EMGOPA). There was no difference in abundance among the four combinations, but differences in their community structures were observed. Populations of bacterial feeders were higher in no-tillage fields than in traditional tillage fields, whereas in fungal feeders it was the opposite. Populations of plant parasites were higher in cv. Cristalina than in cv. EMGOPA, but not different between the two planting systems. Predators and omnivores showed higher populations in the final period of soybean cultivation and also higher in cv. EMGOPA than in cv. Cristalina. Nematode diversity measured by generic richness, Shannon–Weaver's and Simpson's diversity indices were higher in the tillage system with cv. Cristalina, not with cv. EMGOPA. The MI, the mMI and the PPI indicated that soils were less disturbed in the no-tillage system than in the tillage system. Preference of some nematodes was shown: *Dorylaimellus* in no-tillage system with both cultivars, *Cephalobus* in no-tillage system with cv. EMGOPA, *Helicotylenchus* in no-tillage system with cv. Cristalina and *Aphelenchus* in tillage system with cv. Cristalina. The mosaic results mentioned above indicated that nematode communities were influenced significantly not only by two planting systems, but also by two soybean cultivars. The differences in nematode community structure resulted from the interaction of the four components.

Temporal and spatial samplings

Gomes *et al.* (2003) made temporal samplings (three fields sampled monthly for 12 months with a total of 180 samples), and identified 55 nematode taxa, as compared with 41 from the spatial sampling (23 fields within 1 month with a total of 115 samples). In general, the populations of the plant-parasitic nematodes *Helicotylenchus* spp., *Meloidogyne* spp. and *Pratylenchus* spp. increased during the soybean-growing season (April to June), declined by the end of the soybean cycle (July) and stabilized at low levels during the fallow period (August and September). Low populations were found during the soybean seedling period (November to January). When the soybean plants grew bigger in February, nematode populations increased again. Contrary to the plant feeders ($r^2 = -0.766$), the populations of fungal feeders became higher only by the end of the soybean-growing season (June to July), coinciding with the root decomposition period after harvest. Populations of bacterial feeders remained relatively stable for the whole year, only increasing in the rainfall period (December and January). Similarly, these populations of predators and omnivores also remained at low and stable levels during the whole year. Predators slightly increased their populations in July, corresponding to the population increase of fungal feeders. Thirteen ecological indices were applied to assess nematode communities in soybean plantations (Gomes *et al.*, 2003). The coefficients of variation of the values of seven indices (Simpson's and Shannon's indices, both evenness indices, MI, PPI and mMI) were less than 10% in both temporal and spatial samplings, indicating that these index values can characterize the nematode community in soybean fields. The indices of trophic diversity and species richness showed moderate variations (< 25%). The highest coefficients of variation (25–100%) were from the relative abundance of the Dorylaimida, and from the ratio of fungal feeders/bacterial feeders, which indicate different levels of soil disturbance and different pathways of biomass decomposition in different soybean fields, respectively. In this work, only one individual of *Criconemella* sp. was counted, confirming that criconematid nematodes are sensitive to agricultural practices, and hardly survive in intensively managed soybean fields.

Edaphic factors

Having studied the influence of soil type on communities of plant-parasitic nematodes associated with 'cerrado *sensu stricto*', 'cerradão', gallery forest, cultivated perennial and annual plants in the Brazilian central region, Huang and Cares (1995) found 28, 26, 19, 19 and 17 genera in dark-red latosols, yellowish red latosols, acrisols, gleysols and arenosols, respectively. The highest abundance occurred in acrisols, attributed to its better capacity of water retention, as compared with the two latosols and arenosols. The gleysols possessed a high capacity of water retention, but had very low levels of oxygen. Based on nematode density and frequency in cluster analysis, the arenosols, the yellowish red latosols and the dark-red latosols composed one group close to gleysols and distant to acrisols.

Nematode parameters have been significantly related ($P < 0.05$) to some edaphic factors. Cares and Huang (1991) studied six nematode genera (*Tylenchus*, *Coslenchus*, *Ditylenchus*, *Discocriconemella*, *Ecphyadophora* and *Trophotylenchulus*) in a savannah native 'cerrado' system, and found that there were high correlation coefficients ($r = 0.74$–0.82) between the curves of soil humidity and the curves of population fluctuation. The highest abundance was found during the rainy season from October to April, and the lowest during the dry season from May to September, indicating a great dependence of plant-parasitic nematodes on soil water content. In another study on savannah native and cultivated vegetation (Mattos, 1999; Mattos *et al.*, 2000a,b), soil water contents were related negatively to fungal feeders in coffee plantations ($r = -0.54$), but positively to plant parasites in tomato fields ($r = 0.40$). The genera *Dorylaimellus* in gallery forest and *Rotylenchus* in tomato fields were related positively to soil water contents ($r = 0.45$ and 0.60, respectively), whereas *Helicotylenchus* in gallery forest, *Rhabditis* in coffee and *Aphelenchus* in tomato plantations were related negatively to soil water ($r = -0.67$, -0.69 and -0.78, respectively). There was a negative relationship between generic richness and soil moisture ($r = -0.40$) and between amount of organic matter and abundance of fungal feeders ($r = -0.40$). The MI was related negatively to bacterial feeders in gallery forest ($r = -0.64$) and on *Eucalyptus* ($r = -0.84$), but positively to omnivores in cerradão ($r = 0.76$).

Silva (2000) studied the influence of phosphate (90 and 180 kg/ha of phosphate in the form of apatite and 90 kg/ha of superphosphate) on the nematode communities in cow grazed/non-grazed pasture of *Brachiaria decumbens* by itself, and associated with *Stylosanthes guianensis*. He found that these ecological indices differentiated all treatments with phosphate from the control without phosphate application, but they were not sensitive to the different levels and forms of phosphate. Nematode abundance was higher and generic richness was lower in the grazed area than in the non-grazed one. There were no differences in the other indices between the two areas. The application of phosphate increased the populations of bacterial feeders, omnivores and predators, and decreased those of plant parasites and fungal feeders, as compared with the control. In the total of 48 samples, there were positive correlations between soil pH in H_2O ($r = 5.1$–6.2) with the four diversity indices (Simpson's and Shannon's indices and their two evenness indices), and a negative correlation between Bongers' MI and the soil pH in H_2O, and between the MI and the concentrations of copper and of iron.

Jorge *et al.* (1998) investigated the influence of fire on nematode communities in native cerrado *sensu stricto* with four periods of soil sampling: 1 day before fire, 13, 48 and 90 days after fire. There were no differences in nematode abundance in the four sampling periods between fire and no-fire treatments. Before fire, plant parasites dominated nematode communities, followed by bacterial and fungal feeders, while predators and omnivores were low in number. After fire, the populations of plant parasites drastically decreased, whereas bacterial and fungal feeders significantly increased their populations, possibly due to the decrease of available root resources for plant parasites resulting from the effects of fire, and to the increase of microbes resulting from the elevation of soil water content, 48 and 90 days after fire, in the period of the rainy season.

Discussion

The higher diversity of plant-parasitic nematodes in 'terra firme' than in 'várzea' lands (Cares, 1984; Cares and Huang, 1991) is possibly related to plant diversity that, in native systems, was higher in 'terra firme' than in 'várzea'. Similarly, comparing the diversity of plant-parasitic nematodes between the sites in 'terra firme', nematode diversity was higher in the site with native forest and tropical fruit trees than in the site cultivated with annual and perennial crops, indicating that agricultural disturbance decreases nematode diversity. The nematode communities under undisturbed soil covered with natural vegetation, including the plant-parasitic taxa, mostly show high species diversity, contrasting with a low nematode abundance. Sustainability may be negatively affected when agricultural practices such as slash-and-burn and soil cultivation are introduced, since loss of plant diversity and increased soil disturbance contribute to a reduction in nematode diversity and to the dominance of remaining indigenous species or of new ones of agricultural importance introduced. But in the 'várzea', low nematode diversity in native vegetation, as compared with that in annual and perennial cropping fields, is related to low plant diversity due to flooding for several months each year. On the other hand, the sampled crop fields were in the upper parts of 'várzea', covered by a diversity of cultivated and weed plants, and not frequently reached by flood waters. The results of correspondence analysis (Lebardt *et al.*, 1982) indicated that nematode diversity attributed to differences between native and cultivated vegetation was responsible for more than 50% of total variations between native and cultivated vegetations, followed by the variations between the two ecotypes, 'várzea' versus 'terra firme', and last by the variations in nematode frequency in each type of sample (roots vs. soil, and upper layer vs lower layer). The even distribution of nematode species through the soil profile was attributed to the uniformity in soil moisture and root distribution in both soil profiles resulting from abundant rainfall in the sampled region.

Based on the results of two works (Mattos, 1999; Huang and Cares, 2000b), it is valuable to compare the differences in nematode communities between the two major Brazilian native ecosystems: the Amazonian forest and the savannah cerrado *sensu stricto*. Nematode abundance was higher in the Amazonian forest, whereas nematode biodiversity (measured by generic richness, Simpson's index, Shannon–Weaver's index and trophic diversity) was higher in cerrado *sensu stricto*. In terms of functional groups, the nematode community showed a higher abundance of plant parasites and a lower abundance of bacterial and fungal feeders in the Amazonian forest than in savannah vegetation. The values of MI, PPI and mMI were higher in forest than in cerrado soils. It is commonly agreed that high diversity leads to more species interactions within the food web; if so, the cerrado soil is expected to have higher sustainability than forest soils do. We suggest that rainfall is the major factor to contribute to these differences in the two ecosystems because it causes the loss of soil fertility. Three indices of soil disturbance indicated that there is less edaphic stress in forest than in cerrado soils. The major stress factor in cerrado is soil water content, which is different between dry and rainy seasons.

In nematode communities of most ecosystems, plant parasites and bacterial feeders are the two largest functional groups in soil, followed by fungal feeders, then predators and omnivores, frequently in small proportions (Freckman and Caswell, 1985). There were big changes in nematode trophic compositions from native to cultivated ecosystems in Amazonian and savannah regions. As compared with these cultivated systems, native forest and cerrado soils showed a high abundance of plant parasites (over 50% of total) and a low abundance of bacterial feeders (19.37% in cerrado and 10.6% in forest) (Mattos, 1999; Huang and Cares, 2000a,b), indicating that plant parasites played the most important role in soil organic decomposition in both native systems. Nematodes in the superfamily Criconematoidea dominated the group of plant parasites, especially *Discocriconemella* in forest, and *Trophotylenchulus* and *Discocriconemella* in cerrado *sensu stricto*. After conversion of native lands, plant parasites and bacterial feeders were about equally abundant, while, in some cases, bacterial feeders were even more abundant than plant parasites in 'cerrado' cultivated soils and comprised over 20% of total nematodes in Amazonian annual crop soils.

Ferris and Ferris (1974) found that there was less nematode diversity and greater abundance in intensively managed agroecosystems than in native ecosystems. In savannah, nematode biodiversity in four native vegetations ('cerrado', 'cerradão', gallery forest and 'campo limpo') was higher than in two annual crops (maize and tomato), but not different in perennial crops (coffee and *Eucalyptus*), whereas the total nematode abundance was much higher in tomato and maize systems than in the four native and two perennial systems (Mattos, 1999). In the Amazonian region, the nematode diversity in forest was higher than in pasture and in annual crops, but lower than in fallow and in agroforestry, indicating that the management of fallow and agroforestry had improved nematode diversity in soil (Huang and Cares, 2000a,b). These results show that nematode diversity decreases from native systems to cultivated systems, and that the changes are related to plant diversity. The changes in nematode abundance were obviously dependent on root mass in soil.

After conversion of native savannah to agricultural use, cultivated perennial plants, such as pine, *Eucalyptus*, coffee, and annual crops, such as maize, rice, soybean, horticultural plants, some plant parasites, such as *Helicotylenchus*, *Meloidogyne*, *Pratylenchus*, could adapt well, but the others, such as criconematids, could not.

Nematode communities in the two soybean planting systems were greatly differentiated by the abundance of bacterial feeders that was higher in the no-tillage system than in the tillage system. The value of Shannon–Weaver's diversity index was slightly higher in the no-tillage system than in the tillage system (2.20 vs. 2.15, respectively), but both showed no difference in Simpson's diversity index. The major differences between the two soybean cultivars were a high total abundance and abundance of bacterial feeders in cv. EMGOPA, but high abundance of plant parasites in cv. Cristalina. There were a few changes in nematode diversity and soil disturbance in the two sequences of rotation. Even though some results showed statistical differences in these ecological indices, the differences in these values were small. In the intensively managed lands, agricultural practices such as rotation, planting systems and cultivars influence nematode abundance and trophic structure more, but nematode biodiversity and soil disturbance indices less.

The nematode community structure is influenced not only by plant diversity, but also by soil chemical and physical factors. S.P. Huang and J.E. Cares (unpublished data) found that soil water content was related negatively to nematode diversity measured by Simpson's and Shannon–Weaver's indices. Bacterial feeders were abundant in the rainy season and fungal feeders in the dry season and after the final crop-growing cycle (Jorge, 1999; Mattos, 1999; Gomes *et al.*, 2003). The size of nematode populations was related to water drainage and soil oxygen, which were different in soils with different textures (Huang and Cares, 1995). In pasture, Silva (2000) found a positive relation between soil pH in H_2O (in the range 5.1–6.2) and the four diversity indices (Simpson's and Shannon-Weaver's indices and both evenness indices). Calcium was positively related to the abundance of bacterial feeders and omnivores, whereas pH-H_2O and copper were negatively related to the population of plant parasites. In an ecosystem, the more heterogeneity the soil shows, the more diversity the nematode community possesses. The influence of soil edaphic factors on nematode communities is still beyond our understanding.

Conclusions

Plant-parasitic nematodes were sensitive to agricultural practices in periodically flooded plains ('várzea') and non-flooded lands ('terra firme') in Amazon forest. Nematode diversity was related to vegetation diversity. In Amazonian forest and savannah native vegetation, plant parasites were the most important functional group in nematode communities. After the conversion of native lands, bacterial feeders increased their populations. In cerrado soils, nematodes were more abundant in annual cropping systems (tomato and maize), but more diverse in the native vegetation. Rotation (maize–fallow–soybean and soybean–fallow–maize) in savannah soils changed nematode abundance, but did little to nematode diversity and hardly reflected soil disturbance level. The combination of two planting systems (tillage and no-tillage) and two soybean cultivars (cv. Cristalina and cv. EMGOPA) did change trophic structure (which reflected the soil disturbance level), but not the nematode total abundance. Temporal samplings detected more rare nematodes than spatial ones did in soybean fields. In around a year, the period for population increases of plant parasites, fungal feeders and bacterial feeders frequently coincided with the period for plant growth, final plant cycle, and rainy season, respectively.

References

Adámoli, J., Macedo, J., Azevedo, L.G. and Netto, J.M. (1986) Caracterização da região dos cerrados. In: Goedert, W.J. (ed.) *Solos dos cerrado, tecnologia e estratégias de manejo*. Planaltina, CPAC/EMBRAPA, Nobel, pp. 33–74.

Azevedo, L.G. and Adámoli, J. (1988) Avaliação agroecológica dos recursos naturais da região dos cerrados. In: *Simpósio sobre o cerrado savanas: alimento e energia (6)*, Brasília, DF. 1982. Planaltina, CPAC/EMBRAPA, pp. 729–761.

Bernard, E.C. (1992) Soil nematode biodiversity. *Biology and Fertility of Soils* 14, 99–103.

Bongers, T. (1987) *De nematoden van Nederlands*. Pirola Schoorl Natuurhistorische Bibliotheek KNNV 46, Wageningen Agricultural University, The Netherlands.

Bongers, T. (1990) The maturity index: an ecological measure of environmental disturbance based on nematode species composition. *Oecologia* 83, 14–19.

Bongers, T. and Bongers, M. (1998) Functional diversity of nematodes. *Applied Soil Ecology* 10, 239–251.

Cares, J.E. (1984) Fauna fitonematológica de várzea e terra firme nas proximidades de Manaus, AM. Tese MSc. Universidade de Brasília, Brasília, Brazil.

Cares, J.H. (1990) Fauna fitonematológica dos cerrados virgem e cultivado. Tese MSc. Universidade de Brasília, Brasília, Brazil.

Cares, J.H. and Huang, S.P. (1991) Nematode fauna in natural and cultivated cerrados of central Brazil. *Fitopatologia Brasileira* 16, 199–209.

Cares, J.E. and Huang, S.P. (2000) Taxonomia atual de fitonematóides: chave sistemática simplificada para gêneros-Parte I. *Revisão Anual de Patologia de Plantas* 8, 185–223.

Coleman, D.C., Cole, C.V. and Elliott, E.T. (1984) Decomposition, organic matter turnover, and nutrient dynamics in agroecosystems. In: Lowrance, R., Stinner, B.R. and House, G.J. (eds) *Agricultural Ecosystems Unifying Concepts*, pp. 83–104.

Eiten, G. (1978) Delimitation of the cerrado concept. *Vegetation* 36, 169–178.

Eiten, G. (1979) Formas fisionômicas do cerrado. *Revista Brasileira de Botânica* 2, 139–148.

Eiten, G. (1984) Vegetation of Brasília. *Phytocoenologia* 12, 217–292.

Elliot, C.A. (1990) Diversity indices. In: *Principles of Managing Forests for Biological Diversity*. Prentice-Hall, Englewood Cliffs, New Jersey.

Ettema, C.H. and Bongers, T. (1993) Characterization of nematode colonization and succession in disturbed soil using the maturity index. *Biology and Fertility of Soils* 16, 79–85.
Ferris, V.R. and Ferris, J.M. (1974) Inter-relationships between nematode and plant communities in agricultural ecosystems. *Agroecosystems* 1, 275–299.
Ferris, H., Venette, R.C. and Lau, S.S. (1996) Dynamics of nematode communities in tomatoes grown in conventional and organic farming systems, and their impact on soil fertility. *Applied Soil Ecology* 3, 161–175.
Ferris, H., Bongers, T. and Goede, R.G.M. (2001) A framework for soil food web diagnostics: extension of the nematode faunal analysis concept. *Applied Soil Ecology* 18, 13–29.
Fortuner, R., Geraert, E., Luc, M., Maggenti, A.R. and Raski, D.J. (1988) A reappraisal of tylenchina (Nemata). *Extraction of Revue of Nematologie*, ORSTOM, France 10, 127–232, 409–444; 11, 159–188.
Freckman, D.W. and Caswell, E.P. (1985) The ecology of nematodes in agroecosystems. *Annual Review of Phytopathology* 23, 275–296.
Freckman, D.W. and Ettema, C.H. (1993) Assessing nematode communities in agroecosystems of varying human intervention. *Agriculture, Ecosystems and Environment* 45, 239–261.
Freckman, D.W. and Huang, S.P. (1998) Response of the soil nematode community in a shortgrass steppe to long-term and short-term grazing. *Applied Soil Ecology* 9, 39–44.
Garrido, W.E., Azevedo, L.G. and Jarreta-Junior, M. (1982) *O clima da região dos cerrados em relação à agricultura.* Planaltina, EMBRAPA – CPAC, Circular Técnica 9.
Gomes, G.S., Huang, S.P. and Cares, J.E. (2003) Nematode community, trophic structure and population fluctuation in soybean fields. *Fitopatologia Brasileira* 28, 258–266.
Goodland, R.J.A. (1970) Plants of the cerrado vegetation of Brazil. *Phytologia* 20, 57–80.
Goseco, C.G., Ferris, V.R. and Ferris, J.M. (1974a) Revisions in Leptonchoidea (Nematoda: Dorylaimida) Leptonchus, Proleptonchus, Funaria and Meylis n. gen. in Leptonchidae, Leptonchinae. Purdue University Experimental Station Research Bulletin 911.
Goseco, C.G., Ferris, V.R. and Ferris, J.M. (1974b) Revisions in Leptonchoidea (Nematoda: Dorylaimida) Dorlaimoides in Dorylaimoididae, Dorylaimoidinae; Calolaimus and Timmus n. gen. in Dorylaimoididae, Calolaimidnae; Miranema in Miranematidae. Purdue University Experimental Station Research Bulletin 941.
Huang, S.P. and Cares, J.H. (1995) Community composition of plant-parasitic nematodes in native and cultivated cerrados of central Brazil. *Journal of Nematology* 27, 237–243.
Huang, S.P. and Cares, J.E. (2000a) Comparação da comunidade de nematóides em cinco sistemas de uso da terra na região amazônica. *Fitopatologia Brasileira* 25, 337 (abstract).
Huang, S.P. and Cares, J.E. (2000b) The nematode community as a bioindicator to characterize five different land use systems in two Brazilian tropical states, Rondônia and Acre. Tropical Soil Biology and Fertility Programme, TSBF Report 1997–1998, Nairobi, pp. 88.
Huang, S.P. and Mattos, J.K.A. (2000) Composição comunidades de nematóides em cerrado da região central do Brasil. *Fitopatologia Brasileira* 25, 337–338.
Huang, S.P., Pereira, A.C., Dristig, M.C.G. and Souza, R.M. (1991) Ocorrência de *Meloidogyne javanica* e *M. arenaria* em árvores silvestres no cerrado virgem do Brasil central. *Fitopatologia Brasileira* 16, 37 (abstract).
Huang, S.P., Freire, H.C.A. and Cares, J.E. (1996) Grupos composicionais e tróficos dos nematóides associados à sucupira branca (*Pterodon pubescens*) em cerrado nativo. *Fitopatologia Brasileira* 21, 156–160.
Huang, S.P., Cares, J.E. and Vivas, J.P. (1998) Nematode biodiversity of five different land use systems in two Brazilian tropical states, Rondônia and Acre. *Fitopatologia Brasileira* 23, 305 (abstract).
Hyvönen, R. and Persson, T. (1990) Effects of acidification and liming on feeding groups of nematodes in coniferous forest soils. *Biology and Fertility of Soils* 9, 205–210.
Jairajpuri, M.S. and Ahmad, W. (1992) *Dorylaimida: Free Living, Predaceous and Plant-Parasitic Nematodes.* Oxford & IBH/E.J. Brill, New York.
Jorge, C.L. (1999) Comunidades de nematóides em sistemas de plantios direto e convencional de soja, e em rotação com soja e milho. Tese MSc. Universidade de Brasília, Brasília, Brazil.
Jorge, C.L., Huang, S.P. and Cares, J.E. (1998) Influência do fogo na população e na estrutura trófica de nematóides do cerrado nativo. *Fitopatologia Brasileira* 23, 305 (abstract).
Korthals, G.W., Goede, R.G.M., Kammenga, J.E. and Bongers, T. (1996) The maturity index as an instrument for quick assessment of soil pollution. In: van Straalen, N.M. and Krivolutsky, D.A. (eds) *Bioindicator Systems for Soil Pollution.* Kluwer, Dordrecht, The Netherlands, pp. 85–93.
Lebardt, L., Morineau, A. and Fenelon, J.P. (1982) *Traitment des données estatistiques,methodes et programe*, 2nd edn. Dunod, Paris.

Magurran, A.E. (1988) *Ecological Diversity and Its Measurement*. Cambridge University Press, Cambridge, UK.

Mattos, J.K.A. (1999) Caracterização das comunidades de nematóides em oito sistemas de uso da terra nos cerrados do Brasil central. Tese DSc. Universidade de Brasília, Brasília, Brazil.

Mattos, J.K.A., Huang, S.P. and Pimentel, C.M. (2000a) Avaliação ecológica em comunidades de nematóides em oito sistemas de vegetação nos cerrados do Brasil central. *Fitopatologia Brasileira* 25, 338 (abstract).

Mattos, J.K.A., Huang, S.P. and Pimentel, C.M. (2000b) Correlações entre índices descritores da comunidade de nematóides e em relação à umidade de solo nos cerrados do Brasil central. *Fitopatologia Brasileira* 25, 338 (abstract).

McSorley, R. (1997) Relationship of crop and rainfall to soil nematode community structure in perennial agroecosystems. *Applied Soil Ecology* 6, 147–159.

Neher, D.A. and Campbell, C.L. (1996) Sampling for regional monitoring of nematode communities in agricultural soils. *Journal of Nematology* 28, 196–208.

Niblack, T.L. and Bernard, E.C. (1985) Nematode community structure in dogwood, maple and peach nurseries in Tennessee. *Journal of Nematology* 17, 126–131.

Norton, D.C. (1978) *Ecology of Plant Parasitic Nematodes*. John Wiley & Sons, New York.

Pielou, P.C. (1977) *Mathematical Ecology*. John Wiley & Sons, New York.

Scerne, R.M.C., Santos, A.O.S., Santos, M.M. and Neto, F.A. (1996) *Aspectos agroclimáticos da região de Ouro Preto D'Oeste, RO*. CEPLAC/SUPOR, Ministério da Agricultura e do Abastecimento, Belém, Pará, Brazil, Boletim Técnico 13.

Schmitt, D.P. and Norton, D.C. (1972) Relationships of plant parasitic nematodes to sites in native Iowa prairies. *Journal of Nematology* 4, 200–206.

Silva, R.O.C. (2000) Influência dos fatores físicos e químicos do solo nas comunidades de nematóides em campo de pastagem. Relatório de pesquisa. Universidade de Brasília, Brasília, Brazil.

Souza, R.M., Dolinski, C.M. and Huang, S.P. (1994) Survey of *Meloidogyne* spp. in native cerrado of Distrito Federal, Brazil. *Fitopatologia Brasileira* 19, 463–465.

Todd, T.C. (1996) Effects of management practices on nematode community structure in tallgrass prairie. *Applied Soil Ecology* 3, 235–246.

Wasilewska, L. (1989) Impact of human activities on nematode communities in terrestrial ecosystems. In: Clarholm, M. and Bergstroem, L. (eds) *Ecology of Arable Land*. Kluwer, Dordrecht, The Netherlands, pp. 123–132.

Wasilewska, L. (1994) The effect of age of meadows on succession and diversity in soil nematode communities. *Pedobiologia* 38, 1–11.

Yeates, G.W. (1979) Soil nematodes in terrestrial ecosystems. *Journal of Nematology* 11, 213–229.

Yeates, G.W. (1994) Modification and quantification of the nematode maturity index. *Pedobiologia* 38, 97–101.

Yeates, G.W., Bongers, T., Goede, R.G.M., Freckman, D.W. and Georgieva, S.S. (1993) Feeding habits in soil nematode families and genera – an outline for soil ecologists. *Journal of Nematology* 25, 315–331.

9 Diversity of Microfungi in Tropical Soils

L.H. Pfenning and L.M. de Abreu
Departamento de Fitopatologia, Universidade Federal de Lavras, 37200-000 Lavras MG, Brazil, e-mail: ludwig@ufla.br

Introduction

Destruction of natural habitats in the tropics by expanding agriculture and by uncoordinated exploitation of natural resources like timber is increasing. A more reasonable use of agricultural land and the regeneration of degraded areas represents one of the most imperative challenges for the protection of natural resources, including the diversity of microorganisms, their ecological relationships and processes they mediate. Information on structure and composition of fungal populations such as the identity and frequency of plant-pathogenic fungi and their antagonists can provide insight into the soil community stability or level of interference on soil biota in original forest and disturbed areas.

This chapter reviews current knowledge of diversity of soil microfungi in the tropics by focusing on Brazilian ecosystems and evaluates the possible influence of land use on fungal communities. As genetic resource collections provide access to strains required for teaching, research and industrial purposes, the lack of depository authorities for microbial germplasm is identified as one of the major constraints for scientific and technological development of tropical mycology. Methods used in the past and present for the assessment of soil microfungi diversity are discussed, since techniques are not well established. Assessments focusing on predictor sets, which include phytopathogens and their antagonists, are indicated as a reasonable alternative for estimating diversity of soil microfungi if a long-term study is not feasible. Moreover, applications of molecular tools for studies of soil fungal communities are also considered. It is becoming clear that in the future one must rely on methodological approaches coupling culture-based techniques and molecular taxonomic assays to study the major biogeochemical processes occurring in soil ecosystems in order to gain better understanding of fungal communities in soil environments.

Information about microbial interactions in the soil and rhizosphere in tropical ecosystems is still quite sparse, though their importance for ecological equilibrium is uncontested. Amongst the overall microbial components of ecosystems we can find a highly diverse community of different groups of fungi, including the so-called microfungi. Microfungi is only a practical working term, widely used by mycologists, without any taxonomic or phylogenetic implications. This group comprises, by convenience, zygomycetes, ascomycetes with fruiting bodies smaller than 2 mm and conidial states of ascomycetes, formerly called fungi imperfecti or deuteromycetes. Other functional groups of soil-inhabiting

 Soil Biodiversity in Amazonian and Other Brazilian Ecosystems (eds F.M.S. Moreira *et al.*)

fungi such as macromycetes or mycorrhiza-forming fungi require distinct methodologies for their study. One group, the arbuscular mycorrhiza fungi, is covered in Chapter 10, this volume.

Soil microfungi represent the major functional groups responsible for plant diseases, decomposition of organic material and recycling of plant nutrients (Table 9.1). Information on structure and composition of fungal communities and on the identity and frequency of pathogenic fungi and their antagonists is crucial in agriculture and plant disease management (Kennedy and Smith, 1995; Giller *et al.*, 1997; Altieri, 1999). The relevance of species lists obtained from sites under different vegetation types and different land use was thoroughly discussed during the 1960s and 1970s. If they are based on reliable identification of species and provide information on frequency, they may turn out to be a valuable tool for the characterization and monitoring of fungal communities and their influence in the main biogeochemical processes (Christensen, 1981; Kjøller and Struwe, 1982).

Fungi are one of the most species-rich of all groups of organisms, except insects. The working figure of 1.5 million species of fungi is based on several lines of evidence, now generally accepted (Hawksworth, 1991, 2001; May, 1991; Hammond, 1992). The diversity of fungi in tropical habitats is generally poorly explored. Most of the studies are pioneer work, usually regionally and thematically oriented. Therefore, assessments of the diversity of tropical soil microfungi and investigations concerning their importance for sustainable agriculture are likely to receive much attention over the next decades (Hawksworth and Rossman, 1997; Hyde, 1997a,b; Mueller *et al.*, 2004).

Microfungi are also sought after for the production of novel bioactive metabolites and genes for biotechnological development. In bioprospection programmes, soil represents one of the most important sources of genetic diversity, a fact of outstanding importance for ecological studies and technological aspects (Bull *et al.*, 1992; Fox, 1993; Bills, 1995; Wildman, 1997; Bills *et al.*, 2002). Moreover, there are many basic scientific aspects in the exploitation of biodiversity and the discovery of new fungal species, especially in ecosystems scarcely studied. Studies with such an emphasis will add valuable information to systematics, phylogeny and anamorph–teleomorph relationships in the kingdom Fungi (Hennebert, 1995).

Soil Microfungi – More than a Functional Group

Soil must be defined as a habitat or an ecosystem rather than a substrate. This fact brings about problems with definition and methodology, since soil represents a complex mixture of inorganic and organic fractions with water and living organisms. The

Table 9.1. Main groups of soil microfungi.

Plant pathogens	Saprotrophs and antagonists	Insect pathogens
Ascomycetes		
Cylindrocladium, Fusarium, Lasiodiplodia, Sclerotinia, Verticillium	*Clonostachys* (*Gliocadium*), *Coniothyrium, Penicillium, Talaromyces, Trichoderma*	*Beauveria, Metarhizium, Paecilomyces, Verticillium*
Basidiomycetes		
Rhizoctonia, Sclerotium		
Straminipiles (Oomycota)		
Pythium, Phytophthora	*Pythium oligandrum*	

organic fractions are composed of fresh and decaying plant material in different stages of decomposition, living roots, exudates and microorganisms, small invertebrates and their gut contents. For this reason the soil harbours a considerable part of the fungal diversity and no sound estimate of numbers of soil fungal species exists (Hawksworth, 1991; Hawksworth and Rossman, 1997).

Soil microfungi play a key role in decomposition processes that mineralize and recycle plant nutrients (Wainwright, 1988; Lodge, 1993; Beare *et al.*, 1997). In the soil environment, fungi interact with a complex microbial community, including bacteria, actinomycetes and small invertebrates. Saprophytes have a limited specificity for substrates, for instance, zygomycetes that use simple carbohydrates or ascomycetes that may decompose cellulose and hemicellulose (Domsch *et al.*, 1980; Zak and Visser, 1996; Lodge, 1997).

In agroecosystems, plant pathogens and their antagonists are specifically important. Plant pathogens act in the soil, rhizosphere and plant shoots, causing yield losses. They may be specific, but most of them attack a wide range of host plants. Suppressiveness of soils to plant pathogens may be intrinsic, but can also be maintained or augmented by specific agricultural practices like incorporation of organic matter, cover plants and crop diversity. Biological elements have been identified as primary factors in disease suppression (Chet and Baker, 1980; Schneider, 1984; Mazzola, 2002, 2004). It has been experimentally shown that introduction of specific antagonists like *Trichoderma* spp. or *Coniothyrium minitans* can reduce the incidence of a variety of soil-borne diseases (Whipps *et al.*, 1993) and the antagonistic properties of the soil microfungal community is as yet poorly exploited.

Considering the decomposing role, it is important to mention that fungi are responsible for the degradation of xenobiotics and organic pollutants introduced in the soil (Bordjiba *et al.*, 2001; Barratt *et al.*, 2003; Da Silva *et al.*, 2003). Fungi are also an important part of the food chain within the soil environment, mainly for the soil-inhabiting mesofauna (Bonkowski *et al.*, 2000). Maintenance of soil microfungi diversity should therefore directly benefit sustainable agricultural production by providing available nutrients, better physical structure of the soil, and natural biocontrol of soil-borne plant pathogens.

Current Knowledge of Diversity of Soil Microfungi in the Tropics

At present, there are no investigations on soil microfungi with exact species identification that would permit the drawing of a reliable picture of communities present in forest and agricultural soil. The use of the dilution plate method or Warcup plates in most studies conducted to date introduces a considerable bias by favouring the heavily sporulating and rapidly growing species. For this reason, a comparative interpretation of results is hardly possible or useful, but this technique is still widely used.

In his classical study on tropical soil microfungi, Farrow (1954) reported 135 fungal species, most of them ascomycetous anamorphs, which were recovered from 31 soil samples from Costa Rica and Panama using the dilution plate technique. The most frequently isolated genera were: *Penicillium*, *Aspergillus*, *Fusarium*, *Trichoderma*, *Chaetomium* and *Cunninghamella*. From India, preliminary results are available from soil fungi isolated from forest soils (Rama Rao, 1970; Agarwal and Chauhan, 1988) and cultivated soils (Jabbar Miah *et al.*, 1980; Joshi and Chauhan, 1982). A comparative study of species groups in soils under forest and an oil palm plantation was conducted in Malaysia. Species of the genera *Trichoderma* and *Aspergillus* were found to be most common, whereas under the plantation an increase in the frequency of *Fusarium* species was observed (Varghese, 1972). Studies on soil fungi under planted forest in South Africa (Eicker, 1969), pasture (Ogbonna and Pugh, 1982) groundnut (McDonald, 1969) and cowpea (Odunfa and Oso 1979) in Nigeria indicate that certain species have a common occurrence in several geographical regions.

Cunninghamella spp., *Gongronella butleri*, *Chaetomium* spp., *Aspergillus* spp., *Fusarium* spp. or *Paecilomyces lilacinus* were reported from almost all investigated sites.

Gochenaur (1970) studied soil microfungi in about 30 different soil samples from Peru using dilution plates. Most common were species of the genera *Absidia*, *Mucor*, *Ulocladium*, *Trichoderma* and *Fusarium*, but the relative frequency of these fungi was not reported. In the Bahamas, under *Palm* and *Casuarina* trees, 60 and 20 species, respectively, were recovered and identified; *Aspergillus niger*, *Penicillium chrysogenum* and *Cladosporium cladosporioides* were the most frequent ones (Gochenaur, 1975). In the South American subcontinent, but outside the subtropical area, diversity of soil microfungi was investigated in Argentina and Uruguay using the dilution plate technique. Under an undisturbed and a disturbed forest site, 49 and 37 species, respectively, were detected (Cabello and Arambarri, 2002). The most common species isolated were: *Acremonium* sp., *Aspergillus ustus*, *Doratomyces stemonitis*, *Fusarium oxysporum*, *Fusarium solani*, *Clonostachys rosea* and *P. lilacinus* (cited as '*Penicillium lilacinus*'). In Uruguay, preliminary surveys revealed that under primary forest and in different wetland sites, the most common genera were: *Aspergillus*, *Penicillium*, *Eupenicillium*, *Talaromyces*, *Fusarium* and *Gongronella* (Bettucci and Roquebert, 1995; Bettucci *et al.*, 2002). In Costa Rica, an All Taxa Biodiversity Inventory (ATBI) initiative has been ongoing since the 1990s with the objective of assessing diversity of all groups of fungi (Rossman *et al.*, 1998).

In Brazil, some information on microfungi from soil and litter is available. A survey of litter fungi in the region of Manaus-AM was conducted by Katz (1981). During investigations on the infestation of soils in the surroundings of Manaus with *Pythium* spp., Lourd *et al.* (1986) found that these species occur in soil under primary forest with almost the same incidence as in cultivated soils. Crop plants cultivated in forest soils, nevertheless, showed lower disease incidence in biotests where inoculation with the pathogen was made. This indicates that microbial interactions are responsible for the bioprotection phenomenon (Lourd and Bouhot, 1987). The occurrence of the genera *Aspergillus* and *Penicillium* in northern states of Brazil, Maranhão and Pará has been documented by Batista *et al.* (1967a,b). A more comprehensive review on studies of soil-, rhizosphere- and litter-inhabiting microfungi conducted in Brazil can be found in Pfenning (1996, 1997). In an investigation of populations of microfungi in soil and the rhizosphere in the eastern Amazon, 134 species were identified from a total of 830 isolates. Ascomycetous anamorphs were represented by 96 species, followed by sexual ascomycetes with 20 species and 17 species of zygomycetes. The proportion of singletons, species that were isolated only once, was relatively high, representing 38% of the species isolated from the forest stand, and between 30% and 45% of the species recovered from cultivated areas (Pfenning, 1993). A preliminary evaluation of the results from the Atlantic rainforest in south-east Brazil suggests that the fungal community in the rhizosphere shows a low specificity with regard to host plants, similar to the results obtained from the soils in the Amazon region. It seems that the hypothesis of the specificity of rhizosphere fungi cannot be confirmed as otherwise stated (Pfenning, 1997).

In the tropics probably only one site has been thoroughly studied with regard to soil microfungi. From an experimental area investigated since 1979 in the Tai National Park in the Ivory Coast, about 250 species of microfungi were reported from soil and about 300 species from litter samples (Rambelli *et al.*, 1983, 1984; Maggi *et al.*, 1990; Maggi and Persiani, 1992). One of the conclusions is that disturbances due to shifting cultivation may cause sudden changes in frequency of certain soil fungi. However, fallow leads to a relatively rapid recovery of the pre-existing community (Persiani *et al.*, 1998). Investigations from the Ivory Coast and Brazil reached the same conclusions as to some frequent species and the observation that the impact on soil fungal communities due to slash-and-burn is more quantitative than qualitative. Some of the frequent species include the morphotype *Acremonium strictum*, *Gliocephalotrichum bulbilium*, *Gonytrichum macrocladum*, *Metarhizium anisopliae*, *P. lilacinus*, species of *Penicillium*, *Trichoderma harzianum*, *Trichoderma hamatum* and the zygomycetes *Cunninghamella elegans* and *G. butleri*. It is worth mentioning that several of the rare species could be found in both countries, e.g. *Aspergillus candidus*, *Aspergillus janus*, *Farrowia longicollea*, *Scopulariopsis carbonaria*, *Thozetella tocklaiensis*, *Triangularia batistae* and the zygomycetes *Absidia cylindrospora* and *Absidia spinulosa*.

Soil fungal communities both in the tropics and in the temperate regions comprise many cosmopolitan species. Additionally, species with typically tropical distribution occur with lower frequencies. These rare species may play an important role in the characterization of fungal communities, depending on local biotic and abiotic properties of soils. There is strong evidence that more than half of the fungi isolated from soil are actually litter-decomposing fungi, being introduced by activities of the insects, earthworms and other components of the mesofauna. Still, in many tropical forest soils the litter layer and the humus horizon are not well differentiated. As already discussed, this fact brings about restraints in defining what soil fungi are. Although it was frequently stated that *Aspergillus* spp. are common in tropical soils (Christensen and Tuthill, 1985), investigations using soil-washing and particle filter techniques do not confirm this observation (Pfenning, 1997, and unpublished results). In the northern hemisphere it may be possible to predict the vegetation type and region from where samples have been taken in order to isolate the most frequent fungi in the soil (Bills *et al.*, 2004). This affirmation is based on the assumption that a strong correlation exists between the fungal community and the vegetation type and the edapho-climatic conditions. The same is certainly not true for tropical regions. What has already been achieved with regard to soil fungal communities in regions with temperate climates has still to be done for tropical regions. As tropical forests are more species-rich than temperate climate vegetation types, several areas must still be investigated in the tropics with the use of adequate isolation techniques. Only in Brazil, three of the main biomes, the savannah-like vegetation ('cerrado'), the Atlantic rainforest and the forest in the central Amazon region, are characterized by a quite different plant species composition and edapho-climatic conditions.

Diversity of Microfungi and Land Use

For regions with a temperate climate, a series of classical studies are available that compare species diversity and frequency in soils under different land uses and human impact, like fire (Widden and Parkinson, 1973; Christensen, 1981; Widden, 1986; Brodie *et al.*, 2003). The vegetation types investigated, like pine forests or grasslands, are generally uniform and the soil horizons clearly defined. As the type of litter influences the composition process by fungi it is difficult to predict a specific community in soils under tropical forests (Kjøller and Struwe, 1982). The diversity of substrate may contribute to the fungal diversity in tropical regions.

With regard to cultivated soils in the American tropics, areas planted with sugarcane and bananas received more attention. In central America, communities of microfungi associated with the soil rhizosphere of bananas are characterized by a high frequency of *Aspergillus terreus*, *Cunninghamella* spp., *F. solani*, *P. lilacinus*, *Penicillium purpurogenum*, and others (Goos, 1960, 1963; Goos and Timonin, 1962). Studies on soil and rhizosphere fungi of sugarcane are known from Jamaica (Robison, 1970), Trinidad (Mills and Vlitos, 1967) and Brazil (Sanhueza and Balmer, 1985; Santos *et al.*, 1989). Species of the genera *Aspergillus*, *Fusarium*, *Penicillium* and *Trichoderma* were the most common in these plantations. A comparative survey of the diversity of soil microfungi was conducted in an area virtually representing the real land use in the eastern Amazon region. Diversity of soil fungi was investigated in a plantation of cacao, annual crops followed by fallow, pasture and a relic area of primary forest, using a soil-washing technique. The composition of dominant species varied considerably in all investigated stands. In the cultivated sites the proportion of dominant species in relation to the total number of isolates varied between 60% and 70%. Only two of the dominating species were recovered from all sites: *G. butleri* and *T. hamatum*. The proportion of dominant species in the forest stand was considerably lower than in the cultivated areas, indicating higher species diversity in this site. The introduction of agricultural plants in forest-cleared areas resulted in a specific increase of potential plant-parasitic species like *Fusarium*. The data showed that the sites differ primarily in a quantitative aspect composition of species and less in the qualitative composition of species (Pfenning, 1993, 1995, 1997; Table 9.2).

The impact of agricultural practices on some soil-borne plant pathogens in Indonesia and the importance of antagonists were reviewed by Gafur and Darmono (1999). In a review of soil-borne and rhizosphere plant pathogens in Mexico, Rodrigues-Guzman (2001) reported *Phytophthora*, *Pythium*, *Rhizoctonia*, *Fusarium*, *Verticillium* and *Botrytis* (cited as *Phymatotrichum*) as the most frequent genera. More comprehensive studies including the pathogen, the host and the environment are imperative for more rational and sustainable disease management. Management practices like crop rotation and incorporation of organic matter can enhance natural suppressiveness of agricultural soils and are therefore an important component of environmentally sustainable systems for plant production (Lodge, 1997; Mazzola, 2004).

Methods for the Assessment of Soil Fungal Communities: Culture-based Procedures

The classical microbiological procedures for studying soil fungi are based on isolation of microbial propagules or actively growing hyphae from soil and growing them in axenic culture media for further identification and quantification. Other methodologies focus on the analysis of fungal activity and their role in biogeochemical processes in soil environments. Soil and litter fungi, known as one of the most important groups of decomposer organisms, are often still referred to as microbial biomass (Swift and Bignell, 2001). For these purposes, methods that rely on the analysis of soil microbial biomass, soil respiration, nitrogen cycling and fungal fatty acids analysis, or direct observations of actively growing mycelia on soil particles, have been applied (Widden and Parkinson, 1973; Anderson and Ingram, 1989; Houston *et al.*, 1998; Brodie *et al.*, 2003).

Table 9.2. Species composition of soil microfungi communities under different land use in eastern Amazon.

	Forest	Cacao	Fallow	Pasture
Number of species	76	58	60	47
Number of species found exclusively at one site	35	17	14	14
Percentage of the 12 most common species	50	63	60	70
Shannon index	3.9	3.5	3.6	3.3

Source: Pfenning (1993).

However, these methods alone give no information about the fungal species involved in these processes. Isolation and identification procedures are generally required for a better understanding of soil fungal community structure and function (Brodie *et al.*, 2003). Methods used for the isolation of soil, rhizosphere and rhizoplane microfungi are fully revised by several authors (Frankland *et al.*, 1990; Gray, 1990; Gams, 1992; Cannon, 1996; Davet and Rouxel, 2000; Bills *et al.*, 2004). A general overview on methods for studying soilborne plant-pathogenic fungi was given by Singleton *et al.* (1992). Since the use of soil plates became widespread, considerable progress was made using washing techniques, less selective culture media and additives that reduce growth of certain groups of fungi. Some of the common methods for soil fungi assessment based on isolation procedures are presented and discussed. Moreover, principles and applications of molecular tools for soil fungal community studies are also considered.

Soil dilution plate method

The soil dilution plate method is most commonly used as a generic method for isolation and quantitative estimation of both bacteria and fungi. The technique is very simple and several modifications have been described. Basically, a known amount of soil is suspended in sterilized water, making a 10% suspension, which is then agitated for a few minutes. From this suspension a series of tenfold dilutions is prepared until the desired final dilution is achieved. A final dilution factor of 10^{-4} or 10^{-5} has been considered suitable for isolation of fungi (Dhingra and Sinclair, 1985). Aliquots of the final dilution are evenly distributed onto Petri dishes containing agar media, generally amended with antibiotics such as cloramphenicol, streptomycin or penicillin for inhibition of bacterial growth. Quantitative measurements can be achieved by multiplying the mean of colony forming units (CFU) plate count by the dilution factor employed, which gives the estimate of the number of fungal propagules per gram of soil.

It is of general concern that the dilution plate method shows a bias towards fungi that are capable of producing large amounts of spores and grow very fast on rich culture media. Therefore the diversity of fungi that usually exist as actively growing mycelia in the soil and have a low ability to compete with fast-growing species when in axenic media is underestimated by this technique (Tsao *et al.*, 1983; Bååth, 1988). However, when a soil fungal community is subjected to stress such as drought, the changes in community population dynamics, with the selection towards slow-growing melanous ascomycetes, can be assessed by the soil dilution plate (Grishkan *et al.*, 2003). The results can be improved by the use of several different carbon sources in isolation media, coupled with strategies that permit the recovery of low-sporulating fungi. The latter could be accomplished by the use of more viscous dilution media, which permit the transport of mycelium-containing soil particles through the dilution series along with the spores, or coupling the soil-washing procedure with the dilution plate technique (Dhingra and Sinclair, 1985; Petrovic *et al.*, 2000).

Soil-washing technique

The basic purpose of this technique is to eliminate the excess of dormant spores from soil samples, favouring the isolation of low-sporulating, actively growing mycelia. The soil can be washed several times with sterilized distilled water in glass recipients, always discharging the supernatant that contains large amounts of spores. After washing, soil is subjected to the soil dilution plate method as described above (Dhingra and Sinclair, 1985).

In another more common approach, the soil is washed and the component particles are separated in a series of sieves (Fig. 9.1). The particles retained on the smallest mesh sieve are transferred to Petri dishes (Thorn

Fig. 9.1. Methods for isolation of soil microfungi: (a–c) combined soil-washing and particle filter technique; (d) baiting technique for isolation of *Pythium*.

et al., 1996; Tiunov and Scheu, 2000). The size of soil particles is inversely proportional to the number of isolates yielded by each particle; thus the use of sieves with smaller mesh openings, and consequently the plating of smaller soil particles on to culture media, may favour the recovery of slow-growing fungi from those one-fungus-yielding soil particles (Bååth, 1988). The same finding was made for endophytic fungi (Gamboa *et al.*, 2002). A soil-washing apparatus containing a series of nylon meshes can also be used instead of sieves (Widden and Parkinson, 1973; Gams *et al.*, 1998).

Selective media and baiting techniques

For an improved isolation of target groups or species of fungi from soil, selective media are routinely used. The selective media can contain carbon sources that are preferably metabolized by some physiological groups of target fungi or they may be amended with chemicals that inhibit the growth of undesirable organisms. A vast number of selective media have been developed for the isolation of several genera of ascomycetes, basidiomycetes and oomycetes from soil, in particular those that contain plant-pathogenic species (Masago *et al.*, 1977; Tsao *et al.*, 1983; Dhingra and Sinclair, 1985; Sneh *et al.*, 1991; Thorn *et al.*, 1996).

The selective isolation of soil fungi can also be accomplished by the use of baits that are primarily colonized by specific physiological groups of fungi, facilitating their isolation. In most cases the baiting tissue is incubated with a soil sample for a few days and then transferred to a selective agar medium for the isolation of desired fungi; the colonization of the bait by a target fungus can also be accomplished by direct microscopic observations (Gams *et al.*, 1998). Examples of baits used for selective isolation of soil fungi are plant tissues for plant pathogens, paper strips for cellulolytic species, polyester polyurethane for plastic degraders, hair pieces for keratinophilic species, chitin for chitinase producers and insect larvae for entomopathogenic fungi (Marks and Mitchell, 1970; Papavizas *et al.*, 1975; Sneh *et al.*, 1991; Gams *et al.*, 1998; Edena *et al.*, 2000; Gonçalves, 2000; Pettitt *et al.*, 2002; Barratt *et al.*, 2003; Wellington *et al.*, 2003).

DNA-targeted Techniques for the Assessment of Soil Microfungi Diversity

The methodologies of isolation and culturing of fungi from complex ecosystems such as soil show inherent limitations due to the fastidious nature of several species coupled with the inability of culture media to mimic soil habitats (Tsao *et al.*, 1983; Muyzer *et al.*, 1993; Bridge and Spooner, 2001). Even the analysis of relative abundance of culturable species recovered from soil may not represent the dynamics of soil communities well, since culture media impose new selective conditions and can introduce bias to the analyses (Liu *et al.*, 1997). A reliable measure of soil fungal communities without bias requires a laborious programme of systematic isolations with different culture media and isolation strategies covering idiosyncrasies from the diverse taxonomic and physiological groups of fungi occurring in soil. However, some fungal groups such as *Basidiomycota* are difficult to isolate and may fail to sporulate when in axenic media (Thorn *et al.*, 1996). Also, fungi belonging to the phylum *Glomeromycota* that form the arbuscular mycorrhiza are obligate biotrophic fungi that do not grow in the absence of their host plant (Moreira and Siqueira, 2002; Siqueira and Stürmer, Chapter 10, this volume). Another interesting fact is that only a small fraction of the estimated number of fungal species is known (Hawksworth, 2001). Since a considerable part of these species may occur in soil in some part of their life cycle, other methodologies should be used to complement the traditional approaches for a better understanding of diversity and dynamics of soil fungi (Bridge and Spooner, 2001) and these methodologies may include DNA target techniques.

Molecular techniques

Molecular tools based on DNA analysis have been successfully applied for the study of complex bacterial and, recently, fungal assemblages from environmental samples. The majority of molecular approaches are based on the amplification of specific DNA sequences from samples by PCR, and further characterization of the DNA polymorphisms that may represent different taxa. The gene cluster for ribosomal RNA molecules 18S, 8.5S and 28S is thought to be highly conserved among eukaryotic organisms and is usually employed as a molecular marker. DNA length polymorphisms and base sequence variations can be used to group organisms according to their origin and evolutionary relationship. The identification of unknown DNA sequences can also be accomplished by comparing them with nucleotide sequence databases from putatively known taxa. Several copy numbers of the rDNA clusters are found in the eukaryotic genome, facilitating the amplification from very small DNA samples. The rDNA cluster also contains spacers between the coding regions known as internal transcribed spacers (ITSs), which have less conserved base sequences and can be used for differentiation among related species or to assess infraspecific genetic divergences (Baayen *et al.*, 2000; Viaud *et al.*, 2000; Down, 2002).

Since the soil contains a vast number of organisms, obtaining specific fungal rDNA primers is a critical step in PCR amplification. The primers must permit amplification of a broad range of fungal species without losing the specificity to this target group. Based on the ribosomal RNA gene sequences available on specialized databases for a vast number of fungal species, primers that are specific for PCR amplification of fungal DNA from complex soil samples have been developed (Smit *et al.*,

1999; Borneman and Hartin, 2000). However, some of these specific primers have been shown to amplify non-fungal DNA or show bias towards the amplification of particular taxonomic groups inside kingdom Fungi (Smit *et al.*, 1999; Borneman and Hartin, 2000; Anderson *et al.*, 2003). Another crucial step prior to PCR reaction is the DNA extraction directly from soil. The fungal cells present in the soil sample even as mycelia or spores have to be correctly lysed and the obtained DNA thoroughly purified to eliminate humic and phenolic substances that can interfere with the PCR reaction. These requirements are generally met through intricate protocols combining heat, chemical lyses and bead beating, with the crude DNA being subjected to additional purification in commercial DNA purification kits (Liu *et al.*, 1997; Viaud *et al.*, 2000; Bridge and Spooner, 2001). Some fungi present in very low densities in natural soils, such as *Phytophthora* species, are difficult to detect by molecular means since PCR tends to amplify DNA molecules that are dominant among the total extract DNA. For these fungi, molecular detection from soil samples can be improved through baiting techniques coupled with PCR using genus- or species-specific primers (Nechwatal *et al.*, 2001).

DNA sequencing and molecular fingerprint techniques

In order to obtain the identities of the PCR-amplified fungal DNA, the amplicons of correct size can be separated in agarose gels, excised from the gel matrix, purified, connected to plasmid vectors and cloned into bacterial cells. The cloned DNA is sequenced and compared with databases containing fungal rDNA oligonucleotide sequences via software analyses (Borneman and Hartin, 2000; Viaud *et al.*, 2000; Anderson *et al.*, 2003). Sequencing of the 18S rDNA from samples obtained from the rhizoplane of a single plant species, the grass *Arrhenatherum elatius*, revealed 49 distinct phylotypes. Only seven of them could be identified by already deposited sequences of known fungal species (Vandenkoornhuyse *et al.*, 2002). However, these procedures are expensive and time-consuming and not suitable for the assessment of complex environmental samples. Therefore, simpler molecular community fingerprint techniques have been developed. Initially used for assessment of bacterial assemblages from environmental samples, molecular fingerprint techniques such as thermal gradient gel electrophoresis (TGGE) and denaturing gradient gel electrophoresis (DGGE), single-strand-conformation polymorphism (SSCP) and terminal restriction fragment length polymorphism (TRFLP) have been successfully applied for the 18S rDNA analysis of soil fungal communities (Muyzer *et al.*, 1993; Lee *et al.*, 1996; Liu *et al.*, 1997; Smit *et al.*, 1999; Van Elsas *et al.*, 2000; Lowell and Klein, 2001; Brodie *et al.*, 2003).

In TGGE or DGGE, complex soil fungal communities can be monitored by the evaluation of the band patterns presented by the PCR-amplified 18S DNA molecules when migrating in a gel matrix containing a linear thermal (TGGE) or denaturing (DGGE) gradient under electrophoretic conditions. Under these denaturing conditions, DNA stretches of the same length but with different base-pair sequences partially melt due to the differential denaturing of less stable domains in the molecules called melting domains. Partially melted molecules halt their movement in the gel. Therefore, DNA molecules with different base-pair sequences have different melting behaviour and migrate to distinct points in the gel (Muyzer *et al.*, 1993). The fungal community structure differences between the soil bulk and rhizosphere soil of wheat, plus temporal changes inside these communities, were assessed by Smit *et al.* (1999) using TGGE analyses coupled with 18S DNA sequencing. Van Elsas *et al.* (2000) monitored the fate of two fungal strains, *T. harzianum* and *Arthrobotrys oligospora*, inoculated in soil samples by spores and mycelial fragments, respectively, using DGGE. These authors also utilized DGGE to assess the changes in soil microbial community when it was treated with crude oil and could detect significant alterations in band patterns with decrease in band numbers and increase in intensity of some dominant bands in the soil that received crude oil.

Another molecular fingerprint approach uses the differential migration of single-stranded DNA molecules in a gel to study complex microbial communities. In SSCP, the amplified DNA is fully denatured before being submitted to an electrophoresis run. Single-stranded DNA molecules acquire unique folded structures dictated by their nucleotide sequences and migrate to specific points in the gel. Consequently, DNA fragments with the same length but dissimilar in base-pair sequence can be readily separated in SSCP by their differential migration in the gel when in single-strand conformation (Lee *et al.*, 1996). Alterations in the fungal community in cultivated and uncultivated soils after nitrogen application were assessed by Lowell and Klein (2001) with an SSCP approach. From a total of 589 samples, 312 different band migration patterns or haplotypes were detected in this study, and the differences of dominant haplotypes between the nitrogen amended and the control soils were also distinguished.

Individuals in fungal communities can be further characterized not only by migration patterns of amplified DNA in a gel but also by a sequence of several DNA hybridization tests with different oligonucleotide probes. In the oligonucleotide fingerprinting of rRNA genes (OFRG) technique, the fungal rDNA is directly extracted from soil, PCR-amplified and separated in a gel. Subsequently, the DNA fragments are transferred from the gel to a nylon membrane or first cloned into bacterial cells and then transported to the membrane. Once the arrays of the DNA fragments are fixed on a nylon membrane, they are subjected to several hybridization tests, each with a single ^{33}P-labelled oligonucleotide probe. After the experiments, the hybridization pattern for each rDNA applied to the membrane is analysed via software. The OFRG was applied for the assessment of fungal communities in an agricultural soil. After PCR amplification of the fungal DNA, it was cloned and subjected to 27 hybridization experiments. Based on their hybridization fingerprints, the clones were clustered in a phylogenetic tree and the identities of representatives of each cluster were assessed by DNA sequencing (Valinsky *et al.*, 2002). Even if this is a robust and reliable technique, the necessity of several hybridization tests makes the OFRG time-consuming, which can prevent its application for complex soil sample analysis (Lee *et al.*, 1996).

The low variability of rDNA may introduce undesirable artefacts in molecular fingerprint techniques such as TGGE and DGGE, like comigration of different DNA molecules to the same point in the gel. To increase band polymorphism and improve resolution, enzymatic restriction of the amplified DNA prior to electrophoresis can be used. Recently, RFLP of PCR-amplified ITS DNA was employed for the assessment of soil fungal communities by comparing the RFLP band patterns of fungal DNA directly extracted from soil and from fungal colonies previously isolated from the same soil by the soil dilution plate technique. From a total of 58 RFLP groups detected, only one was coincident between the DNA from cultivable fungi and the DNA isolated from the soil environment (Viaud *et al.*, 2000). This result shows the incompleteness of molecular and culture-based methods alone and reinforces the need of a polyphasic approach for a better understanding of soil fungal communities (Van Elsas *et al.*, 2000). Related fungal species can be identified by their typical RFLP patterns after PCR amplification of DNA directly extracted from soil using genus-specific primers (Nechwatal *et al.*, 2001).

Looking for a molecular method that could provide phylogenetic information about the dominant microbial groups in environmental samples and yet serve as a quantitative molecular approach, Liu *et al.* (1997) developed an extended version of the PCR-RFLP called terminal restriction fragment length polymorphism (TRFLP). In TRFLP, the DNA, originally the bacterial 16S rDNA, is PCR-amplified with one of the two primers labelled with fluorescence. After amplification the DNA is digested with restriction enzymes and the restriction fragments can be separated in a gel according to their size. As the terminal restriction fragments (TRFs) are labelled with fluorescence they can be detected and quantified in an automated sequencer. By searching in rDNA databases the TRFs of several fungal species already sequenced can be predicted and used for the identification of unknown TRFs originating from soil samples. This strategy was used by Brodie *et al.* (2003) in a study of soil fungal communities in temperate upland grassland soil. However, it was unsatisfactory since some TRFs did not match any fungal species from the databank and others matched as many as 23 different species of ascomycetes and basidiomycetes. The differences of soil fungal communities amongst three contrasting soils in France were assessed by Edel-Hermann *et al.* (2004) using TRFLP. They also detected shifts in fungal populations in soils amended with manure or mushroom compost.

As discussed before, the ITS region in the rDNA cluster has a higher variability than the 18S and 28S regions. Therefore this natural polymorphism in ITS among fungal species can be used for the evaluation of soil fungi. In automated ribosomal intergenic spacer analysis (ARISA), the fungal DNA is extracted from soil and PCR-amplified using specific primers for the ITS region with one of the primers fluorescently labelled. Following the amplification, the different ITSs can be separated in a gel, detected and quantified in an automated sequencer. The fungal ARISA was employed by Ranjard *et al.* (2001) to evaluate the soil fungi from five different soils. It proved to be a highly sensitive method and could detect as much as 118 ITS types from a single soil sample. This method, however, cannot provide a clear separation between phylogenetic groups of fungi since a single fungal species can exhibit ITS polymorphism and even a single heterokaryotic mycelium can contain more than one ITS type (O'Donnell and Cigelnik, 1997; Viaud *et al.*, 2000).

Quantitative PCR

Some studies, especially with soil-borne phytopathogenic fungi, focus on only one or a few fungal species present in the soil. For such studies, very specific PCR primers must be designed for correct amplification of DNA from the target species, and other molecular markers can be employed such as the sequences of β-tubulin gene or the translation elongation factor EF 1-α gene (Baayen *et al.*, 2000; Mauchline *et al.*, 2002; Filion *et al.*, 2003; Li and Hartman, 2003). Besides molecular detection, the quantification of fungal DNA in soil is generally required in epidemiological and ecological studies. However, the conventional PCR is not suitable for quantitative approaches since small variations during the exponential phase of the amplification reaction can drastically alter the amounts of PCR products. For quantitative assays, modifications of conventional PCR as competitive PCR (cPCR) and real-time PCR have been developed.

In cPCR, DNA fragments containing the same primer sites as the sample DNA are added in known amounts to the PCR reactions and co-amplified with the target DNA. After amplification, the distinct PCR products are separately quantified by their relative band intensities in agarose gels. By adding different amounts of competitor DNA to a standard amount of sample DNA in a series of PCR reactions, and monitoring the quantities of competitor PCR products yielded in each reaction, the amount of DNA present in the sample can be calculated (Siebert and Larrick, 1992). The correct detection and quantification of a genetically modified strain of *Trichoderma virens*, capable of degrading organophosphate pesticides, was accomplished by Baek and Kenerley (1998) with a cPCR approach. Mauchline *et al.* (2002) employed specific primers and cPCR for a quantitative assessment of the nematophagous fungus *Verticillium chlamydosporium*. The population dynamics, host specificity and saprophytic growth of *V. chlamydosporium* in sterilized and non-sterilized soil could be assessed by this quantitative molecular approach. The authors also could correlate the numbers of chlamydospores added to soil samples to the amounts of cPCR products.

Another method employed in quantitative approaches is real-time PCR. Fluorogenic probes or dyes added in the PCR reactions permit the correct monitoring of products yielded during the amplification. The fluorescent markers added with the other reagents of PCR are capable of emitting fluorescence in the presence of double-strand DNA; therefore the increase of DNA molecules during amplification leads to an increase in the intensity of fluorescent emission that can be measured. A standard curve can be constructed correlating known amounts of DNA to the fluorescent intensities after a defined number of PCR cycles, and used to directly quantify DNA from unknown samples. This technique is faster and simpler than cPCR since there is no need for the construction of competitor DNA and no post-PCR analysis is required. Real-time PCR has been successfully employed for quantitative assessments of the plant pathogens *Colletotrichum coccodes*, *F. solani* f. sp. *phaseoli* and *Rhizoctonia solani* from soil samples (Cullen *et al.*, 2002; Lees *et al.*, 2002; Filion *et al.*, 2003).

Application of molecular tools for soil fungal community assays is in its infancy. These approaches were derived mainly from studies on bacterial diversity and presuppose limitations of isolation techniques for fungi (Selenska and Klingmüller, 1992; Tsai and Olson, 1992; O'Donnell *et al.*, 1994). However, in the case of fungi the quantity of non-culturable fungal species is surely overemphasized. What can be true for strictly symbiotic organisms like most N-fixing bacteria and arbuscular mycorrhizal fungi does not apply to saprophytic fungi. Even most of the known plant-pathogenic fungi are not obligate parasites. The main constraints have been overcome with the advent of more sophisticated isolation techniques like soil-washing and particle filtration methods, as already discussed. A large number of different techniques have been published; however, there are only a handful of publications on each methodological approach concerning soil fungi. Community fingerprint techniques can monitor complex fungal assemblages from soil but a better characterization and phylogenetic analysis of dominant taxa invariably requires sequencing and comparison with public DNA databanks. However, fungal taxonomic approaches based only on molecular data may present some problems due to the incompleteness of databanks and the presence of misidentified DNA sequences deposited, originating from incorrect morphological identifications, amplification and sequencing of fungal contaminants or even sequencing of chimeras (Crous, 2002; Bridge *et al.*, 2003; Hawksworth, 2004b).

Another limitation of molecular approaches based on DNA amplification is the lack of knowledge concerning actively growing or functional groups of soil fungi in the total DNA extract from soil. As already discussed, the DNA extraction protocols are capable of lysing and extracting nucleic acids either from actively growing mycelia or from dormant spores, making the differentiation of functional groups based only on PCR product analyses difficult. Recent developments based on RT-PCR from mRNA extracted from soil and stable isotope probing (SIP) showed good potential for distinguishing active bacterial groups in complex ecosystems (Wellington *et al.*, 2003), and these procedures will probably be described for soil fungi soon. For now, a more complete understanding of fungal communities in soil should rely on polyphasic approaches coupling culture-based and molecular taxonomic assays with the study of major biogeochemical processes occurring in soil.

Can we rapidly measure diversity of soil microfungi?

To obtain a confident assessment of diversity of soil fungi, there are two basic constraints. Investigations should be designed as long-term studies involving teamwork of specialists in different taxonomic groups and limitations with regard to the methodology used must be overcome (Hyde and

Hawksworth, 1997). The construction of rarefaction curves aids the evidence about whether real diversity was achieved or whether the limit of the employed methodologies has been reached. Broadly accepted standard methods for cataloging fungal diversity are not yet available (Cannon, 1997). A generally accepted method should permit taxonomic and functional characterization of soil microfungi as well as monitoring changes in space and time across a land use intensity gradient or changes due to other physical or anthropogenic impacts. An outlook on the ATBI initiative, given by Rossman (1994), presents valuable information on sampling and isolation strategies. An overview on specific protocols was also published by Rossman *et al.* (1998) and Mueller *et al.* (2004). A comprehensive overview on sampling design, isolation techniques and how to document soil microfungi diversity was published recently (Bills *et al.*, 2004).

When a long-term study involving several specialists is not feasible because of limitations of time and resources, the use of recognizable taxonomic units (RTUs) is a good choice. This approach is useful for species-rich groups of insects, where morphotypes can easily be distinguished, but may not give useful information in the case of fungi (Cannon, 1997, 1999; Hyde and Hawksworth, 1997). There are examples where the use of indicator organisms or target groups give a good picture of disturbance of ecosystems. Changes in fungal communities can be assessed by the study of target groups of fungi that have putatively known functions in soil ecosystems. A limited group of known organisms can be a good indicator of environmental disturbance (Hyde and Hawksworth, 1997). The concept of predictor sets, developed by entomologists (Kitching, 1993), may be the most promising alternative for a rapid estimation of the diversity of fungi (Hyde, 1997b; Hyde and Hawksworth, 1997). A predictor set can also be a valuable tool when the degree of impact due to agricultural practices has to be evaluated. Soil-borne plant pathogens and their natural antagonists putatively represent a good option for a predictor set of species since they comprise a well-studied group of fungi whose distribution and dominance changes can be related to human intervention on soil by agricultural practices. Three major groups of soil-borne fungi covering distinct taxonomic and physiological groups are suitable for a rapid assessment of fungal diversity of tropical soils, subjected to different levels of human interference and are selected for use as a predictor set. The set includes plant-pathogenic oomycetes like *Phytophthora* and *Pythium*; the plant-pathogenic anamorphic basidiomycete *Rhizoctonia* and the ascomycetes including the potentially plant-pathogenic genera *Cylindrocarpon*, *Cylindrocladium*, *Fusarium*, *Lasiodiplodia* and *Verticillium*, and their respective antagonists such as *Clonostachys*, *Coniothyrium*, *Talaromyces*, and *Trichoderma* (Fig. 9.2). Occurrence and relative frequency of species is assessed in sampling points representing different degrees of disturbance. Diversity of other species is registered as far as possible.

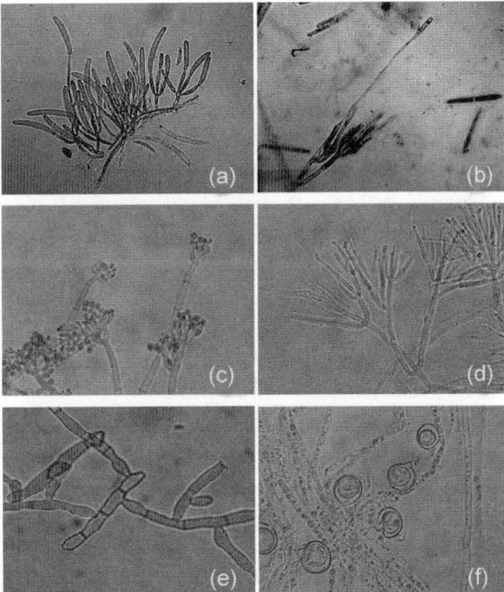

Fig. 9.2. Examples of common soil microfungi: (a) *Fusarium solani*, young sporodochium; (b) *Cylindrocladium clavatum*, conidiophore with conidiogenic cells; (c) *Trichoderma virens*, conidiophores and conidia; (d) *Clonostachys rosea*, penicilliate conidiophores and conidia; (e) *Rhizoctonia* sp.; (f) *Pythium* sp., oospores.

This specific predictor is being tested by analysing about 100 sampling points in the Brazilian benchmark site of the UNEP-GEF Project 'Conservation and Sustainable Management of Below-Ground Biodiversity'.

Isolation of target species

Besides a number of saprophytes belonging to the phylum Oomycota distributed among humid habitats, two genera of oomycetes, *Phytophthora* and *Pythium*, are important soil-borne plant pathogens and deserve a place amongst the target groups. Detection and isolation of *Pythium* species for a high number of soil samples can be achieved using susceptible plants as baits. An excellent review on the genus is now available (Lévesque and Cock, 2004). Despite their active growth in soil and litter, the macro- and microfungal species of basiodimycetes are not regularly recovered from soil due to their specific nutritional requirements, low rate of growth and incapability of sporulating in axenic cultures. Representing the phylum Basidiomycota, a target group can be designated with the basidiomycetous anamorph genus *Rhizoctonia*, a fully studied genus of plant-pathogenic fungi that cause root rot and damping-off in a number of crops. Several methods for isolation and quantification of *Rhizoctonia* species from soil have been developed and compiled in works such as Sneh *et al.* (1991) and Dhingra and Sinclair (1985). However, the majority of research relies on isolation of *Rhizoctonia* from agricultural soils where the fungus is causing symptoms of disease and can be detected more readily. Ascomycetes are a major group of soil microfungi including strictly saprophytic, entomopathogenic, plant-pathogenic and antagonistic species. However, almost all these species exhibit a saprophytic phase in the soil and can be isolated by a unified soil-plating procedure. A combined soil-washing and particle filter technique is proposed here as a suitable method for the study of plant pathogens and their antagonists.

Preservation of Genetic Resources

Inventories of biodiversity provide useful information and a scientific base for the preservation of habitats and sustainable land use and therefore need to be done. Genetic resource collections are now requested in all parts of the world with the aim of *ex situ* preservation of species and to supply research institutions and industry with authentic material. Reference collections must be supported by countries with a strong policy in science and technology because they contain information on geographical and host distribution and provide basic working material for those studying characteristics and variation, as well as practical and economic applications of species (Hawksworth, 1993; Kirsop, 1996). The actual situation of fungal genetic resource collections and the challenges to support the needs of fungal genomics, molecular biology and conservation are very timely and have received considerable attention (Hawksworth, 2004a). In the future, long-term preservation of voucher specimens and cultures will be essential to validate entries with accurate sequence and genomic databases, as is already the case for publishing and naming new species (Agerer *et al.*, 2003).

There are remarkably few microbial culture collections in tropical countries. In Brazil, although about 40 minor specialized collections are registered, spread over the country, there is no official depository collection for reference strains and microbial genetic resources, except for some special groups like N_2-fixing bacteria or arbuscular mycorrhizal fungi (WFCC, 1999). In the field of plant pathology, a traditional claim for depositing reference material is now of high priority due to the demand for characterization of emerging plant pathogens, development of diagnostic kits and plant breeding programmes looking for pathogen resistance (Smith and Waller, 1992). The need for genetic resource collections with an industrial and a conservation and biodiversity perspective has been widely emphasized (Kirsop and Hawksworth, 1994; Kelley,

1995; Hawksworth, 1996, 2004a; Canhos and Manfio, 2000; Ryan and Smith, 2004) and should be a strong policy in developing countries with rich biodiversity.

While the knowledge of taxonomy, distribution and ecology is still poor, there is high demand for support and training of local scientists. As more fungal genomes are analysed and put in databases, the same must be done with tropical fungal diversity. As there are still many practical constraints in tropical mycology, expansion and improvement of genetic resource collections in a particular region will be of great importance and value, providing easy access to strains required for teaching, research and industrial purposes.

Concluding Remarks

Tropical ecosystems exhibit a very high diversity of different groups of organisms, including soil fungi, that are poorly studied. Advances in tropical mycology with ever-improved modern techniques will lead to the discovery of many of the still unknown taxa. Also, much more work is needed on the geographical distribution of fungal taxa. The discovering of new fungal genomes is of great importance for systematics and biotechnology. A crucial point would be to pay more attention to specific habitats and substrates and, above all, to the development of novel isolation and identification techniques. In no case can the whole diversity of fungi be easily assessed.

The most important constraint on comparative analyses of the soil fungi is methodology. At present, it is not possible to accept the hypothesis that specific communities of soil microfungi exist for certain sites with a certain vegetation type. Therefore, it would be highly desirable to discuss and establish a widely accepted strategy for isolation of this group of fungi, probably using a sophisticated particle filter technique and some specific culture media. This would provide a powerful tool for assessing diversity and, at the same time, quantitative and qualitative changes in communities under different land use. This information could be used for recommendations for management of soil-borne plant diseases and maintenance of fertility and therefore contribute to more sustainable agriculture. Considering that we have a reasonable knowledge of the role of soil microfungi, adoption of strategies that emphasize the maintenance of fungal diversity may enhance sustainability of agroecosystems by providing balanced biological interactions between plant pathogens and antagonists. Furthermore, knowledge about diversity in certain areas and habitats is imperative for both conservation and exploitation of biological resources.

References

Agarwal, A.K. and Chauhan, R.K.S. (1988) Fungal communities and seasonal succession of microfungi in the forest soil of Chandpata, Shivpuri, MP. *Acta Botanica Indica* 16, 204–209.

Agerer, R., Ammirati, J., Blanz, P., Courtecuisse, R., Dejardin, D.E., Gams, W., Hallenberg, N., Halling, R., Hawksworth, D.L., Horak, E., Korf, R.P., Mueller, G.M., Oberwinkler, F., Rambold, G., Summerbell, R.C., Triebel, D. and Watling, R. (2003). Always deposit vouchers. *Mycological Research* 104, 642–644.

Altieri, M.A. (1999) The ecological role of biodiversity in agroecosystems. *Agriculture Ecosystems and Environment* 74, 19–31.

Anderson, J.M. and Ingram, J.S.I. (1989) *Tropical Soil Biology and Fertility, a Handbook of Methods*. CAB International, Wallingford, UK, 171 pp.

Anderson, I.C., Campbell, C.D. and Prosser, J.I. (2003) Potential bias of fungal 18S rDNA and internal transcribed spacer polymerase chain reaction primers for estimating fungal biodiversity in soil. *Environmental Microbiology* 5, 36–47.

Bååth, E. (1988) A critical examination of the soil washing technique with special reference to the effect of the size of the soil particles. *Canadian Journal of Botany* 66, 1566–1569.

Baayen, R.P., O'Donnell, K., Bonants, P.J.M., Cigelnik, E., Kroon, L.P.N., Roebroeck, E.J.A. and Waalwijk, C. (2000) Gene genealogies and AFLP analyses in the *Fusarium oxysporum* complex identify monophyletic and nonmonophyletic formae speciales causing wilt and rot disease. *Phytopathology* 90, 891–900.
Baek, J. and Kenerley, C.M. (1998) Detection and enumeration of a genetically modified fungus in soil environments by quantitative competitive polymerase chain reaction. *FEMS Microbiology Ecology* 25, 419–428.
Barratt, S.R., Ennos, A.R., Greenhalgh, M., Robson, G.D. and Handley, P.S. (2003) Fungi are the predominant micro-organisms responsible for degradation of soil-buried polyester polyurethane over a range of soil water holding capacities. *Journal of Applied Microbiology* 95, 78–85.
Batista, A.C., Silva, J.O., Maciel, M.J.P., Lima, J.A. and Ramos de Moura, N. (1967a) Micropopulações fúngicas dos solos do Território Federal do Amapa. *Atas Instituto Micologia* 4, 117–121.
Batista, A.C., Silva, J.O., Maciel, M.J.P. and Almeida, A.G. (1967b) Aspergillaceae dos solos das zonas fisiográficas de Bragança e do Baixo Amazonas, Estado do Pará. *Atas Instituto Micologia* 4, 185–229.
Beare, M.H., Vikram Reddy, M., Tian, G. and Srivastava, S.C. (1997) Agricultural intensification, soil biodiversity and agroecosystem function in the tropics, the role of decomposer organisms. *Applied Soil Ecology* 6, 87–108.
Bettucci, L. and Roquebert, M.F. (1995) Microfungi from a tropical rain forest litter and soil, a preliminary study. *Nova Hedwigia* 61, 111–118.
Bettucci, L., Malvarez, I., Dupont, J., Bury, E. and Roquebert, M.F. (2002) Paraná river delta wetlands soil microfungi. *Pedobiologia* 46, 606–623.
Bills, G.F. (1995) Analyses of microfungal diversity from a user's perspective. *Canadian Journal of Botany* 73, S33–S41.
Bills, G.F., Dombrowsky, A., Pelaez, F., Polishook, J.D. and An, Z. (2002) Recent and future discoveries of pharmacologically active metabolites from tropical fungi. In: Watling, R., Frankland, J.C., Ainsworth, A.M., Isaac, S. and Robinson, C.H. (eds) *Tropical Microfungi. Micromycetes*, Vol. 2. CAB International, Wallingford, UK, pp. 165–194.
Bills, G.F., Christensen, M., Powell, M. and Thorn, G. (2004) Saprobic soil fungi. In: Mueller, G.M., Bills, G.F. and Foster, M.S. (eds) *Biodiversity of Fungi. Inventory and Monitoring Methods.* Elsevier, Amsterdam, pp. 271–302.
Bonkowski, M., Griffith, B.S. and Ritz, K. (2000) Food preference of earthworms for soil fungi. *Pedobiologia* 44, 666–676.
Bordjiba, O., Steiman, R., Kadri, M., Semadi, A. and Guiraud, P. (2001) Removal of herbicides from liquid media by fungi isolated from a contaminated soil. *Journal of Environmental Quality* 30, 418–426
Borneman, J. and Hartin, R.J. (2000) PCR primers that amplify fungal fRNA genes from environmental samples. *Applied and Environmental Microbiology* 66, 4356–4360.
Bridge, P. and Spooner, B. (2001) Soil fungi, diversity and detection. *Plant Soil* 232, 147–154.
Bridge, P.D., Roberts, P.J., Spooner, B.M. and Panchal G. (2003) On the unreliability of published DNA sequences. *New Phytologist* 160, 43–48.
Brodie, E., Edwards, S. and Clipson, N. (2003) Soil fungal community structure in a temperate upland grassland soil. *FEMS Microbiology Ecology* 45, 105–114.
Bull, A.T., Goodfellow, M. and Slater, J.H. (1992) Biodiversity as a source of innovation in biotechnology. *Annual Review of Microbiology* 46, 219–252.
Cabello, M. and Arambarri, A. (2002) Diversity in soil fungi from undisturbed and disturbed *Celtis tala* and *Scutia buxifolia* forests in the eastern Buenos Aires province (Argentina). *Microbiological Research* 157, 115–125.
Canhos, V.P. and Manfio, G.P. (2000) Microbial resource centres and *ex-situ* conservation. In: Priest, F.G. and Goodfellow, M. (eds) *Applied Microbial Systematics.* Kluwer, Dordrecht, The Netherlands, pp. 421–446.
Cannon, P.F. (1996) Filamentous fungi. In: Hall, G. (ed.) *Methods for the Examination of Organismal Diversity in Soils and Sediments.* CAB International, Wallingford, UK, pp. 127–145.
Cannon, P.F. (1997) Strategies for rapid assessment of fungal diversity. *Biodiversity and Conservation* 6, 669–680.
Cannon, P.F. (1999) Options and constraints in rapid diversity analysis of fungi in natural systems. *Fungal Diversity* 2, 1–15.
Chet, I. and Baker, K.F. (1980) Induction of suppressiveness to *Rhizoctonia solani* in soil. *Phytopathology* 70, 994–998.
Christensen, M. (1981) Species diversity and dominance in fungal communities. In: Wicklow, D.T. and Carroll, G.C. (eds) *The Fungal Community, Its Organization and Role in Ecosystem.* Marcel Dekker, New York, pp. 201–232.

Christensen, M. and Tuthill, D.E. (1985) *Aspergillus*: an overview. In: Samson, R.A. and Pitt, J.I. (eds) *Advances in* Aspergillus *and* Penicillium *systematics*. Plenum, New York, pp. 195–209.
Crous, P. (2002) Adhering to good cultural practice. *Mycological Research* 106, 1378–1379.
Cullen, D.W., Lees, A.K., Toth, I.K. and Duncan, J.M. (2002) Detection of *Colletotrichum coccodes* from soil and potato tubers by conventional and quantitative real-time PCR. *Plant Pathology* 51, 281–292.
Da Silva, M., Umbuzeiro, G.A., Pfenning, L.H., Canhos, V.P. and Esposito, E. (2003) Filamentous fungi isolated from estuarine sediments contaminated with industrial discharges. *Soil and Sediment Contamination* 12, 345–356.
Davet, P. and Rouxel, F. (2000) *Detection and Isolation of Soil Fungi*. Science Publishers, New Hampshire, 188 pp.
Dhingra, O.D. and Sinclair, J.B. (1985) *Basic Plant Pathology Methods*. CRC Press, Boca Raton, Florida, 355 pp.
Domsch, K.H., Gams, W. and Anderson, T.H. (1980) *Compendium of Soil Fungi*, Vols I and II. Academic Press, London.
Down, G. (2002) Fungal family trees – finding relationships from molecular data. *Mycologist* 16, 51–58.
Edel-Hermann, V., Dreumont, C., Pérez-Piqueres, A. and Steinberg, C. (2004) Terminal restriction fragment length polymorphism analysis of ribosomal RNA genes to assess changes in fungal community structure in soils. *FEMS Microbiology Ecology* 47, 397–404.
Edena, M.A., Hillb, R.A. and Galpothage, M. (2000) An efficient baiting assay for quantification of *Phytophthora cinnamomi* in soil. *Plant Pathology* 49, 515–522.
Eicker, A. (1969) Microfungi from surface soil of forest communities in Zululand. *Transactions of the British Mycological Society* 53, 381–392.
Farrow, W.M. (1954) Tropical soil fungi. *Mycologia* 46, 632–646.
Filion, M., St-Arnaud, M. and Jabaji-Hare, S.H. (2003) Direct quantification of fungal DNA from soil substrate using real-time PCR. *Journal of Microbiological Methods* 53, 67–76.
Fox, F.M. (1993) Tropical fungi, their commercial potential. In: Isaac, S., Frankland, J.C., Watling, R. and Whalley, A.J.S. (eds) *Aspects of Tropical Mycology*. Cambridge University Press, Cambridge, UK, pp. 253–263
Frankland, J., Dighton, J. and Boddy, L. (1990) Methods for studying fungi in soil and forest litter. In: Grigorova, R.G. and Norris, J.R. (eds) *Methods in Microbiology*, Vol. 22. Academic Press, London, pp. 343–404.
Gafur, A. and Darmono, T.W. (1999) Impacts of different land use systems on the abundance of soil-borne pathogens. In: Gafur, A., Susilo, F.X., Utomo, M. and van Noordwijk, M. (eds) Management of Agrobiodiversity in Indonesia for Sustainable Land Use and Global Environmental Benefits. Bogor, Indonesia. ASB Indonesia Report no. 9, pp. 93–102.
Gamboa, M.A., Loureano, S. and Bayman, P. (2002) Measuring diversity of endophytic fungi in leaf fragments: does size matter? *Mycopathologia* 156, 41–45.
Gams, W. (1992) The analysis of communities of saprophytic microfungi with special reference to soil fungi. In: Winterhoff, W. (ed.) *Fungi in Vegetation Science*. Kluwer, Dordrecht, The Netherlands, pp. 183–223.
Gams, W., Hoekstra, E.S. and Aptroot, A. (1998) *CBS Course of Mycology*, 4th edn. Centraalbureau voor Schimmelcultures, Baarn, The Netherlands, 165 pp.
Giller, K.E., Beare, M.H., Lavelle, P., Izac, A.M.N. and Swift, M.J. (1997) Agricultural intensification, soil biodiversity and agroecosystem function. *Applied Soil Ecology* 6, 3–16.
Gochenaur, S.E. (1970) Soil mycoflora of Peru. *Mycopathologia Mycologia Applicata* 42, 259–272.
Gochenaur, S.E. (1975) Distributional patterns of mesophilous and thermophilous microfungi in two Bahamian soils. *Mycopathologia Mycologia Applicata* 57, 155–164.
Gonçalves, R.C. (2000) Iscas para quantificação de Cylindrocladium spp. no solo e flutuação da densidade de inóculo do patógeno em jardim clonal de Eucalyptus spp. Viçosa, UFV, Brazil, 54 pp. (master thesis).
Goos, R.D. (1960) Soil fungi from Costa Rica and Panama. *Mycologia* 52, 877–883.
Goos, R.D. (1963) Further observations on soil fungi in Honduras. *Mycologia* 55, 142–150.
Goos, R.D. and Timonin, M.I. (1962) Fungi from the rhizosphere of banana in Honduras. *Canadian Journal of Botany* 40, 1371–1377.
Gray, T.R.G. (1990) Methods for studying the microbial ecology of soil. In: Grigorova, R.G. and Norris, J.R. (eds) *Methods in Microbiology*, Vol. 22. Academic Press, London, pp. 310–342.
Grishkan, I., Nevo, E., Wasser, S.P. and Beharav, A. (2003) Adaptive spatiotemporal distribution of soil microfungi in 'Evolution Canyon' II, Lower Nahal Keziv, western Upper Galilee, Israel. *Biological Journal of the Linnean Society* 78, 527–539.

Hammond, P.M. (1992) Species inventory. In: Groombridge, B. (ed.) *Global Diversity, Status of the Earth's Living Resources.* Chapman & Hall, London, pp. 17–39.

Hawksworth, D.L. (1991) The fungal dimension of biodiversity, magnitude, significance and conservation. *Mycological Research* 95, 641–655.

Hawksworth, D.L. (1993) The tropical fungal biota, census, pertinence, prophylaxis, and prognosis. In: Isaac, S., Frankland, J.C., Watling, R. and Whalley, A.J.S. (eds) *Aspects of Tropical Mycology.* Cambridge University Press, Cambridge, UK, pp. 265–293.

Hawksworth, D.L. (1996) Microbial collections as a tool in biodiversity and biosystematic research. In: Samson, R.A., Stalpers, J.A., van der Mei, D. and Stouthamer, A.H. (eds) *Culture Collections to Improve Quality of Life.* Proceedings of the 8th International Congress for Culture Collections. CBS, Baarn, pp. 26–35.

Hawksworth, D.L. (2001) The magnitude of fungal diversity, the 1.5 million estimate revisited. *Mycological Research* 105, 1422–1432.

Hawksworth, D.L. (2004a) Fungal diversity and its implications for genetic resource collections. *Studies in Mycology* 50, 9–18.

Hawksworth, D.L. (2004b) 'Misidentifications' in fungal DNA sequence databanks. *New Phytologist* 161, 13–15.

Hawksworth, D.L. and Rossman, A.Y. (1997) Where are all the undescribed fungi? *Phytopathology* 87, 888–891.

Hennebert, G.L. (1995) Fungal diversity in tropical forests. International Biodiversity Seminar ECCO XIV Meeting Gozd Martulek, Slovenia, pp. 75–93.

Houston, A.P.C., Visser, S. and Lautenschlager, R.A. (1998) Microbial processes and fungal community structure in soils from clear-cut and unharvested areas of two mixedwood forests. *Canadian Journal of Botany* 76, 630–640.

Hyde, K.D. (1997a) *Biodiversity of Tropical Microfungi.* Hong Kong University Press, Hong Kong, 412 pp.

Hyde, K.D. (1997b) Can we rapidly measure fungal diversity? *Mycologist* 11, 176–178.

Hyde, K.D. and Hawksworth, D.L. (1997) Measuring and monitoring the biodiversity of microfungi. In: Hyde, K.D. (ed.) *Biodiversity of Tropical Microfungi.* Hong Kong University Press, Hong Kong, pp. 11–28.

Jabbar Miah, M.A., Varshney, J.L. and Sarbhoy, A.K. (1980) Soil fungi of South India. *Proceedings of the Indian National Science Academy*, B 46, 593–602.

Joshi, I.J. and Chauhan, R.K.S. (1982) Investigations into the soil mycoecology of Chambal ravines of India. I. Fungal communities and seasonal succession. *Plant Soil* 66, 329–338.

Katz, B. (1981) Preliminary results of leaf litter decomposing microfungi survey. *Acta Amazônica* 11, 410–411.

Kelley, J. (1995) Microorganisms, indigenous intellectual property rights and the Convention on Biological Diversity. In: Allsopp, D., Colwell, R.R. and Hawksworth, D.L. (eds) *Microbial Diversity and Ecosystem Function.* CAB International, Wallingford, UK, pp. 415–426.

Kennedy, A.C. and Smith, K.L. (1995) Soil microbial diversity and the sustainability of agricultural soils. *Plant Soil* 170, 75–86.

Kirsop, B. (1996) *Access to Ex-situ Microbial Genetic Resources within the Framework to the Convention on Biological Diversity*. World Federation of Culture Collections, WFCC, 25 pp.

Kirsop, B. and Hawksworth, D.L. (1994) *The Biodiversity of Microorganisms and the Role of Microbial Resource Centres.* World Federation of Culture Collections, WFCC, 104 pp.

Kitching, R.L. (1993) Rainforest canopy arthropods: problems for rapid biodiversity assessment. In: Beattie, A. (ed.) *Rapid Biodiversity Assessment, Proceedings of the Biodiversity Assessment Workshop.* Macquarie University, Sydney, pp. 26–30.

Kjøller, A. and Struwe, S. (1982) Microfungi in ecosystems, fungal occurrence and activity in litter and soil. *Oikos* 39, 391–422.

Lee, D., Zo, Y. and Kim, S. (1996) Nonradioactive method to study genetic profiles of natural bacterial communities by PCR single-strand-conformation polymorphism. *Applied and Environmental Microbiology* 62, 3112–3120.

Lees, A.K., Cullen, D.W., Sullivan, L. and Nicholson, M.J. (2002) Development of conventional and quantitative real-time PCR assays for the detection and identification of *Rhizoctonia solani* AG-3 in potato and soil. *Plant Pathology* 51, 293–302.

Lévesque, C.A. and Cock, A.W.A.M. (2004) Molecular phylogeny and taxonomy of the genus *Pythium. Mycological Research* 108, 1363–1383.

Li, S. and Hartman, G.L. (2003) Molecular detection of *Fusarium solani* f. sp. *glycines* in soybean roots and soil. *Plant Pathology* 52, 74–83.

Liu, W., Marsh, T.L., Cheng, H. and Forney, L. (1997) Characterization of microbial diversity by determining terminal restriction fragment length polymorphisms of genes encoding 16S rRNA. *Applied and Environmental Microbiology* 63, 4516–4522.

Lodge, D.J. (1993) Nutrient cycling by fungi in wet tropical forests. In: Isaac, S., Frankland, J.C., Watling, R. and Whalley, A.J.S. (eds) *Aspects of Tropical Mycology.* Cambridge University Press, Cambridge, UK, pp. 37–58.

Lodge, D.J. (1997) Factors related to diversity of decomposer fungi in tropical forests. *Biodiversity and Conservation* 6, 681–688.

Lourd, M. and Bouhot, D. (1987) Recherche et caractérisation de sols résistants aux *Pythium* spp. en Amazonie brésilienne. *Bulletin OEPP/EPPO* 17, 569–575.

Lourd, M., Alves, M.L.B. and Bouhot, D. (1986) Análise qualitativa e quantitativa de espécies de *Pythium* patogênicas dos solos no município de Manaus. I. Solos de terra firme. *Fitopatologia Brasileira* 11, 479–485.

Lowell, J.L. and Klein, D.A. (2001) Comparative single-strand conformation polymorphism (SSCP) and microscopy-based analysis of nitrogen cultivation interactive effects on the fungal community of a semi-arid steppe soil. *FEMS Microbiology Ecology* 36, 85–92.

Maggi, O. and Persiani, A.M. (1992) Etudes comparatives sur les micro-champignons en écosystèmes tropicaux. Rapport final sur les recherches mycologiques du sol. *Mycologia Helvetica* 5, 79–98.

Maggi, O., Persiani, A.M., Casado, M.A. and Pineda, F.D. (1990) Edaphic mycoflora recovery in tropical forests after shifting cultivation. *Acta Oecologica* 11, 337–350.

Marks, G.C. and Mitchell, J.E. (1970) Detection, isolation and pathogenicity of *Phytophthora megasperma* from soils and estimation of inoculum levels. *Phytopathology* 60, 1687–1690.

Masago, H., Yoshikawa, M., Fukada, M. and Nakanishi, N. (1977) Selective inhibition of *Pythium* spp. on a medium for direct isolation of *Phytophtora* spp. from soils and plants. *Phytopathology* 67, 425–428.

Mauchline, T.H., Kerry, B.R. and Hirsch, P.R. (2002) Quantification in soil and the rhizosphere of the nematophagous fungus *Verticillium chlamydosporium* by competitive PCR and comparison with selective plating. *Applied and Environmental Microbiology* 68, 1846–1853.

May, R.M. (1991). A fondness for fungi. *Nature* 352, 475–476.

Mazzola, M. (2002) Mechanisms of natural soil suppressiveness to soilborne diseases. *Antonie van Leeuwenhoek* 81, 557–564.

Mazzola, M. (2004) Assessment and management of soil microbial community structure for disease suppression. *Annual Review of Plant Pathology* 42, 35–59.

McDonald, D. (1969) The influence of the developing groundnut fruit on soil mycoflora. *Transactions of the British Mycological Society* 53, 393–406.

Mills, J.T. and Vlitos, A.J. (1967) Studies on the rhizosphere of sugar cane. *Tropical Agriculture* (Trinidad) 44, 151–157.

Moreira, F.M.S. and Siqueira, J.O. (2002) *Microbiologia e bioquímica do solo.* Editora UFLA, Lavras, Brazil, 626 pp.

Mueller, G.M., Bills, G.F. and Foster, M.S. (2004) *Biodiversity of Fungi. Inventory and monitoring methods.* Elsevier, Amsterdam, 777 pp.

Muyzer, G., Waal, E.C. de and Uitterlinden, A.G. (1993) Profiling of complex microbial populations by denaturing gradient electrophoresis analysis of polymerase chain reaction-amplified genes coding for 16S rRNA. *Applied and Environmental Microbiology* 59, 695–700.

Nechwatal, J., Schlenzig, A., Jung, T., Cooke, D.E.L., Duncan, J.M. and Osswald, W.F. (2001) A combination of baiting and PCR techniques for the detection of *Phytophthora quercina* and *P. citricola* in soil samples from oak stands. *Forest Pathology* 31, 85–97.

O'Donnell, K. and Cigelnik, E. (1997) Two divergent intragenomic rDNA ITS2 types within a monophyletic lineage of the fungus *Fusarium* are nonorthologous. *Molecular Phylogenetics and Evolution* 7, 103–116.

O'Donnell, A.G., Goodfellow, M. and Hawksworth, D.L. (1994) Theoretical and practical aspects of the quantification of biodiversity among microorganisms. *Philosophical Transactions of the Royal Society London. Series B, Biological Sciences* 345, 65–73.

Odunfa, V.S.A. and Oso, B.A. (1979) Fungal populations in the rhizosphere and rhizoplane of cowpea. *Transactions of the British Mycological Society* 73, 21–26.

Ogbonna, C.I.C. and Pugh, G.J.F. (1982) Nigerian soil fungi. *Nova Hedwigia* 36, 795–808.

Papavizas, G.C., Adams, P.B., Lomsden, R.D., Lews, J.A., Dow, R.L., Ayers, W.A. and Kantzer, J.G. (1975) Ecology and epidemiology of *Rhizoctonia solani* in field soil. *Phytopathology* 65, 871–877.

Persiani, A.M. Maggi, O., Casado, M.A. and Pineda, F.D. (1998) Diversity and variability in soil fungi from a disturbed tropical rain forest. *Mycologia* 90, 206–214.

Petrovic, U., Gunde-Cimerman, N. and Zalar, P. (2000) Xerotolerant mycobiota from high altitude Anapurna soil, Nepal. *FEMS Microbiology Letters* 182, 339–342.
Pettitt, T.R., Wakeham, A.J., Wainwright, M.F. and White, J.G. (2002) Comparison of serological, culture, and bait methods for detection of *Pythium* and *Phytophthora* zoospores in water. *Plant Pathology* 51, 720–727.
Pfenning, L.H. (1993) Mikroskopische Bodenpilze des Ostamazonischen Regenwaldes (Brasilien). PhD thesis, Universität Tübingen, Germany, 192 pp.
Pfenning, L.H. (1995) Rhizosphere microfungi in tropical rain forest ecosystems, diversity and practical importance. ISME-7, Santos SP, Brazil, Abstracts, p. 64.
Pfenning, L.H. (1996) Diversity of microfungi. In: Bicudo, C.E.M. and Menezes, N.A. (eds) *Biodiversity in Brazil, a First Approach*. CNPq, São Paulo, Brazil, pp. 65–80.
Pfenning, L.H. (1997) Soil and rhizosphere microfungi from Brazilian tropical forest ecosystems. In: Hyde, K.D. (ed.) *Biodiversity of Tropical Microfungi*. Hong Kong University Press, Hong Kong, pp. 341–365.
Rama Rao, P. (1970) Studies on soil fungi III. Seasonal variation and distribution of microfungi in some soils from Andhra Pradesh (India). *Mycopathatologia Mycologia Applicata* 40, 277–298.
Rambelli, A., Persiani, A.M., Maggi, O., Lunghini, D., Onofri, S., Riess, S., Dowgiallo, G. and Puppi, G. (1983) Comparative studies on microfungi in tropical ecosystems. Mycological studies in southwestern Ivory Coast forest. Report no. 1, MAB, UNESCO, Rome, 102 pp.
Rambelli, A., Persiani, A.M., Maggi, O., Onofri, S., Riess, S., Dowgiallo, G. and Zucconi, L. (1984) Comparative studies in microfungi in tropical ecosystems. Further mycological studies in southwestern Ivory Coast forest. Report no. 2. *Giornale Botanici Italica* 118, 201–243.
Ranjard, L., Poly, F., Lata, J.C., Mouguel, C., Thioulouse, J. and Nazaret, S. (2001) Characterization of bacterial and fungal soil communities by automated ribosomal intergenic spacer analysis fingerprints, biological and methodological variability. *Applied and Environmental Microbiology* 67, 4479–4487.
Robison, B.M. (1970) Micro-fungi of sugar-cane roots and soil in Jamaica. *Tropical Agriculture (Trinidad)* 47, 23–29.
Rodrigues-Guzman, M.P. (2001) Biodiversidad de los hongos fitopatógenos del suelo de Mexico. *Acta Zoológica Mexicana*, numero especial 1, 53–78.
Rossman, A.Y. (1994) A strategy for an all-taxa inventory of fungal biodiversity. In: Peng, C.I. and Chou, C.H. (eds) *Biodiversity and Terrestrial Ecosystems*. Institute of Botany, *Academia Sinica Monograph Series* no. 14, 169–194.
Rossman, A.Y., Tulloss, R.E., O'Dell, T.E. and Thorn, R.G. (1998) *Protocols for an All-Taxa Biodiversity Inventory of Fungi in a Costa Rican Conservation Area*. Parkway Publishers, Boone, North Carolina, 195 pp.
Ryan, M.J. and Smith, D. (2004) Fungal genetic resource centres and the genomic challenge. *Mycological Research* 108, 1351–1362.
Sanhueza, R.M.V. and Balmer, E. (1985) Levantamento de fungos associados à podridão de raízes de cana-de-açúcar na região de Campos, RJ. *Fitopatologia Brasileira* 10, 505–513.
Santos, A.C. dos, Cavalcanti, M.A. and Santos Fernandes, M.J. dos (1989) Fungos isolados da rizosfera de cana-de-açúcar da Zona da Mata de Pernambuco. *Revista Brasileira de Botânica* 12, 23–29.
Schneider, R.W. (1984) *Suppressive Soils and Plant Disease*. APS Press, St Paul, Minnesota.
Selenska, S. and Klingmüller, W. (1992) Direct recovery and molecular analysis of DNA and RNA from soil. *Microbial Releases* 1, 41–46.
Siebert, P.D. and Larrick, J.W. (1992) Competitive PCR. *Nature* 359, 557–558.
Singleton, L.L., Mihail, J.D. and Rush, C.M. (1992) *Methods for Research on Soilborne Phytopathogenic Fungi*. APS Press, St Paul, Minnesota, 266 pp.
Smit, E., Leeflang, P., Glandorf, B. and Van Elsas, J.D. (1999) Analysis of fungal diversity in the wheat rhizosphere by sequencing of cloned PCR-amplified genes encoding 18S rRNA and temperature gradient gel electrophoresis. *Applied and Environmental Microbiology* 66, 2614–2621.
Smith, D. and Waller, J.M. (1992) Culture collections of microorganisms, their importance in tropical plant pathology. *Fitopatologia Brasileira* 17, 5–12.
Sneh, B., Burpee, L. and Ogoshi, A. (1991) *Identification of Rhizoctonia species*. APS Press, St Paul, Minnesota, 133 pp.
Swift, M. and Bignell, D. (2001) Standard methods for assessment of soil biodiversity and land use practice. ICRA, Bogor, ASB Lecture Note 6B, 34 pp.
Thorn, R.G., Reddy, C.A., Harris, D. and Paul, E.A. (1996) Isolation of saprophytic basidiomycetes from soil. *Applied and Environmental Microbiology* 62, 4288–4292.

Tiunov, A.V. and Scheu, S. (2000) Microfungal communities in soil, litter and casts of *Lumbricus terrestris* L. (Lumbricidae), a laboratory experiment. *Applied Soil Ecology* 14, 17–26.
Tsai, Y.L. and Olson, B.H. (1992) Rapid method for direct extraction of DNA from soil and sediments. *Applied and Environmental Microbiology* 58, 1070–1074.
Tsao, P.H., Erwin, D.C. and Bartnicki-Garcia, S. (1983) *Phytophthora. Its Biology, Taxonomy, Ecology and Pathology.* APS Press, St Paul, Minnesota, 392 pp.
Valinsky, L., Vedova, G.D., Jiang, T. and Borneman, J. (2002) Oligonucleotide fingerprinting of rRNA genes for analysis of fungal community composition. *Applied Environmental Microbiology* 68, 5999–6004.
Vandenkoornhuyse, P., Baldauf, S., Leyval, C., Straczek, J. and Young, J.P.W. (2002) Extensive fungal diversity in plant roots. *Science* 295, 2051.
Van Elsas, J.D., Duarte, G.F., Keijzer-Wolters, A. and Smit, E. (2000) Analysis of the dynamics of fungal communities in soil via fungal-specific PCR of soil DNA followed by denaturing gradient gel electrophoresis. *Journal of Microbiological Methods* 43, 133–151.
Varghese, G. (1972) Soil microflora of plantations and natural rain forest of west Malaysia. *Mycopathologia Mycologia Applicata* 48, 43–61.
Viaud, M., Pasquier, A. and Brygoo, Y. (2000) Diversity of soil fungi studied by PCR-RFLP of ITS. *Mycological Research* 104, 1027–1032.
Wainwright, M. (1988) Metabolic diversity of fungi in relation to growth and mineral cycling in soil – a review. *Transactions of the British Mycological Society* 90, 159–170.
Wellington, E.M.H., Berry, A. and Krsek, M. (2003) Resolving functional diversity in relation to microbial community structure in soil, exploiting genomics and stable isotope probing. *Current Opinion in Microbiology* 6, 295–301.
Whipps, J.M., McQuilken, M.P. and Budge, S.P. (1993) Use of fungal antagonists for biocontrol of damping-off and Sclerotinia disease. *Pesticide Science* 37, 309–317.
Widden, P. (1986) Microfungal community structure from forest soils in southern Quebec, using discriminant function and factor analysis. *Canadian Journal of Botany* 64, 1402–1412.
Widden, P. and Parkinson, D. (1973) Fungi from Canadian coniferous forest soils. *Canadian Journal of Botany* 51, 2275–2290.
Wildman, H.G. (1997) Potential of tropical microfungi within the pharmaceutical industry. In: Hyde, K.D. (ed.) *Diversity of Tropical Microfungi.* Hong Kong University Press, Hong Kong, pp. 29–46.
World Federation of Culture Collections (WFCC) (1999) *Guidelines for the Establishment and Operation of Collections of Cultures of Microorganisms,* 2nd edn. WFCC, 44 pp.
Zak, J.C. and Visser, S. (1996) An appraisal of soil fungal diversity: the crossroads between taxonomic and functional biodiversity. *Biodiversity and Conservation* 5, 169–183.

10 Diversity of Arbuscular Mycorrhizal Fungi in Brazilian Ecosystems

S.L. Stürmer[1] and J.O. Siqueira[2]
[1]Departamento de Ciências Naturais (DCN), Universidade Regional de Blumenau (FURB), Cx.P. 1507, 89010-971 Blumenau, SC, Brazil, e-mail: sturmer@furb.br;
[2]Departamento de Ciência do Solo (DCS), Universidade Federal de Lavras (UFLA), Cx.P. 37. 37200-000 Lavras, MG, Brazil, e-mail: siqueira@ufla.br

Introduction

Diversity has two main components. The first is the number of species or species richness, which is generally related to environmental parameters such as climate, latitude, ecosystem productivity and land use disturbance. The second component is evenness (or equitability) that measures the relative abundance of each species. This measure is important since species richness weighs taxa equally without considering whether it is rare or abundant within the community (Magurran, 1988; Schluter and Ricklefs, 1993). Nevertheless, diversity analyses can be carried out at taxonomic levels higher than species (i.e. genera, family), revealing distribution patterns that in turn can be related to ecosystem productivity. The major goal of biodiversity studies is the compilation of inventories of living organisms for conservation purposes to serve as references for monitoring natural resources, to maintain genetic diversity and to exploit biochemical or functional potential (Palleroni, 1994; Beare *et al.*, 1995). Especially, the importance of biodiversity to ecosystem function, including biogeochemical cycling and ecosystem stability and productivity, has received much attention (Beare *et al.*, 1995; Wardle *et al.*, 2004). This has brought the vast biodiversity of soil organisms (e.g. fungi, bacteria, soil fauna) that perform certain biogeochemical transformations within ecosystems into focus.

Among soil organisms, members of the kingdom Fungi represent a major component of total soil biomass. They are involved in ecosystem processes such as plant litter decomposition (Cromack and Caldwell, 1992), ammonification of organic nitrogen and nitrification (Read *et al.*, 1989), weathering of soil minerals through the excretion of organic acids and Fe siderophores (Mehta *et al.*, 1979) and the influencing of soil structure by filamentous (hyphal) growth and protein production (Miller and Jastrow, 1992; Wright and Upadhyaya, 1998). Despite their roles, fungi have received little attention in discussions of biodiversity, even though they are considered the second largest group of organisms after insects (Hawksworth, 1991). On the basis of Ainsworth and Bisby's *Dictionary of the Fungi* (Hawksworth *et al.*, 1983) and the *Index of Fungi* of the International Mycological Institute, the number of known species of fungi is about 69,000. However, estimates based on vascular plant:fungus ratios in different regions of the world raise this number to over 1.5 million species. This estimate can be considered conservative since fungi associated with insects were not fully represented and plant:fungus ratios were derived primarily from the northern temperate regions (Hawksworth, 1991). Some parts of the world are virtually unsampled for fungi and tropical regions are expected to be a rich source of new species. Among fungi associated with vascular plants, arbuscular mycorrhizal fungi (AMF) are among the most ubiquitous organisms in all terrestrial ecosystems, forming the arbuscular endomycorrhizal associations with plant roots.

In this chapter, we first provide a brief description of this association and the current systematics of its fungal component. Thereafter, we present an analysis of AMF diversity in different Brazilian ecosystems, providing a list of species so far detected in surveys and establishing some patterns of distribution for species and families. Then, we review how fungal diversity is related to plant diversity and ecosystem productivity and discuss management practices affecting mycorrhizal fungal communities.

Arbuscular Mycorrhizal Fungi

Amongst the diverse parasitic and mutualistic symbioses formed by plants and microorganisms, the one established by plant roots and AMF is by far the most common mutualistic association found in natural ecosystems and agroecosystems. Evidence from the fossil record (Pirozynski, 1981), molecular biology (Simon *et al.*, 1993) and phylogenetic analyses (Morton, 2000) place the origin of these fungi and, therefore, the symbiosis back to 353–462 millions years ago during the Devonian period. This long evolutionary history, associated with the lack of host specificity, resulted in a pattern of distribution of these fungi throughout plant communities rivalled only by bacteria. Therefore, this association can be detected in the roots of Pteridophytes, Gymnosperms and Magnoliophytas, and in most natural ecosystems such as sand dunes, tropical forests, deserts, savannahs, grasslands as well as in agroecosystems (e.g. fruit crops, pastures, field crops) and degraded lands.

AMF are obligate symbionts because for completion of their life cycle, they must be associated with a living plant root that provides them with carbon and all the necessary factors for development and sporulation (Siqueira *et al.*, 1985). Within the root cortex and expanding to the bulk soil, AMF form meristic structures with distinct symbiotic functions. Arbuscules are the characteristic structures unique to AMF (Fig. 10.1a); they consist of highly branched hyphae that develop between the cell wall and the plasma membrane of plant cells and are responsible for the exchange of nutrients between symbionts. Vesicles are globose, elliptical or knobby structures (Fig. 10.1b) containing lipids and glycogen granules serving as a storage organ for the fungus. They can be formed within or in between cortex cells and are differentiated only by members of the families Acaulosporaceae and Glomeraceae (Smith and Read, 1997). Auxiliary cells (Fig. 10.1c) also serve as storage organs but they are formed outside the roots only by members of Gigasporaceae. Intra- and extraradical hyphae and mycelia are important for establishing new mycorrhizal associations (Fig. 10.1d) and for scavenging and uptaking nutrients from the soil and enhancing soil aggregation. Asexual spores formed by AMF (Fig. 10.1e and f) are related to fungal dispersal and survival. They are the largest soil fungal spores (diameter 45–700 μm), with colours ranging from hyaline and pale yellow to brownish, reddish and even black. The spore wall structure, which can be smooth or ornamented, together with spore ontogeny form the basis for current taxonomy and systematics (Morton and Benny, 1990).

The main effect of AMF on the plant host is the increased nutrient uptake from the soil, which results in improvement of plant growth, survival and yield (Smith and Read, 1997). Phosphorus is the most important nutrient that mycorrhizal fungi help plants to absorb because of its low availability in most soils and slow diffusion rate through soil. Extraradical hyphae of mycorrhizal fungi

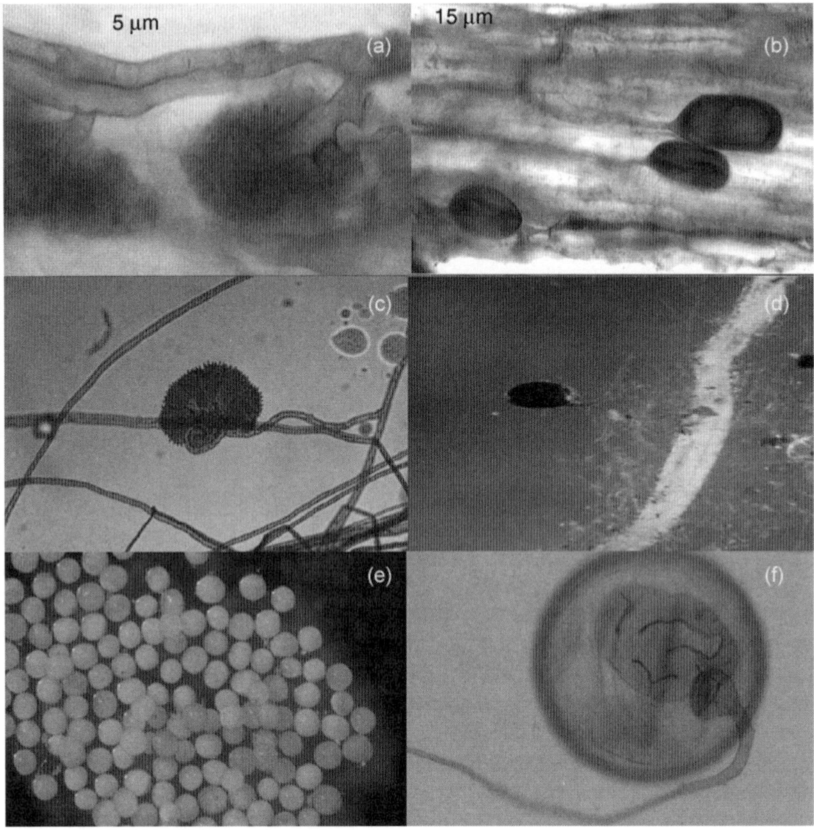

Fig. 10.1. Vegetative and reproductive structures differentiated by arbuscular mycorrhizal fungi: (a) arbuscule of *Glomus* within a root cortex cell, (b) vesicles formed in the root cortex, (c) auxiliary cell differentiated by a species of *Gigaspora,* (d) extraradical hyphae growing into the soil and attached to the root, (e) spores produced by a member of Gigasporaceae and (f) spores of *Scutellospora* sp. showing the germination shield. Data on arbuscules and vesicles available at http://invam.caf.wvu.edu

grow behind the P-depletion zone formed around actively absorbing plant roots, scavenge and take up P ions from the bulk soil, thereby increasing the soil volume explored for this nutrient (Bolan, 1991). Besides P nutrition, mycorrhizal fungi have been implied in plant–water relationships improving drought resistance (Sánchez-Diaz *et al.*, 1990), protection against pathogens like nematodes and root rot fungi (Newsham *et al.*, 1995a) and alleviating environmental stress (Diaz *et al.*, 1996). Extraradical hyphae also influence soil structure by increasing aggregate stability and formation (Miller and Jastrow, 1992; Wright *et al.*, 1996). Based on the multitude of roles that AMF play in plant–soil systems, the endomycorrhizal association has been viewed recently as being multifunctional (Newsham *et al.*, 1995b) and an important indicator for the assessment of soil quality (Kling and Jakobsen, 1998). Moreover, since the association represents an important link between the biotic portion of ecosystems and the geochemical matrix, arbuscular mycorrhizae can affect higher-level processes within ecosystems and the AMF species represent a keystone group of soil organisms (O'Neill *et al.*, 1991) that can potentially affect plant productivity, above-ground diversity and soil biotic and abiotic characteristics related to ecosystem sustainability.

Taxonomy and Phylogeny

Analysis of biodiversity and species distribution patterns for any group of organisms needs primarily the establishment of a taxonomic framework. The systematics of glomalean fungi have changed dramatically over the last decade, especially in the past few years. Studies on spore ontogeny and on the use of molecular tools associated with morphological characters have resulted in new taxa. The historical background of glomalean systematics has been reviewed elsewhere (Stürmer, 1999) and is not discussed here, but major aspects of the taxonomic framework are briefly described in this chapter. The first Linnean classification of the AMF established by Gerdemann and Trappe (1974) included all described species in the family Endogonaceae (division Zygomycota, order Endogonales). AMF identification and species description were based on spore size and colour and on the analysis of phenotypically distinct walls of asexual spores (e.g. Walker, 1983; Morton, 1986). Morton and Benny (1990) proposed a new classification of glomalean fungi based on cladistic analysis of 57 AMF species using morphological characters of spores and mycorrhizal morphology. In this classification, all AMF species formed a monophyletic group defined by the establishment of mutualistic symbiosis with land plants and the production of highly branched intraradical arbuscules. The order Glomerales was then erected to comprise all AMF species in three distinct families, Acaulosporaceae (genera *Acaulospora* and *Entrophospora*), Glomeraceae (genera *Glomus* and *Sclerocystis*) and Gigasporaceae (genera *Gigaspora* and *Scutellospora*). Further studies were undertaken to determine the origin and individuality of spores' subcellular characters used to limit and to identify the species. This resulted in a research programme that defined distinct stages in spore ontogeny of *Scutellospora* (Franke and Morton, 1994), *Gigaspora* (Bentivenga and Morton, 1995), *Glomus* (Stürmer and Morton, 1997), *Acaulospora* and *Entrophospora* (Stürmer and Morton, 1999). Developmental stages are illustrated for spores of *Glomus etunicatum* (Fig. 10.2a–c) and *Acaulospora spinosa* (Fig. 10.2d–f).

More recently, molecular tools including the analysis of ribosomal DNA of selected species resulted in a completely new picture of the AMF systematics at the genus, family and higher levels of the taxonomic hierarchy. Redecker *et al.* (2000a), on the basis of a phylogenetic analysis of the 18S ribosomal subunit, demonstrated that *Glomus sinuosum* and *Sclerocystis*

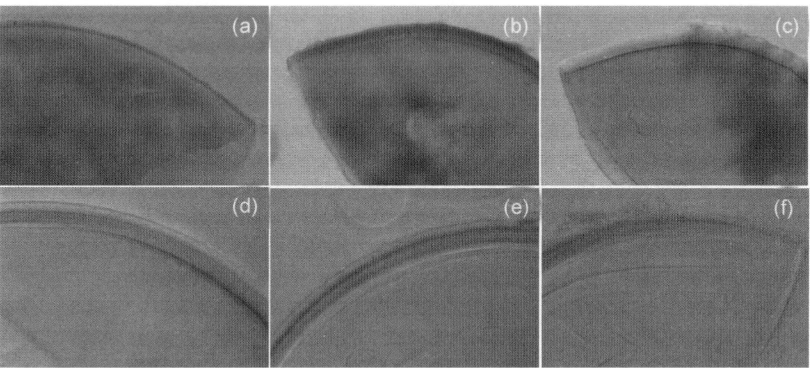

Fig. 10.2. Spore development stages of (a–c) *G. etunicatum* and (d–f) *A. spinosa*: (a) spore wall formed only by a mucilagenous layer, (b) spore wall differentiating laminated layers, (c) mature spore with laminated layer fully developed and mucilagenous layer sloughing, (d) spore wall formed by a mucilagenous layer and a laminated layer, (e) spore differentiating inner germinal walls and (f) spore fully mature with spore wall and inner germinal walls.

coremioides were closely related and both inserted within a larger monophyletic clade formed by well-established species like *Glomus mosseae* and *Glomus intraradices*. Their observation resulted in the transfer of the monospecific genus *Sclerocystis* into the genus *Glomus*. Ribosomal DNA sequences also revealed two ancient clades with a large phylogenetic distance between each other and with all other three Glomerales families (Morton and Redecker, 2001). This distance suggested that the new clades deserved above-genus ranking within the taxonomic hierarchy and the families Archaeosporaceae and Paraglomeraceae were proposed. Former species of *Acaulospora* and *Glomus* were transferred to the newly erected genera *Archaeospora* (F. Archaeosporaceae) and *Paraglomus* (F. Paraglomeraceae). Both families were also distinguished from the others based on fatty acid content, mycorrhizal morphology and staining as well as immunological reactions against monoclonal antibodies.

Classification of AMF experienced major changes after the molecular phylogenetic analyses based on the SSU rRNA sequences carried out by Schüßler *et al.* (2001). As a result, the AMF were removed from the polyphyletic Zygomycota and placed in a newly erected monophyletic group, the Glomeromycota. This change placed this group of organisms at the same level as the classical groups of Basidiomycota and Ascomycota. Schüßler *et al.* (2001) also proposed three new orders and several families separated from the former Glomerales. Congruence of morphological, biochemical and molecular data sets may provide a more refined understanding of the phylogenetic relationships among AMF. We adopt the recognition of a new division as given by Schüßler *et al.* (2001), but we consider AMF still pertaining to the order Glomerales, resulting in the classification scheme followed throughout this chapter (Fig. 10.3, Table 10.1). Recent evidence indicates that the AMF are highly divergent descendants of allelic nucleotide sequences that evolved asexually into a multigenomic (multinucleous) organism with high within-individual genetic variation (Kuhn *et al.*,

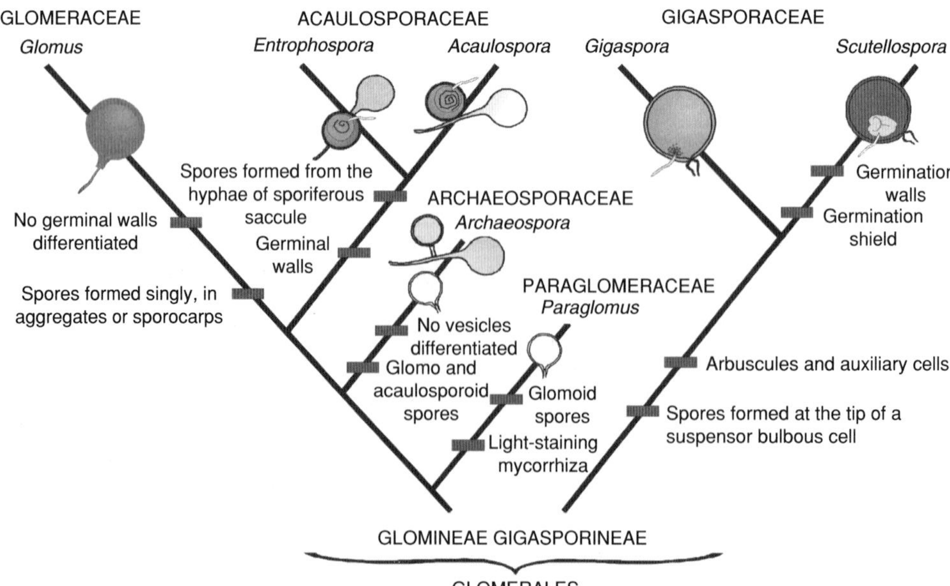

Fig. 10.3. Classification of AMF indicating main characters defining families within the order Glomerales. Based on and modified from http://invam.caf.wvu.edu

Table 10.1. Classification of AMF and major morphological characters defining genera in Glomerales.

Division Glomeromycota Schüβler, Scharzott & Walker
Order Glomerales Morton & Benny
Suborder Glomineae Morton & Benny
Family Glomeraceae Pirozysnki & Dalpé
Glomus Tulasne & Tulasne (85 species)

Spores formed blastically on a subtending hyphae, singly, in loose aggregates or in a sporocarp. Vesicles are thin-walled and ellipsoid. Intraradical hyphae rarely coiled, with cross-connecting branch hyphae. Mycorrhizae stain darkly. Arbuscules with flared or cylindrical trunks with incremental narrowing of branch hyphae. Spores with spore wall formed by a variable number of layers all originating from the subtending hyphae, no germinal walls differentiated. Germination through the lumen of the subtending hyphae or through the spore wall.

Family Acaulosporaceae Morton & Benny
Acaulospora Gerd. & Trappe emend. Berch (31 species)

Spores formed laterally from the neck of a sporiferous saccule, which leaves one scar on the spore surface. Vesicles vary in shape with knobs and concavities. Intraradical hyphae straight or coiled near the entry points. Mycorrhizae stain weakly. Arbuscules with flared or cylindrical trunks with incremental narrowing of branch hyphae. Spores with spore wall formed by three layers and two inner germinal walls, each with two thin layers that can be adherent. The innermost germinal wall has a beaded surface. Germination through a flexible, plate-like germination orb.

Entrophospora Ames & Schneider (4 species)

Spores formed within the neck of a sporiferous saccule, which leaves two scars on the spore surface. Vesicles, arbuscules, intraradical hyphae and mycorrhizae staining as in *Acaulospora*. Spores with spore wall formed by two layers. Other spore subcellular structures and germination identical to that in *Acaulospora*.

Family Archaeosporaceae Morton & Redecker
Archaeospora Morton & Redecker (3 species)

Spores formed terminally from a subtending hyphae or as a branch from a structure resembling a sporiferous saccule. Arbuscules and intraradical hyphae stain lightly. Vesicles and auxiliary cells are not differentiated. Spores with spore wall formed by 3–4 layers and no true bilayered germinal wall formed. Dimorphic species found forming acaulosporoid and glomoid spores.

Family Paraglomeraceae Morton & Redecker
Paraglomus Morton & Redecker (2 species)

Spores formed terminally from a subtending hyphae (like in *Glomus*). Arbuscules and intraradical hyphae stain lightly. Vesicles and auxiliary cells are not differentiated. Spore subcellular structures and germination like in *Glomus*.

Suborder Gigasporineae Morton & Benny
Family Gigasporaceae Morton & Benny
Gigaspora Gerd. & Trappe (5 species)

Spores formed terminally on a bulbous sporogenous cell; auxiliary cells finely papillate or echinulate. No vesicles produced. Intraradical hyphae frequently coiled, especially near entry points, often knobby or with projections. Arbuscules with swollen trunks with abrupt narrowing of branch hyphae. Spores with spore wall formed by two permanent layers, no inner germinal walls differentiated. At germination, a thin layer interspersed with warts differentiates and a germ tube grows through the spore wall.

Scutellospora Walker & Sanders (30 species)

Spores formed terminally on a bulbous sporogenous cell; auxiliary cells almost smooth to knobby. No vesicles produced. Arbuscules and intraradical hyphae similar in morphology to organisms of *Gigaspora*. Spores with spore wall formed by two permanent layers and 1–3 inner germinal walls, each with two layers. Germ tube grows from flexible, plate-like germination shield that differentiates on the surface of the last germinal wall.

Sources: compiled from Morton and Benny (1990) and http://invam.caf.wvu.edu

2001). This fact may limit achieving a correct and appropriate taxonomical framework, and therefore represents a major drawback for accurate diversity studies of these fungi.

Diversity and Distribution

Species richness

The presence and identification of spores in the rhizosphere or bulk soil is the most common and simplest method to estimate AMF species abundance and richness in plant communities. Spore-based identification is necessary because spores are the only fungal stages that possess morphological characters to define species in this group of organisms (Morton *et al.*, 1995). Species separation based on mycorrhizal roots, however, is possible when the root colonization pattern is well established (Abbott and Gazey, 1994). To assess the status of the Brazilian inventory of AMF species, we analysed 28 scientific publications appearing in national and international journals over the last 20 years, mainly covering six different ecosystems with quite distinctive plant communities and management (Table 10.2). These surveys include agroecosystems (annual crops, fruit crops, pastures, etc), coffee plantations, a degraded land resulting from heavy-metal pollution and four natural ecosystems (maritime sand dunes, tropical and temperate forest (Atlantic and *Araucaria* forests) and the Brazilian tropical savannah known as 'cerrado'). Although coffee plantations fit within agroecosystems, we considered them separately since coffee represents the best-studied crop for mycorrhizal associations in Brazil. Ten studies were carried out in agroecosystems, three to five in natural ecosystems and two in degraded areas (Table 10.2).

In Brazilian ecosystems, a total of 79 AMF have been identified to species level (Table 10.3), which represents approximately 50% of the total number of AMF species (160) formally described in the literature (available at http://invam.caf.wvu.edu). These data indicate that Brazilian ecosystems are an important source of AMF diversity, deserving further studies and a defined germplasm conservation policy. This species richness is relatively high considering that fungal inventories in Brazil are mainly concentrated in some regions, especially in those with researchers with taxonomic expertise, resulting in less than 35% of all politically defined states in Brazil being surveyed for AMF (Fig. 10.4). Most of the studies were carried out in the states of São Paulo and Minas Gerais, both in the south-eastern region of the country. Only four studies were done in the south region and only three of the nine states of the north-eastern region have been sampled for AMF. Despite the presence of representative and relevant ecosystems, in several states mostly in the north and the western central part, the presence of AMF has been completely ignored. For instance, in the Amazon region where the largest area of the Amazon tropical forest is located, an extensive survey of AMF diversity was not found in the literature. Finding out mycorrhizal status and AMF occurrence in the pristine forest and in any of the diverse land use systems (e.g. pastures, annual crops) in that region is urgent. The Atlantic forest, which represents one of the most diverse ecosystems in the world, has been surveyed twice, but only in the São Paulo state.

The relatively high diversity of the AMF recorded in agroecosystems (Table 10.2) can be explained by the high frequency of surveys carried out in these systems. A large number of species, 23–49, were reported in the other ecosystems, although in bauxite mine soils under rehabilitation, only six species were recovered (Melloni *et al.*, 2003). It is interesting to note the high proportion of undescribed species that also represents part of AMF diversity. In coffee plantations, 26% of the species (12 species) recorded could not be assigned to any known species and in agroecosystems and sand dunes, 20% and 21%, respectively, of the species were not identified (Table 10.2). Coffee plantations and cerrado had the largest average number of AMF species per

Table 10.2. Total number of identified (ID) and unidentified AMF species and average number of species recorded in independent surveys in different Brazilian ecosystems.

System/ecosystem	AMF species total/non-ID[a]	AMF species average per study (range)[b]	Reference/code
Agrosystem (Ag)	79/16	19 (9–25)	Trufem & Bononi (1985) (Ag1); Siqueira *et al.* (1989) (Ag2); Schenck *et al.* (1989) (Ag3); Maia & Trufem (1990) (Ag4); Paula *et al.* (1993) (Ag5); Weber & Oliveira (1994) (Ag6); Melo *et al.* (1997) (Ag7); Carrenho *et al.* (1998) (Ag8); Trindade (1998) (Ag9); Carrenho *et al.* (2001b) (Ag10)
Cerrado (C)	44/4	22 (17–28)	Bononi & Trufem (1983) (C1); Schenck *et al.* (1989) (C2); Siqueira *et al.* (1989) (C3)
Coffee plantations (Cf)	46/12	22 (8–32)	Lopes *et al.* (1983) (Cf1); Fernandes & Siqueira (1989) (Cf2); Oliveira *et al.* (1990) (Cf3); Colozzi-Filho & Cardoso (2000) (Cf4)
Forest (F)	49/7	17 (13–22)	Trufem & Viriato (1990) (F1); Gomes & Trufem (1998) (F2); Breuninger *et al.* (2000) (F3); Carrenho *et al.* (2001a) (F4); Zandavalli (2001) (F5)
Sand dunes (Sd)	37/8	16 (12–24)	Trufem *et al.* (1989) (Sd1); Trufem *et al.* (1994) (Sd2); Stürmer & Bellei (1994) (Sd3); Cordoba *et al.* (2001) (Sd4)
Degraded areas (Da)	23/5	13 (6–21)	Klauberg-Filho *et al.* (2002) (Da1); Melloni *et al.* (2003) (Da2)

[a]Numbers indicate total number of AMF species recovered from each ecosystem/number of non-identified species.
[b]Range indicates the minimum and the maximum number of species recovered in each ecosystem considering all surveys.

study, a total of 22, followed by agroecosystems with 19 species. The largest range in the number of AMF species recorded in studies within an ecosystem was in coffee plantations (24) where in one study only 8 AMF species were recorded, while in another study 32 species were detected (Table 10.2). The range for all other ecosystems lies between 9 and 15.

Distribution of genera and families

Diversity analyses carried out at genera and family levels may provide important clues on species diversity patterns as long as the taxonomic level reveals the historical development of the biota (Schluter and Ricklefs, 1993). Considering the origin and long period of evolution of the glomalean fungi since their appearance as a taxonomic group (Redecker *et al.*, 2000b) and the long time they had for dispersal and distribution over the ecosystems (Heckman *et al.*, 2001), it can be assumed that species occurrence in a given area reflects historical and recent occupation of that area. The seven Glomerales genera are represented in all the ecosystems analysed, except *Entrophospora*, which was not recorded so far in sand dunes (Fig. 10.5a). Considering

Table 10.3. List of AMF species and their occurrence in Brazilian ecosystems.

Family/AMF species	Ecosystems					
	Ag	C	Cf	F	Sd	Da
Acaulosporaceae						
Acaulospora bireticulata Rothwell & Trappe	x			x		
A. delicata Walker, Pfeiffer & Bloss	x					
A. dilatata Morton	x					
A. elegans Trappe & Gerdemann	x					
A. foveata Trappe & Janos	x			x		x
A. lacunosa Morton				x		
A. laevis Gerdemann & Trappe	x	x	x	x	x	
A. longula Spain & Schenck	x		x	x		
A. mellea Spain & Schenck	x	x	x	x		x
A. morrowiae Spain & Schenck	x	x	x	x		x
A. myriocarpa Spain, Sieverding & Schenck	x	x	x			
A. rehmii Sieverding & Toro	x			x		
A. rugosa Morton	x					
A. scrobiculata Trappe	x	x	x	x	x	x
A. spinosa Walker & Trappe	x	x	x	x	x	
A. tuberculata Janos & Trappe	x			x	x	x
Entrophospora colombiana Spain & Schenck	x	x	x	x		x
E. infrequens (Hall) Ames & Schneider	x		x	x		
E. kentinensis Wu & Liu				x		
Glomeraceae						
Glomus aggregatum Schenck & Smith			x	x	x	
G. albidum Walker & Rhodes	x	x	x			
G. ambisporum Smith & Schenck				x	x	
G. claroideum Schenck & Smith	x			x		
G. clarum Nicol. & Schenck	x	x	x	x		x
G. clavisporum (Trappe) Almeida & Schenck	x	x	x			
G. coremioides (Berk. & Broome) Redecker & Morton	x	x	x			
G. deserticola Trappe, Bloss & Menge	x					
G. diaphanum Morton & Walker	x	x	x			
G. etunicatum Becker & Gerdemann	x	x	x	x	x	
G. fasciculatum (Thaxter) Gerd. & Trappe emend. Walker & Koske	x	x	x	x	x	
G. fulvum (Berk & Broome) Trappe & Gerd.	x	x		x		
G. geosporum (Nicol. & Gerd.) Walker	x	x		x		
G. globiferum Koske & Walker	x			x	x	
G. intraradices Schenck & Smith	x	x	x			x
G. invermaium Hall	x			x		
G. macrocarpum Tulasne & Tulasne	x	x	x	x		
G. maculosum Miller & Walker	x					
G. microaggregatum Koske, Gemma & Olexia				x		x
G. microcarpum Tulasne & Tulasne	x	x	x	x		
G. monosporum Gerdemann & Trappe	x	x			x	
G. mosseae (Nicol. & Gerd.) Gerd. & Trappe	x	x	x			
G. pansihalos Berch & Koske				x		
G. rubiforme (Gerd. & Trappe) Almeida & Schenck	x					
G. sinuosum (Gerd. & Bakshi) Almedia & Schenck	x	x	x	x	x	
G. scintillans Rose & Trappe		x				
G. tenue (Greenhall) Hall					x	

Continued

Table 10.3. List of AMF species and their occurrence in Brazilian ecosystems. – cont'd

Family/AMF species	Ecosystems					
	Ag	C	Cf	F	Sd	Da
G. tortuosum Schenck & Smith	x		x	x		
Gigasporaceae						
Gigaspora albida Schenck & Smith	x				x	x
Gigaspora decipiens Hall & Abbott	x	x	x	x	x	
Gigaspora gigantea (Nicol. & Gerd.) Gerd. & Trappe	x	x	x	x	x	
Gigaspora margarita Becker & Hall	x	x	x	x	x	x
Gigaspora rosea Nicol. & Schenck		x	x			
Scutellospora auriglоba (Hall) Walker & Sanders	x	x		x		
S. biornata Spain, Sieverding & Toro	x			x		
S. calospora (Nicol. & Gerd.) Walker & Sanders	x	x		x	x	
S. castanea Walker						x
S. cerradensis Spain & Miranda	x					
S. coralloidea (Trappe, Gerd. & Ho) Walker & Sanders	x				x	
S. dipapillosa (Walker & Koske) Walker & Sanders	x	x	x			
S. erythropa (Koske & Walker) Walker & Sanders	x			x		
S. fulgida Koske & Walker	x			x	x	x
S. gilmorei (Trappe & Gerd.) Walker & Sanders	x	x	x		x	
S. gregaria (Schenck & Nicol.) Walker & Sanders	x				x	
S. hawaiiensis Koske & Gemma					x	
S. heterogama (Nicol. & Gerd.) Walker & Sanders	x	x	x	x		x
S. minuta (Ferr. & Herr.) Walker & Sanders	x					
S. nigra (Redhead) Walker & Sanders	x	x				
S. pellucida (Nicol & Schenck) Walker & Sanders	x	x	x	x	x	x
S. persica (Koske & Walker) Walker & Sanders	x				x	x
S. reticulata (Koske, Miller & Walker) Walker & Sanders		x				
S. scutata Walker & Diederichs					x	
S. tricalyptra (Herr. & Ferr.) Walker & Sanders		x				
S. verrucosa (Koske & Walker) Walker & Sanders		x	x	x	x	
S. weresubiae Koske & Walker					x	
Archaeosporaceae						
Archaeospora leptoticha (Schenck & Smith) Morton & Redecker	x	x	x		x	x
A. gerdemanni (Rose, Daniels & Trappe) Morton & Redecker	x	x	x	x		
A. trappei (Ames & Linderman) Morton & Redecker	x					
Paraglomeraceae						
Paraglomus brasilianum (Spain & Miranda) Morton & Redecker	x					
P. occultum (Walker) Morton & Redecker	x	x	x	x	x	x

the number of species recorded per genus, *Glomus* is predominant in agroecosystems, cerrado, coffee plantations and forest. Only in sand dunes *Scutellospora* predominates followed by *Glomus* and in a heavy-metal-contaminated area, *Acaulospora* and *Scutellospora* codominated (Fig. 10.5a). However, the pattern depicted in Fig. 10.5a can be misleading when the total number of species formally described within each genus is considered. *Glomus* is the largest genus with 85 species, followed by *Acaulospora*

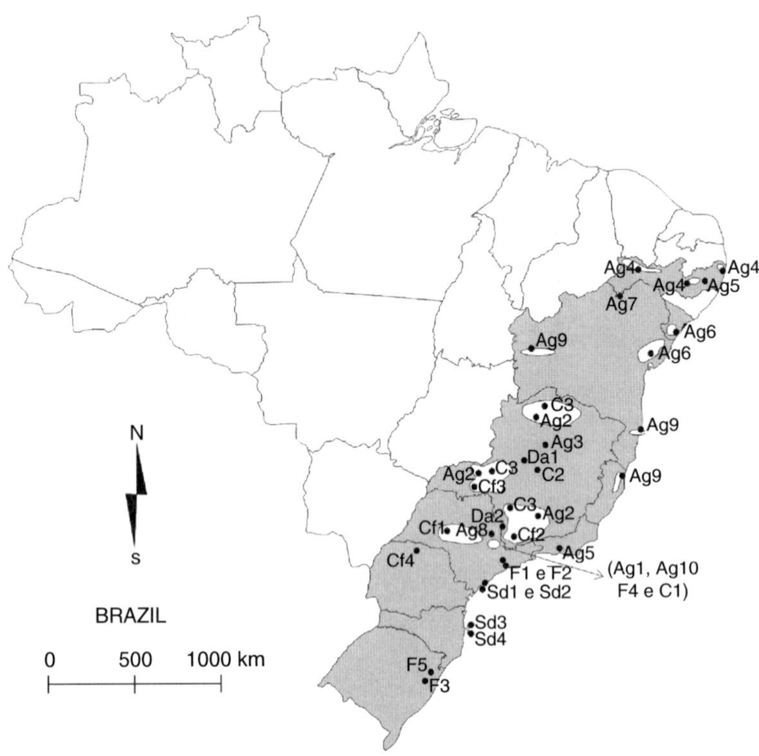

Fig. 10.4. Map of Brazil indicating the location of the 28 studies analysed in this chapter. Note that some studies were carried out in more than one locality and are indicated as white areas within grey-filled states. See Table 10.2 for explanation of codes.

and *Scutellospora* with 31 and 30 species, respectively. When the number of species within a genus in each ecosystem is considered as a proportion of the total species number described for that genus, a completely new picture arises (Fig. 10.5b). In agroecosystems and degraded areas, the genera *Acaulospora* and *Scutellospora* predominate instead of *Glomus*. *Scutellospora* is also predominant in cerrado and sand dunes while *Acaulospora* predominates in forest and coffee plantations. We excluded *Gigaspora*, *Entrophospora*, *Archaeospora* and *Paraglomus* from this analysis because of the low number of species (< 5) described in each of these genera.

Family distribution within ecosystems closely follows genera distribution. This relationship is expected considering that each family is composed of one or two genera and some genera have a low number of species. Considering absolute numbers of species within each family, the Glomeraceae are dominant in agroecosystems, cerrado, coffee plantations and forest areas, while the Gigasporaceae dominate in sand dunes (Fig. 10.6a). Acaulosporaceae is the second dominant family in forest ecosystems, being codominant with Gigasporaceae in coffee plantations and degraded areas. Considering the proportion of total species described within each family, a similar pattern to that detected for genera emerges. Comparing Fig. 10.6b with Fig. 10.5b, it can be seen that the most striking difference between genus and family distribution is that Acaulosporaceae and Gigasporaceae codominate in coffee plantations, whereas *Acaulospora* is dominant at the genus level in this ecosystem.

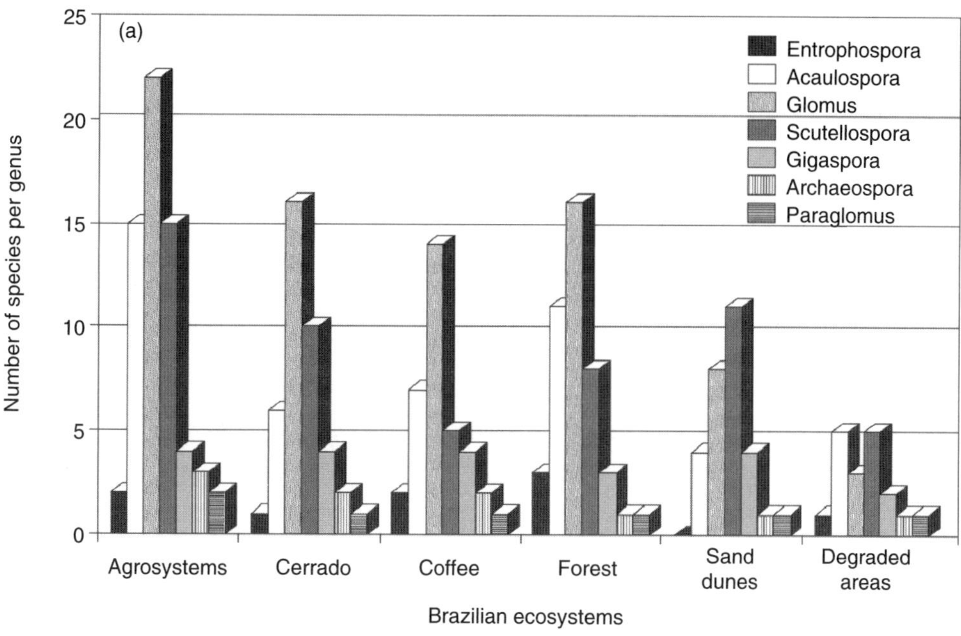

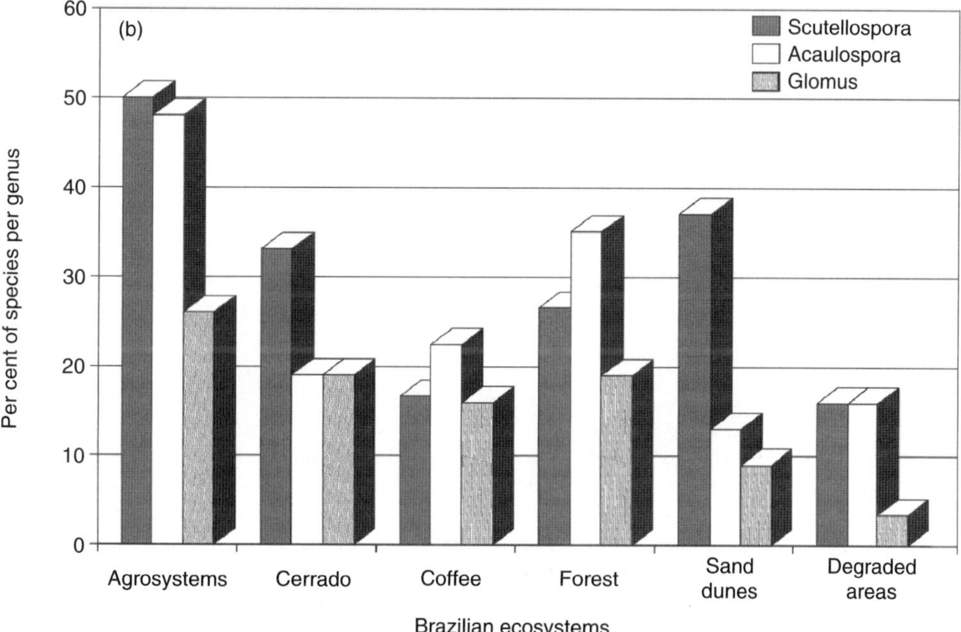

Fig. 10.5. Number of species and per cent of species per genera of AMF occurring in different Brazilian ecosystems.

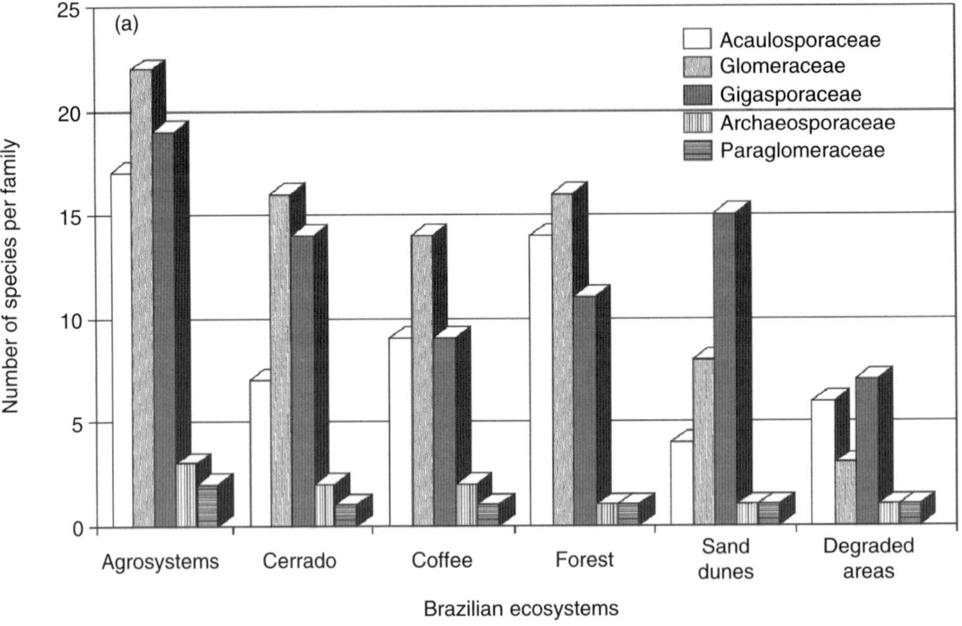

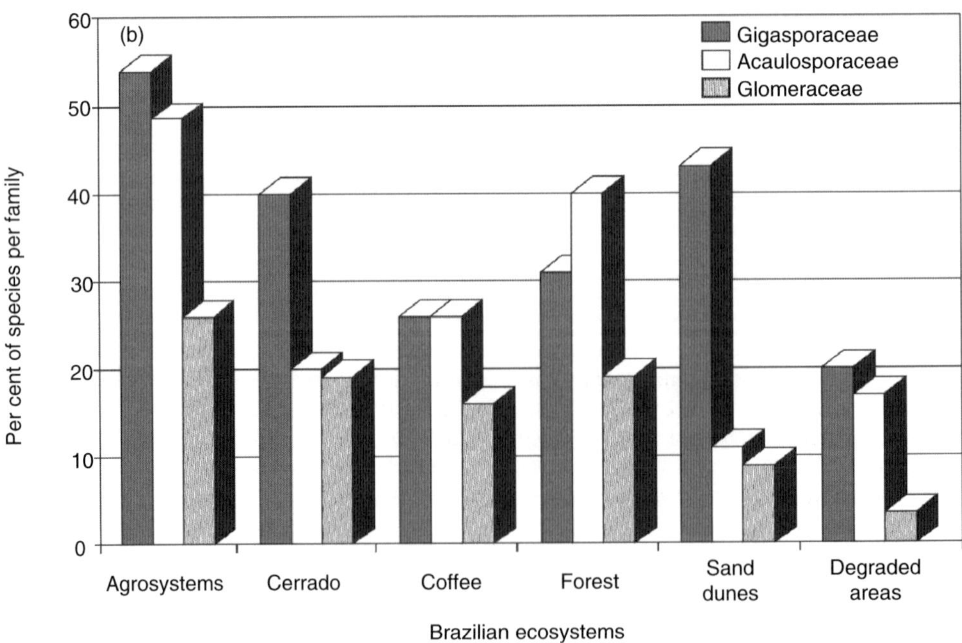

Fig. 10.6. Number of species and per cent of species per family of AMF occurring in different Brazilian ecosystems.

AMF species distribution

Diversity studies in any ecosystem often include calculations of species frequency of occurrence. Frequency provides a measure of whether a species is rare or common within an ecosystem and whether it is directly or indirectly related to fungal sporulation (a measure of abundance) (Stürmer and Bellei, 1994; Saggin-Júnior and Siqueira, 1996) or not. It also gives some clues on how adapted a species is to a range of soil and environmental conditions. Our analyses of frequent species in Brazilian ecosystems are based on only 15 publications (out of 28) that provided data on frequency of occurrence; the remaining papers only mentioned which species were the most common without any occurrence value associated with them.

Based on the five most common AMF species detected in each study, we observe that no prediction can be made on the distribution of a given species in an ecosystem, as illustrated by several examples depicted in Fig. 10.7. For instance, no overlap was detected among the five most frequent species between the two studies carried out in sand dune ecosystems. In sand dunes, species of *Gigaspora* were always the most dominant: *Gigaspora albida* in Santa Catarina state and *Gigaspora gigantea* in São Paulo state. A similar picture emerges in the cerrado ecosystem: in two studies carried out by the same group of researchers, different species were detected in each study as being the most frequent. Siqueira *et al.* (1989) found *Entrophospora colombiana* and *Scutellospora pellucida* in 32% and 29%, respectively, of the samples they analysed, while Schenck *et al.* (1989) detected *S. pellucida* in 100% of the samples, followed by *Gigaspora margarita* and *Glomus diaphanum*, both with a frequency of 83%. Nevertheless, *S. pellucida* and *A. spinosa* were among the five most frequent species recovered in both studies. Among the six surveys in agroecosystems, *Acaulospora scrobiculata* was the most frequent organism recovered in three of them and appeared among the five most frequent ones in two other studies, followed by *A. spinosa* (Fig. 10.7). *A. scrobiculata* was also among the most common fungi recovered from coffee plantations where *A. spinosa*, *Acaulospora morrowiae* and *G. etunicatum* were detected in at least two of the three studies analysed. The frequency data for these 15 studies indicate that *A. scrobiculata* is the most frequent fungus recovered from Brazilian ecosystems, being present in nine studies and detected in all ecosystems considered herein. *A. spinosa* and *G. etunicatum* were among the most common fungi in seven studies, although none of them were detected in degraded lands. Other commonly found species were *S. pellucida* (five studies), *A. morrowiae* (four studies) and *G. margarita* (four studies).

Geographical distribution of these six AMF species in Brazilian states and regions was evaluated based on their citation in each of the 28 surveys (Table 10.2). Note that, as shown in Fig. 10.4, some surveys carried out in different ecosystems by different researchers were done in the same geographical region and that in some cases more than one region was sampled within the same survey (e.g. Ag 4). Further analyses indicate that *A. scrobiculata* is the most widespread species found in the southernmost region, associated with *Araucaria* forest in the state of Rio Grande do Sul to Pernambuco state in the north-eastern region associated with several crops in humid and semiarid habitats (Fig. 10.8). *A. spinosa* was detected mainly in the states of São Paulo and Minas Gerais and was detected in two surveys in other regions. *G. etunicatum* shows a wider distribution than *A. spinosa* and similar to that of *A. scrobiculata*. Both *A. morrowiae* and *G. margarita* were not detected so far in the southern part of Brazil. The former was mostly detected in Minas Gerais and Bahia states, the latter in Minas Gerais and São Paulo states (Fig. 10.8).

Most of the 28 surveys analysed reported data on spore abundance at the sampling time for each of the species recovered in a given ecosystem. Spore abundance of a given species may not reflect its functional importance or its relative abundance within the community as a whole (Morton *et al.*, 1995; Douds and Millner,

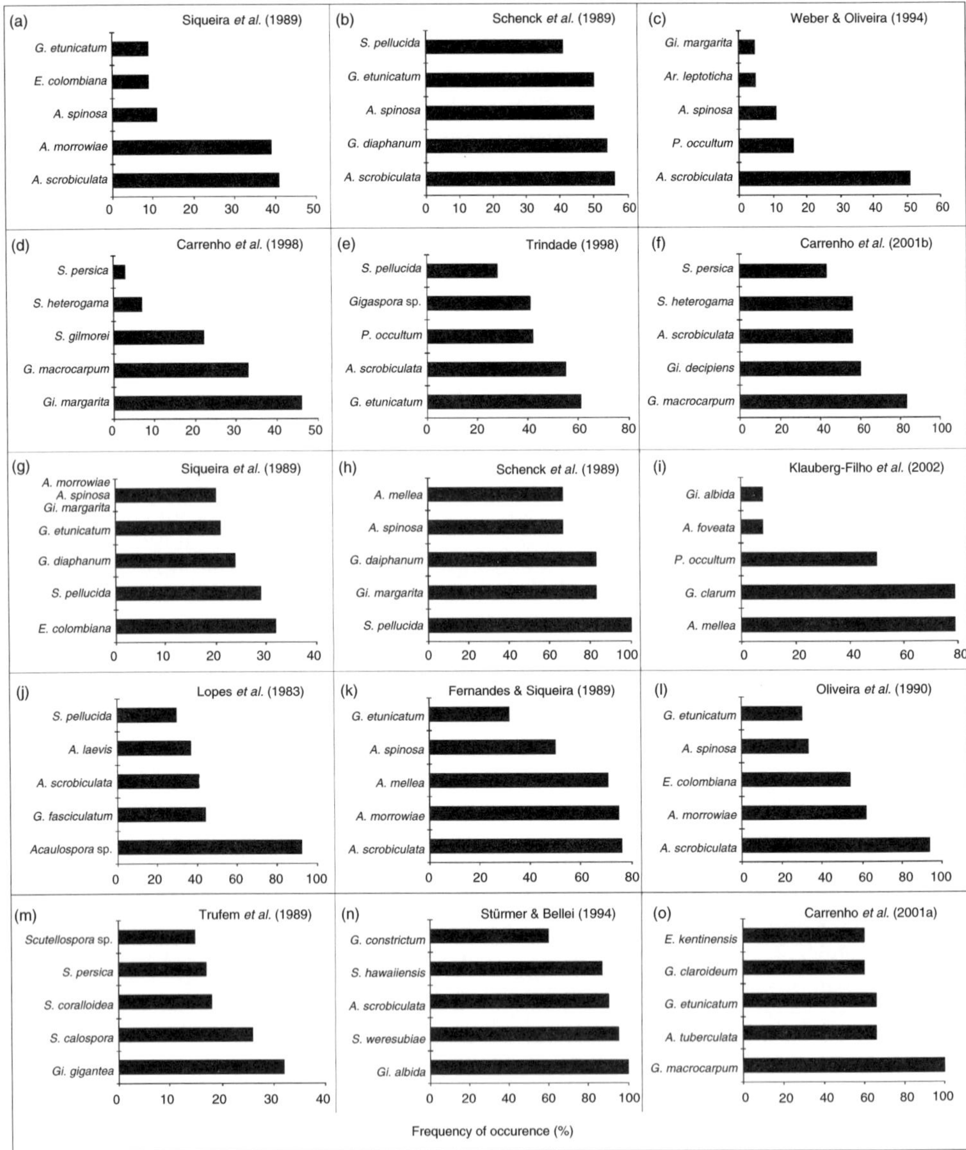

Fig. 10.7. The most frequent AMF species detected in several field studies on selected Brazilian ecosystems: (a–f) agroecosystems, (g,h) cerrado, (i) degraded areas, (j–l) coffee plantations, (m,n) sand dunes and (o) forest. Note differences of scales between graphics.

1999). Nevertheless, counting the number of spores of each AMF species is used to determine the abundance of an AMF species within a community and spore number or spore biovolume can be used as an assessment of the community structure such as the dominant species within a mycorrhizal community (Morton *et al.*, 1995). We chose five studies representing the ecosystems studied, except cerrado, to picture the mycorrhizal community structure based on the relative spore abundance of AMF species. In each study, those species that had a relative abundance >10% were

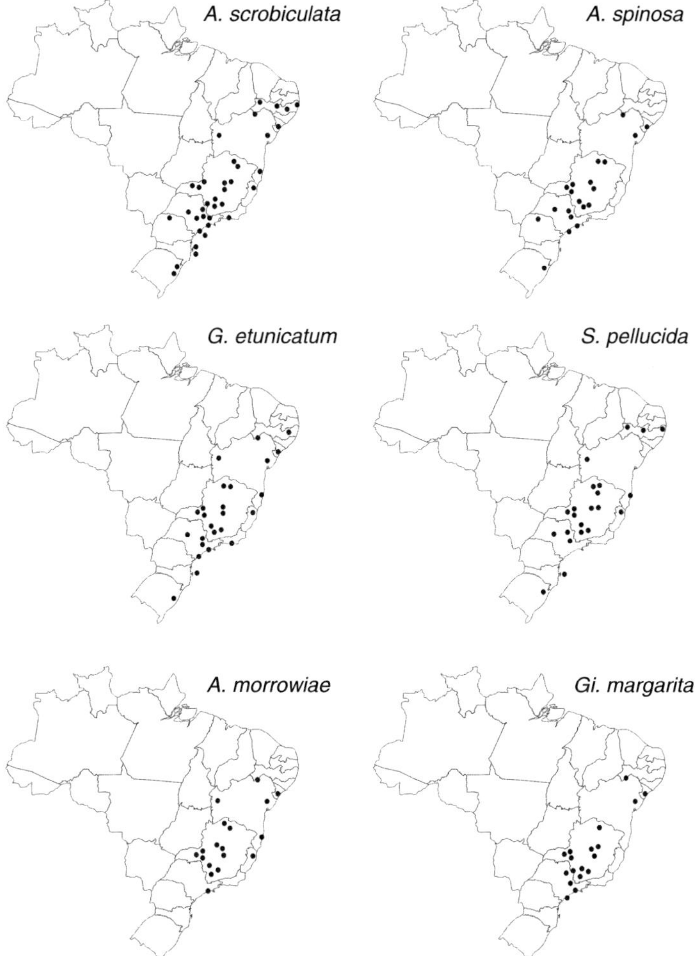

Fig. 10.8. Geographical distribution of the most frequent species of AMF in different Brazilian ecosystems. Dots indicate locations where the species were observed.

identified while the others were lumped. In all ecosystems, the mycorrhizal community was dominated by two to three species, which together accounted for 50–96% of the spores produced (Fig. 10.9). In coffee plantations, only two species, *Acaulospora* sp. and *A. scrobiculata*, accounted for 50% of the total spores produced (Fig. 10.9c). *A. scrobiculata* was also the main sporulator in maritime sand dunes in south Brazil, accounting alone for 51% of spore production (Fig. 10.9e). The most extreme situation was found in heavy-metal-degraded sites where Klauberg-Filho *et al.* (2002) found that *Glomus clarum*, *Acaulospora mellea* and *Paraglomus occultum* accounted for 96% of all spores recovered from field samples. This suggests indirect evidence of carbon allocation to sporulation of each species (Gazey *et al.*, 1992) that is taking place in the rhizosphere during sampling. This may represent an important link between above- and below-ground processes.

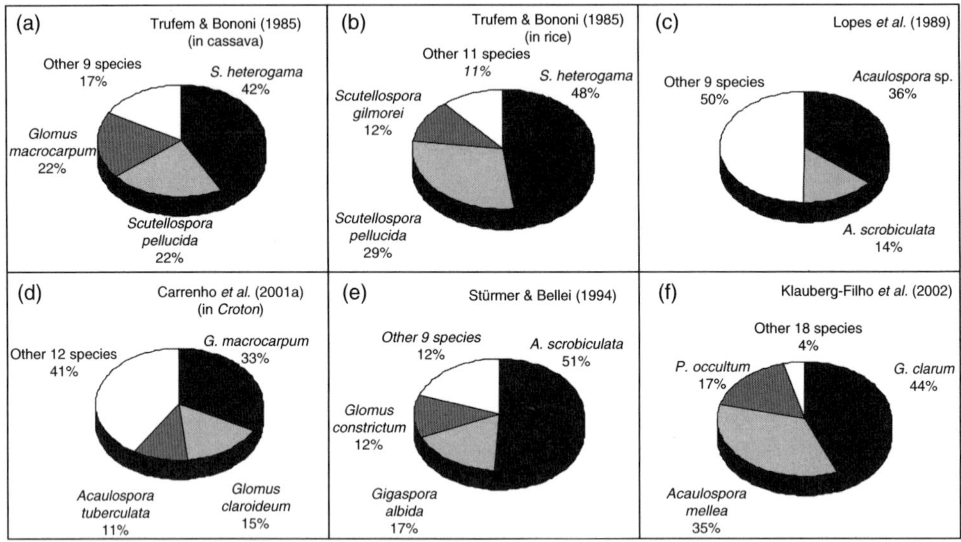

Fig. 10.9. Relative density of spores (percent of total recovered) in the AMF community under (a,b) agroecosystems, (c) coffee plantations, (d) forest, (e) sand dunes and (f) area degraded by heavy-metal contamination.

Determinant factors of AMF diversity

Most of the studies on occurrence of AMF tend to relate fungal species diversity with physical (e.g. soil texture, soil moisture) or chemical (e.g. nutrient levels, pH) soil properties, although authors commonly recognize the difficulty in establishing a clear relationship between soil and environmental variables with AMF occurrence and diversity. Siqueira *et al.* (1989) observed that *Acaulospora* spp. are favoured in soils with pH 6.5 while species of *Gigaspora* were less frequent in soils with pH > 6.5 and available (Mehlich I) P > 6 mg/kg soil. In coffee plantations, Lopes *et al.* (1983) found that in soils with pH ranging from 4.7 to 6.9 there was no relationship of this variable and the occurrence of *Acaulospora* species, whereas *Glomus* species were less frequent in soil with pH 5.0. Fernandes and Siqueira (1989), also working with coffee plantations, observed that spore density of *A. spinosa* was negatively correlated with soil Zn levels and plant Fe levels, both of these nutrients being associated with higher sporulation of *Glomus tortuosum* and *Glomus microcarpum*. Relating soil characteristics to AMF species occurrence represents an attempt to explain the effects of local/ecological processes upon local fungal diversity (number of species within a small homogenous habitat). Other processes that influence AMF diversity at a local level include soil physical disturbance, competition between fungal species, predation by several components of soil fauna, hyperparasitism and plant community structure.

Local processes that tend to reduce diversity are balanced by regional processes that act on a long-term basis and facilitate coexistence of species and addition of new species at a local level (Ricklefs, 1989). Regional processes and regional diversity, however, are rarely considered in AMF studies, particularly by Brazilian researchers. Maia and Trufem (1990), in the north-east of Brazil, observed that diversity tended to decrease from the more humid east region (15 species) to inland with a semiarid climate (8 species). In the USA, Koske (1987) observed an effect of temperature on fungal diversity in sand dunes along a latitudinal gradient. Comparing diversity among ecologically similar, but geographically distant, localities provides a way to demonstrate the influence of regional/historical factors on fungal species diversity (Ricklefs, 1989). A hierarchical cluster analysis (average method) was performed considering the presence of AMF species in order to observe whether studies performed within the same ecosystem, and therefore ecologically similar, could be grouped together. The dendrogram (Fig. 10.10) shows that the mycorrhizal community associated with papaya in Bahia and Espirito Santo states (Ag9) and with field crops in Pernambuco (Ag4) formed the most distinct groups. Among surveys in the same ecosystems, only those carried out in sand dunes formed a distinct cluster, both studies in São Paulo state (Sd1 and Sd2) clustered together and joined one of the studies in Santa Catarina (Sd3). Conversely, none of the studies performed on the forest ecosystems clustered together or were grouped with studies on agroecosystems and coffee plantations. Even studies F3 and F5, both analysing AMF in

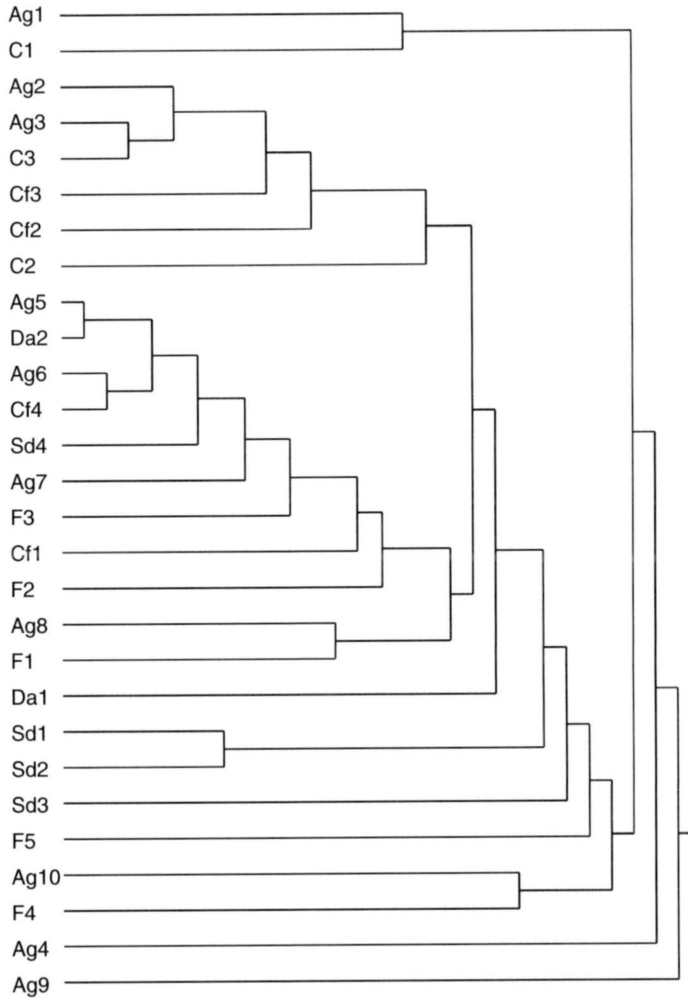

Fig. 10.10. Average-link dendrogram for the 28 surveys on AMF diversity in different Brazilian ecosystems. See Table 10.2 for explanation of codes.

Araucaria forests, clustered with agroecosystems and formed a distinct group remotely linked with the majority of other studies. Most of the studies in agroecosystems also clustered with studies in other ecosystems (Fig. 10.10). Sørenson's similarity index (Table 10.4) shows that coffee plantations and cerrado share a large proportion of species, with an index value of 0.83. AMF species compositions in sand dunes and degraded areas are relatively different from those detected in other ecosystems, with maximum similarity achieved when contrasted with forest ecosystems with values of 0.51 and 0.47, respectively. In contrast, species compositions of agroecosystems, cerrado, coffee plantations and forest areas were moderate to highly similar with indices ranging from 0.59 to 0.83.

What are cluster analyses and similarity indices telling us about processes governing AMF species diversity in Brazilian ecosystems? Lack of consistent grouping among the several studies developed in each ecosystem, as revealed by the cluster analyses, indicates that regional/historical processes should be considered as important determinants of AMF species diversity in natural ecosystems. We are not completely ignoring the influence of ecological factors on species diversity, but if local processes are the main determinant of species diversity, fungal communities developing under similar conditions (e.g. forest, cerrado, coffee plantations) would be expected to be similar with regard to species composition. Nevertheless, that is not really the case. As discussed by Moreira and Siqueira (2002), AMF occurrence varied substantially amongst maize fields in the states of Minas Gerais, Pernambuco and São Paulo in Brazil and Florida in the USA. The number of species found in these independent studies ranged from 9 to 14, similar to spore density and species composition. In spite of the same host and similar agronomic practices, none of the predominant species were common to these four locations. Although historical processes such as sympatric speciation and dispersal of fungal species among habitats and ecosystems (Ricklefs, 1989) could enhance AMF diversity in natural ecosystems, there is no evidence for that in the tropics. The fact that distinct ecosystems (agroecosystems, cerrado, coffee plantations and forest), as reviewed here, shared a high proportion of fungal species (Table 10.4) suggests that occurrence of these fungi may not be directly affected by local processes.

In contrast, we hypothesize that in more extreme conditions, ecological processes might be more important than the historical ones in determining AMF diversity. For instance, in a heavy-metal-polluted site studied by Klauberg-Filho *et al.* (2002), previously covered by cerrado vegetation, a very low similarity index (0.38) was found as compared with the original cerrado. Melloni *et al.* (2003) showed that bauxite mining reduced AMF colonization and propagule density by over 80%, whereas diversity as measured by H' index (Shanon–Weaver) was reduced by 40% compared with a non-mined reference site. In sand dunes, plants and soil microorganisms are subjected to extreme conditions such as high levels of salinity, sandy and unstable soils often poor in nutrients. Indeed, sand dunes and bauxite mine soils shared only

Table 10.4. Sørenson's index of similarity in different Brazilian ecosystems.

Ecosystems	C	Cf	F	Sd	Da[a]
Agrosystems (Ag)	0.68	0.65	0.69	0.48	0.40
Cerrado (C)		0.83	0.59	0.46	0.38
Coffee plantations (Cf)			0.62	0.47	0.42
Forest (F)				0.51	0.47
Sand dunes (Sd)					0.38

[a]Da, degraded areas.

a moderate number of species with all other ecosystems analysed, with indices ranging from 0.38 to 0.51. Additionally, disturbance of the ecosystems usually cause consistent shifts in the AMF community. Cultivation of the cerrado soil reduced AMF species richness and favoured dominance by certain fungal species (Siqueira *et al.*, 1989). The average species richness was 19 for undisturbed cerrado compared with 12 for adjacent agroecosystems. Comparative studies on AMF composition over different habitats within an ecosystem and among ecosystems are desired to understand the role of local versus regional processes in determining species diversity. Besides that, monitoring changes (or its absence) in fungal occurrence in different land use systems established over areas previously covered by a pristine environment represents an effective method to assess the role of local processes in determining diversity of this group of organisms.

Germplasm Banks of AMF

Germplasm banks have an important role in scientific development since they serve as centres for information exchange concerning biological diversity at local or regional levels, allow training of human resources and help to preserve biodiversity (Hawksworth and Mound, 1991). There are *c.* 250,000 fungal strains preserved in germplasm banks in several countries, representing only 17% of the known species and 0.8% of the estimated total number of fungal species (Hawksworth, 1991).

Two germplasm banks of AMF are renowned internationally. In the USA, the 'International Culture Collection of Arbuscular and Vesicular–Arbuscular Mycorrhizal Fungi' (INVAM), whose curator is Dr Joseph B. Morton, represents the largest germplasm bank of AMF, harbouring more than 1000 fungal isolates from different countries (Morton *et al.*, 1993). In France, 'La Banque Européenne des Glomales' (BEG), under the supervision of Dr Vivianne Gianinazzi-Pearson, has approximately 500 isolates. These banks act mainly as repository of fungi utilized in research as experimental isolates and provide fungal inoculum for research purposes. Although these banks have AMF species originating from tropical soils, species from temperate regions represent most of their germplasm (e.g. only 20% of AMF isolates at INVAM are from tropical soils).

Development of germplasm banks must be encouraged in tropical countries, especially in Brazil, which represents one of the ten countries that harbour a biological megadiversity that needs to be maintained in the country to avoid its loss by biopiracy. As already mentioned, in Brazil, 50% of the known AMF species have been detected in disparate geographical regions, in different natural ecosystems and agroecosystems, and in association with a wide range of hosts. However, no national culture collection has been developed to harbour this diversity. The situation in Brazil is that culture collections are maintained by researchers in their own institutions (e.g. Universidade Federal de Lavras, Embrapa-Agrobiologia, Insituto de Botânica de São Paulo, Insituto de Micologia de Recife, Universidade Regional de Blumenau) and fungal isolates are exchanged between researchers without a rigorous control over quality of inoculum and taxonomic identification. Siqueira and Klauberg-Filho (2000) pointed out that for application of AMF on a large scale in Brazil, problems in taxonomy and ecology, in development of procedures to produce inoculants and on availability of fungal isolates to field studies must be overcome. These authors also recognize that AMF research in Brazil must accomplish the development of a Brazilian germplasm bank of Glomerales and training of human resources in specific areas like taxonomy and systematics.

Relationship between AMF Diversity and Plant Communities

Studies on AMF occurrence in natural ecosystems represent a first step for further applications of mycorrhizal fungi by providing a list of species (taxonomic diversity) and ecological measurements of diversity (e.g. species richness, evenness, dominance). It is well known that plant communities (hosts) influence the ecology of plant symbionts. However, relationships between fungal species diversity and aboveground diversity and productivity are rarely assessed or briefly speculated on in these surveys for many reasons. Characters used to assess taxonomic diversity (analysis of spore walls) of AMF are not linked to fungal life-history traits (external hyphae architecture, arbuscule production, absorptive and transport capacity of hyphae) responsible for the functional diversity of the symbiosis (Morton and Bentivenga, 1994). Also, arbuscular mycorrhizal function under field conditions has been assessed in a very limited number of species, mainly in temperate regions (Newsham *et al.*, 1995b). Besides that, research on mycorrhizal ecology has assumed that AMF are functionally redundant since these fungi are assumed to have low specificity towards their hosts (Bever *et al.*, 2001) and because of the assumption that mycorrhizal diversity is relatively poor compared with plant diversity (Allen *et al.*, 1995). In a recent study (Poyu-Rojas, 2002), it was shown that mycotrophic promiscuity of several wild tree species ranged from highly general to very restrictive against ten fungal isolates. By the same token, AMF isolates behave quite differently on distinct hosts. *G. clarum*, for instance, was able to colonize all hosts tested while *A. scrobiculata* colonized only 10% of the hosts, being considered very restrictive. Understanding aspects of host–fungus relationship is crucial for assessing occurrence, diversity and species shifts in the tropics. In other studies, it has been shown that plant susceptibility and responsiveness to AMF are directly related to seed weight and successional stage of tropical wild tree species (Siqueira *et al.*, 1998). Although a very low proportion of the known tropical plant species have been studied for their mycorrhizal status, available data indicate that the vast majority of them are highly mycotrophic, exhibiting high responsiveness but varied mycorrhizal dependency (Siqueira and Saggin-Júnior, 2001).

Plants associated with AMF generally have a higher capacity to access scarce or immobile soil minerals, especially phosphorus, resulting in enhanced nutrition, plant growth rates and development. Such benefits have been widely demonstrated under controlled laboratory and greenhouse conditions, being markedly high and of great consistency in tropical regions where soils are strongly P-limiting. Newsham *et al.* (1995b) suggested that because of inconclusive results on plant P nutrition provided by arbuscular mycorrhizae in certain hosts in temperate regions and the ubiquity of this association in natural environments, AMF can perform a range of functions providing benefits to plants other than improvement of mineral nutrition. The multifunctionality of AMF likely results from individual species within a mycorrhizal community, but the biology of these fungi is very limited (Streitwolf-Engel *et al.*, 1997; Bever *et al.*, 2001). One reason is that AMF species functioning has generally been measured in single fungal isolates originating from different habitats rather than from the same community (Stürmer, 1998). Nevertheless, isolates from different coffee fields exhibited different effectiveness on coffee outplants in a wide range of applied soil P under controlled environments (Saggin-Júnior *et al.*, 1994) and under field situations as well (Siqueira *et al.*, 1998).

Some studies, however, have demonstrated that AMF species from the same community impact plant growth traits differently. As a result, AMF species diversity can strongly influence plant community structure and productivity. Streitwolf-Engel *et al.* (1997) studied the influence of three AMF isolated from the same community upon clonal growth traits of two *Prunella* species. They observed that all three fungal isolates had no effect on P concentration for both plant species, but one *Glomus* isolate (Basle Pi) increased leaf number, stolon branch number and the total length of stolons of *Prunella vulgaris* relative to the other two isolates. Stürmer (1998) analysed the effectiveness of fungal isolates from three distinct communities and showed that at least one isolate from each community increased plant shoot biomass accumulation and foliar P content of soybean. Similar results were found in fungal assemblages for soybean in Brazilian soil (Paula *et al.*, 1988). In Stürmer's study, responses of highly effective or non-effective isolates were maintained when tested on soybean and clover, suggesting that this fungal trait may be inheritable. van der Heijden *et al.* (1998a) studied the influence of four fungal isolates on growth traits of co-occurring grassland species. They observed that the response of *Bromus* did not exhibit as much variation as that of *Hieracium* sp. regarding plant growth variables measured as shoot and root dry mass, shoot and root P concentration and specific root length. Further studies with 11 plant species in microcosms simulating calcareous grassland also demonstrated that biomass accumulation of plant species varied according to different AMF isolates inoculated alone or together (van der Heijden *et al.*, 1998b). On the basis of these results and mycorrhizal dependency of these plants, they hypothesized that AMF species in natural communities have the potential to determine plant diversity and community structure. Flores-Aylas *et al.* (2003) showed that AMF inoculation increased growth and favoured a balanced growth of woody species established by direct seeding in low-fertility soil. Their results showed that the presence of AMF reduced plant dominance under low available soil P. Considering that plant biodiversity has the potential to increase the productivity of ecosystems, as demonstrated for grasslands (Naeem *et al.*, 1994; Tilman *et al.*, 1996), and that plant growth responses are dependent on a particular combination of plant and AMF species, it is reasonable to consider that AMF species may influence ecosystem productivity by influencing the plant community. Conversely, plant diversity and its relationship with fungal diversity have been rarely studied and results are highly inconsistent and conflicting. Johnson *et al.* (1991) found that AMF species richness increased with plant diversity along old-field successional stages. Studying the AMF community along a stabilization gradient in sand dunes, Cordoba *et al.* (2001) found that AMF species richness increased from embryonic dunes dominated by a few halophytes to fixed dunes where plant richness was higher. On the other hand, Johnson and Wedin (1997) showed that conversion of dry tropical forest to grassland has no effect on the Simpson index and species richness of the mycorrhizal community, although forest sites were dominated by at least 14 plant species while the grassland was dominated by the African grass *Hyparrhenia rufa*. Their results were recently corroborated by Picone (2000) in lowland tropical forest and pastures in Nicaragua and Costa Rica. He used species-accumulation curves and demonstrated that local AMF species richness was not significantly affected when forest areas were converted to pastures. Simpson's diversity index of AMF was not statistically different between forest (1.58) and pasture areas (1.87) and both habitats shared the most frequent fungal species. In Africa, Wilson *et al.* (1992) found that plantations of *Terminalia* sp. had a higher AMF species richness than native tropical forest. There is an urgent need to assess the impact of human activities on AMF occurrence and diversity and the relationship of these organisms with plant production and ecosystem sustainability.

A clear picture of the relationship between AMF diversity with above-ground plant diversity has not emerged due to two main reasons. First, plant diversity has not been accurately assessed; most studies only mention the dominant plant species without providing any diversity index or plant species richness measurement. Second, AMF diversity measurement has been based solely on field-collected spores, which can be found in low numbers in the soil and lack important morphological characteristics for an accurate taxonomic identification. Studies on microcosms with known AMF and plant species richness and the use of different methods to detect fungal species may help to better understand the

AMF versus plant diversity relationships. Grime *et al.* (1987) were the first to demonstrate that mycorrhizae increased plant diversity mainly by increasing biomass production of subordinate species relative to canopy species in microcosms mimicking calcareous grassland. van der Heijden *et al.* (1998b) inoculated different numbers of AMF species in macrocosms simulating North American old-field ecosystems. They found that Simpson's diversity of plant community and above-ground productivity increased with increasing AMF species richness. The lowest plant diversity and productivity were calculated in non-inoculated macrocosms while the highest of these parameters were detected when 8–14 AMF species were present. The authors point out a mechanism for their results based on the increase of hyphal length in soil with increased AMF species richness and decrease of soil P concentration, indicating that a more diverse AMF community is more efficient in exploring the resources available in the ecosystem. One additional aspect in determining these relationships is the fact that several fungi may not be sporulating at the sampling time. Single sampling may underestimate fungal diversity, and different approaches such as trapping of field soils is one way to detect these cryptic AMF species (Stutz and Morton, 1996; Bever *et al.*, 1996). In North Carolina, Bever *et al.* (2001) detected 37 AMF species within a 75 m^2 field using different trapping procedures over a 5 year period; this fungal diversity was approximately the same as plant diversity at that site and indicates that the mycorrhizal community comprises a higher number of species than are usually reported in surveys based solely on single samplings.

Management of Arbuscular Mycorrhizal Fungi

Considering that distinct fungal species or isolates within a mycorrhizal community impact their host differently, management of AMF is needed to optimize the benefits of the mycorrhizal association for sustainable plant production (Kling and Jakobsen, 1998). Assessment of AMF species occurrence and diversity is indispensable for assessing the impact of soil and crop management practices on the AMF community (Douds and Millner, 1999). Different isolates comprising the mycorrhizal community can be obtained in pure culture and tested for their effectiveness to plants for P nutrition (Bolan, 1991), limiting heavy-metal uptake (Gildon and Tinker, 1983) and protecting against soil-borne pathogens (Davis and Menge, 1980). Regarding phosphorus, AMF isolates and species behave differently in relation to levels of this nutrient in the soil and maintain a mutualistic symbiosis and symbiotic efficiency in a broad range of available soil P (Saggin-Júnior and Siqueira, 1995). These authors demonstrated that symbiotic efficiency varied among four different geographic isolates of *G. etunicatum* and ranged from 32% (Varginha isolate) to 81% (Patrocinio isolate). In the same study, the symbiotic efficiency of an isolate of *G. margarita* and that of native fungi were 97% and 23%, respectively. Species or fungal isolates, proven to confer advantage on plants, can be used to inoculate seedlings before transplanting to the field or different management practices can be developed to increase beneficial fungal isolates in cropping sites. While the former has proved to be feasible and helps the establishment of plants in the field, such as coffee outplants (Siqueira *et al.*, 1998), the latter needs further studies. Several management practices (e.g. tillage, crop rotation) impact upon the mycorrhizal community by disturbing the external mycelial network (McGonigle *et al.*, 1990), by decreasing species diversity (An *et al.*, 1993) or by selecting for less mutualistic species (Johnson, 1993). For instance, studies in the cerrado of Minas Gerais demonstrated that cultivation practices such as liming, fertilization and changes in plant cover increased AMF propagule density, but caused a shift in the community, favouring species dominance (Schenck *et al.*, 1989). These authors speculate that such shifts in the fungal community may affect crop performance and productivity in the cerrado area.

Studies in tropical and temperate soils have investigated the effect of pre-crops or crop rotations on mycorrhizal fungal populations. Dodd *et al.* (1990) observed that pre-crop treatments with *Sorghum*, cassava, *Pueraria* and *Brachiaria* in two tropical soils in Colombia not only increased spore numbers of native and introduced AMF species differently, but also created completely different communities depending on the plant host. Higher mycorrhizal inoculum potential and a sixfold increase of AMF spore numbers were detected in fescue plots compared with soybean plots by An *et al.* (1990). Hendrix *et al.* (1995) found that continuous cropping with soybean decreased AMF spore numbers relative to those in rotated plots with maize and fescue. Species richness and the Shannon–Weaver diversity index were also significantly lower in continuous soybean than in rotated plots. Besides that, continuous soybean resulted in higher densities of *Gigaspora* spp. while rotated fields were dominated by *Glomus* spp. Miranda and Miranda (1997) reported that pre-cropping with soybean and mucuna resulted in a more than tenfold increase of AMF sporulation associated with *Sorghum* when compared with pre-cropping with cabbage or fallow. Results of these studies emphasize that propagules of a species or of a group of species that experimentally have proven to be effective in promoting growth of a target plant can be increased by growing different hosts in the field before planting the target plant. This procedure bypasses the need to use commercial fungal inoculum and might be more feasible to small farmers to enhance sustainability.

Distinct farming systems, ranging from conventional agriculture to low-input and no-input agriculture, also impact diversity and spore density of mycorrhizal communities. Franke-Snyder *et al.* (2001) analysed AMF diversity associated with maize and soybean on conventional and low-input farming systems and observed that 15 consecutive years of the same farming regime caused no major differences among mycorrhizal communities. In all treatments, the mycorrhizal community was composed of nearly the same fungal species and the general structure of AMF communities, measured by Shannon–Weaver and Simpson's indices, evenness and species richness, was similar in all treatments. Under soybean and maize, AMF spore density averaged 293/g soil in a low-input area compared with 174/g soil in a conventionally managed area (Kurle and Pfleger, 1994). Fertilization of soil has been implicated in selection of AMF species or isolates that are less mutualistic or even parasitic to plants (Johnson, 1993). The author used mycorrhizal communities from 'fertilized' and 'unfertilized' plots to inoculate the grass *Andropogon gerardi*. After 1 month, plants inoculated with fertilized communities were smaller than plants inoculated with unfertilized communities and produced fewer inflorescences after 3 months. The implications of these studies are that converting natural ecosystems into agricultural systems with high fertilizer inputs might decrease plant yield in the long term if populations of less effective mycorrhizal fungi increase and become parasitic. Therefore, sustainability in these systems will be achieved only with high inputs of fertilizers, which are high in cost and may be unhealthy to the environment.

The effects of different tillage practices on the mycorrhizal community and soil structure have been demonstrated by several authors. McGonigle *et al.* (1990) observed that P uptake and mycorrhizal colonization was higher in maize plants growing in no-till or ridge tillage than in mouldboard-ploughed soils. Kabir *et al.* (1997) studied mycorrhizal development under conventional tillage, reduced tillage and no-tillage and observed that the densities of both total and metabolically active hyphae and root colonization were lower in conventional tillage than in the two other tillage practices. Douds *et al.* (1995) observed that tillage systems impacted differently on spore density of AMF species: a no-till regime favoured sporulation by *Glomus occultum* while mouldboard and chisel-disc tillage favoured *Glomus* sp. and *G. etunicatum* sporulation. The impact of tillage on mycorrhizal colonization and nutrient uptake mediated by mycorrhizal fungi occurs as a result of breaking up the extraradical network of AMF mycelia, which is related to nutrient absorption and improvement of soil structure (Kling and Jakobsen, 1998; Douds and Millner, 1999). Disrupting the mycorrhizal mycelium network also decreases soil aggregation as hyphae participate physically in the entanglement of mineral soil particles, creating a skeletal structure that binds micro- and macroaggregates (Miller and Jastrow, 1992) and biochemically by producing an external glycoprotein named 'glomalin', which acts as a cementing agent in aggregation (Wright and Upadhyaya, 1998).

Knowledge of AMF species composition in a site is of foremost importance when studying the impact of any kind of agricultural management practices upon AMF. The logic behind this is that researchers could focus their efforts on increasing the population of those species that prove to be highly effective in increasing plant yield. From a practical point of view, the use of a species with a widespread distribution implies that mycorrhizal inoculum produced with one or many species can potentially be used under different soil and climatic conditions. We contrasted the most common species recorded on surveys (Fig. 10.11) with those commonly used by Brazilian researchers as experimental isolates in plant growth and nutrition studies. Experimental isolates were compiled from abstracts of the Brazilian Conferences on Mycorrhizae from 1985 to 1994 (REBRAM I, II, III, IV and V). This analysis demonstrates that among the most frequent species widespread in natural ecosystems (Fig. 10.11a), only *G. etunicatum* and *G. margarita* appear as the two fungi commonly used as experimental isolates (Fig. 10.11b). *A. scrobiculata* and *A. spinosa*, both with a wide range of distribution, are rarely used as experimental isolates; the former was cited in 19 abstracts (out of 410) and the latter was not cited at all. *G. clarum* represents the third most-often studied AMF species. It is interesting that the two species used by Brazilian researchers as experimental isolates, *Glomus macrocarpum* and *Glomus fasciculatum*, represent species whose taxonomic boundaries are not clearly defined and in many cases that originated from overseas. Moreover, *A. scrobiculata* has very low effectiveness in most of the hosts tested so far and *G. margarita* has a limited distribution and very low spore density in most sites (Fernandes and Siqueira, 1989; Siqueira *et al.*, 1989).

Concluding Remarks

Measurement of taxonomic diversity of AMF in different ecosystems has often been considered by Brazilian mycorrhizologists

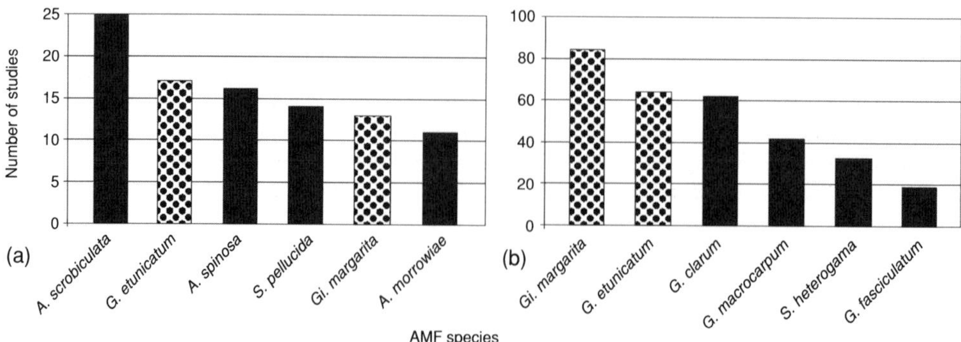

Fig. 10.11. Comparison of the (a) AMF species most frequently recovered in Brazilian ecosystems and (b) AMF species used as experimental isolates by Brazilian researchers. Number of studies represent the 28 papers analysed in this chapter for (a) and 410 abstracts of the REBRAM I to V for (b). Bars for the only two overlapping species are filled with circles. Note differences of scales between graphs.

in their research programmes, providing a considerable amount of information about species distribution, frequency of occurrence, biogeography and major edaphic factors influencing diversity patterns in different ecosystems. We acknowledge that the analysis performed in this review is not thorough, but it represents an attempt to exploit fundamental aspects of AMF ecology and to establish patterns of AMF species distribution in representative ecosystems in Brazil. Regarding AMF diversity in Brazilian agroecosystems and natural ecosystems, we conclude:

1. Research efforts have to be made to sample geographical locations representing important biomes and ecosystems in Brazil, such as the Amazon tropical rainforest and the Pantanal, where studies on the ecology and diversity of AMF have not been conducted. Studies on AMF diversity have been conducted in the eastern part of the country from the southern to northeastern regions and, with few exceptions, most represent specific 'case studies' lacking more intensive sampling within an ecosystem.
2. Despite this, the taxonomic diversity of glomalean fungi can be considered high as 79 formally described species of AMF have been found in Brazilian ecosystems surveyed to date. This represents approximately 50% of the total known diversity at species level. Consistent sampling of more pristine ecosystems would probably increase these numbers and even reveal undescribed species.
3. Relative proportions of genera and families of Glomerales varied markedly according to the ecosystems analysed. Considering the number of described species within each family, Acaulosporaceae and Gigasporaceae, with the genera *Acaulospora* and *Scutellospora*, respectively, are the most representative organisms found in Brazilian ecosystems.
4. Species distribution within a given ecosystem can hardly be predicted when the frequency of occurrence of species are analysed. Different surveys within the same ecosystem showed no consistent pattern of the most frequent species. Overall, *A. scrobiculata*, *A. spinosa* and *G. etunicatum* are the most common glomalean species recovered in Brazilian ecosystems.
5. Historical processes (e.g. dispersal, speciation) should be considered as determinants of glomalean diversity in agroecosystems and natural Brazilian ecosystems, although ecological processes acting nowadays are probably more important in determining diversity in more extreme ecosystems like sand dunes and degraded areas.
6. Relationships between AMF species diversity and plant community diversity are not clear and have been established mainly in microcosms. No study in Brazil has attempted to relate AMF diversity with either above-ground diversity or plant productivity.
7. Management of mycorrhizal fungi and commercial inoculum production should be based on occurrence of the most common fungi within a given ecosystem/geographical region or on the effectiveness of fungal isolates tested under experimental and field conditions.
8. There is an urgent need to strengthen the AMF research programme in Brazil to attain an inventory of this ancient and important component of all terrestrial ecosystems. Brazil is a major centre of AMF biodiversity and therefore deserves a well-defined conservation policy and the establishment of a Brazilian bank of germplasm of Glomerales.

References

Abbott, L.K. and Gazey, C. (1994) An ecological view of the formation of VA mycorrhizas. *Plant and Soil* 159, 69–78.

Allen, E.B., Alllen, M.F., Helm, D.J., Trappe, J.M., Molina, R. and Rincon, E. (1995) Patterns and regulation of mycorrhizal plant and fungal diversity. *Plant and Soil* 170, 47–62.

An, Z.-Q., Grove, J.H., Hendrix, J.W., Hershman, D.W. and Henson, G.T. (1990) Vertical distribution of endogonaceous mycorrhizal fungi associated with soybean as affected by soil fumigation. *Soil Biology and Biochemistry* 22, 715–719.

An, Z.-Q., Hendrix, J.W., Hershman, D.E., Ferris, R.S. and Henson, G.T. (1993) The influence of crop rotation and soil fumigation on a mycorrhizal fungal community associated with soybean. *Mycorrhiza* 3, 171–182.

Beare, M.H., Coleman, D.C., Crossley, D.A. Jr, Hendrix, P.F. and Odum, E.P. (1995) A hierarchical approach to evaluating the significance of soil biodiversity to biogeochemical cycling. *Plant and Soil* 170, 5–22.

Bentivenga, S.P. and Morton, J.B. (1995) A monograph of the genus *Gigaspora* incorporating developmental patterns of morphological characters. *Mycologia* 87, 720–732.

Bever, J.D., Schultz, P.A., Pringle, A. and Morton, J.B. (2001) Arbuscular mycorrhizal fungi: more diverse than meets the eye, and the ecological tale of why. *BioScience* 51, 923–931.

Bolan, N.S. (1991) A critical review on the role of mycorrhizal fungi in the uptake of phosphate by plants. *Plant and Soil* 134, 189–207.

Bononi, V.L.R. and Trufem, S.F.B. (1983) Endomicorrizas vesículo-arbusculares do Cerrado da Reserva Biológica de Moji-Guaçu, SP, Brazil. *Rickia* 10, 55–84.

Breuninger, M., Einig, W., Magel, E., Cardoso, E. and Hampp, R. (2000) Mycorrhiza of Brazil pine (*Araucaria angustifolia* [Bert. O. Ktze.]). *Plant Biology* 2, 4–10.

Carrenho, R., Trufem, S.F.B. and Bononi, V.L.R. (1998) Arbuscular mycorrhizal fungi in *Citrus sinensis/C. limon* treated with fosetyl-al and metalaxyl. *Mycological Research* 102, 677–682.

Carrenho, R., Trufem, S.F.B. and Bononi, V.L.R. (2001a) Fungos micorrízicos arbusculares em rizosferas de três espécies de fitobiontes instaladas em área de mata ciliar revegetada. *Acta Botanica Brasilica* 15, 115–124.

Carrenho, R., Silva, E.S., Trufem, S.F.B. and Bononi, V.L.R. (2001b) Successive cultivation of maize and agricultural practices on root colonization, number of spores and species of arbuscular mycorrhizal fungi. *Brazilian Journal of Microbiology* 32, 262–270.

Colozzi-Filho, A. and Cardoso, E.J.B.N. (2000) Detecção de fungos micorrízicos arbusculares em raízes de cafeeiro e de crotalária cultivada na entrelinha. *Pesquisa Agropecuária Brasileira* 35, 2033–2042.

Cordoba, A.S., Mendonça, M.M., Stürmer, S.L. and Rygiewicz, P.T. (2001) Diversity of arbuscular mycorrhizal fungi along a sand dune stabilization gradient: a case study at Praia da Joaquina, Ilha de Santa Catarina, South Brasil. *Mycoscience* 42, 379–387.

Cromack, K. and Caldwell, B.A. (1992) The role of fungi in litter decomposition and nutrient cycling. In: Carroll, G.C. and Wicklow, D.T. (eds) *The Fungal Community, Its Organization and Role in the Ecosystem*. Marcel Dekker, New York, pp. 601–608.

Davis, R.M. and Menge, J.A. (1980) Influence of *Glomus fasiculatum* and soil phosphorous on *Phytophthora* root rot of citrus. *Phytopathology* 70, 447–452.

Diaz, G., Azconaguilar, C. and Honrubia, M. (1996) Influence of arbuscular mycorrhizae on heavy metal (Zn and Pb) uptake and growth of *Lygeum spartum* and *Anthyllis cytisoides*. *Plant and Soil* 180, 241–249.

Dodd, J.C., Arias, I., Koomen, I. and Hayman, D.S. (1990) The management of populations of vesicular–arbuscular mycorrhizal fungi in acid-infertile soils of a savanna ecosystem II. The effects of pre-crops on the spore populations of native and introduced VAM-fungi. *Plant and Soil* 122, 241–247.

Douds, D.D. and Millner, P.D. (1999) Biodiversity of arbuscular mycorrhizal fungi in agroecosystems. *Agriculture Ecosystems and Environment* 74, 77–93.

Douds, D.D., Galvez, L., Janke, R.R. and Wagoner, P. (1995) Effect of tillage and farming system upon populations and distribution of vesicular–arbuscular mycorrhizal fungi. *Agriculture Ecosystems and Environment* 52, 111–118.

Fernandes, A.B. and Siqueira, J.O. (1989) Micorrizas vesicular–arbusculares em cafeeiros da região Sul do estado de Minas Gerais. *Pesquisa Agropecuária Brasileira* 24, 1489–1498.

Flores-Aylas, W.W., Saggin-Júnior, O.J., Siqueira, J.O. and Davide, A.C. (2003) Efeito de *Glomus etunicatum* e fósforo no crescimento inicial de espécies arbóreas em semeadura direta. *Pesquisa Agropecuária Brasileira* 38, 257–266.

Franke, M. and Morton, J.B. (1994) Ontogenetic comparisons of the endomycorrhizal fungi *Scutellospora heterogama* and *Scutellospora pellucida*: revision of taxonomic character concepts, species descriptions, and phylogenetic hypotheses. *Canadian Journal of Botany* 72, 122–134.

Franke-Snyder, M., Douds, D.D., Galvez, L., Phillips, J.G., Wagoner, P., Drinkwater, L. and Morton, J.B. (2001) Diversity of communities of arbuscular mycorrhizal (AM) fungi present in conventional versus low-input agricultural sites in eastern Pennsylvania, USA. *Applied Soil Ecology* 16, 35–48.

Gazey, C., Abbott, L.K. and Robson, A.D. (1992) The rate of development of mycorrhizas affects the onset of sporulation and production of external hyphae by two species of *Acaulospora*. *Mycological Research* 96, 643–650.
Gerdemann, J.W. and Trappe, J.M. (1974) Endogonaceae in the Pacific Northwest. *Mycologia Memoir* 5, 1–76.
Gildon, A. and Tinker, P.B. (1983) Interactions of vesicular–arbuscular mycorrhizal infections and heavy metals in plants. II. The effects of infection on uptake of copper. *New Phytologist* 95, 263–268.
Gomes, S.P. and Trufem, S.F.B. (1998) Fungos micorrízicos arbusculares (Glomales, Zygomycota) na Ilha dos Eucaliptos, represa do Guarapiranga, São Paulo, SP. *Acta Botânica Brasilica* 12, 393–401.
Grime, J.P., Mackey, J.M.L., Hillier, S.H. and Read, D.J. (1987) Floristic diversity in a model system using experimental microcosms. *Nature* 328, 420–422.
Hawksworth, D.L. (1991) The fungal dimension of biodiversity: magnitude, significance, and conservation. *Mycological Research* 95, 641–655.
Hawksworth, D.L. and Mound, L.A. (1991) Biodiversity database: the crucial significance of collections. In: Hawksworth, D.L. (ed.) *Biodiversity of Microorganisms and Invertebrates and Its role in Sustainable Agriculture*. CAB International, Wallingford, UK, pp. 17–39.
Hawksworth, D.L., Sutton, B.C. and Ainsworth, G.C. (1983) *Ainsworth and Bisby's Dictionary of the Fungi*, 7th edn. Commonwealth Mycological Institute, Kew, UK.
Heckman, D.S., Geiser, D.M., Eidell, B.R., Stanffer, R.L, Kardos, N.L. and Hedges, S.B. (2001) Molecular evidence for the early colonization of land by fungi and plants. *Science* 293, 1129–1133.
Hendrix, J.W., Guo, B.Z. and An, Z.-Q. (1995) Divergence of mycorrhizal fungal communities in crop production systems. In: Collinds, H.P., Robertson, G.P. and Klug, M.J. (eds) *The Significance and Regulation of Soil Biodiversity*. Kluwer, Dordrecht, The Netherlands, pp. 131–140.
Johnson, N.C. (1993) Can fertilization of soil select less mutualistic mycorrhizae? *Ecological Applications* 3, 749–757.
Johnson, N.C. and Wedin, D.A. (1997) Soil carbon, nutrients, and mycorrhizae during conversion of dry tropical forest to grassland. *Ecological Applications* 7, 171–182.
Johnson, N.C., Zak, D.R., Tilman, D. and Pfleger, F.L. (1991) Dynamics of vesicular–arbuscular mycorrhizae during old field succession. *Oecologia* 86, 349–358.
Kabir, Z., O'Halloran, I.P., Fyles, J.W. and Hamel, C. (1997) Seasonal changes of arbuscular mycorrhizal fungi as affected by tillage practices and fertilization: hyphal density and mycorrhizal root colonization. *Plant and Soil* 192, 285–293.
Klauberg-Filho, O., Siqueira, J.O. and Moreira, F.M.S. (2002) Fungos micorrízicos arbusculares em solos de área poluída com metais pesados. *Revista Brasileira de Ciência do Solo* 26, 125–134.
Kling, M. and Jakobsen, I. (1998) Arbuscular mycorrhiza in soil quality assessment. *Ambio* 27, 29–34.
Kuhn, G., Hijri, M. and Sanders, I.R. (2001) Evidence for the evolution of multiple genomes in arbuscular mycorrhizal fungi. *Nature* 414, 74–748.
Kurle, J.E. and Pfleger, F.L. (1994) Arbuscular mycorrhizal fungus spore populations respond to conversions between low-input and conventional management practices in a corn–soybean rotation. *Agronomy Journal* 86, 467–475.
Koske, R.E. (1987) Distribution of VA mycorrhizal fungi along a latitudinal temperature gradient. *Mycologia* 79, 55–68.
Lopes, E.S., Oliveira, E., Dias, R. and Schenck, N.C. (1983) Occurrence and distribution of vesicular–arbuscular mycorrhizal fungi in coffee (*Coffea arabica* L.) plantations in central São Paulo state, Brazil. *Turrialba* 33, 417–422.
Magurran, A.E. (1988) *Ecological Diversity and Its Measurement*. Princeton University Press, Princeton, New Jersey.
Maia, L.C. and Trufem, S.F.B. (1990) Fungos micorrízicos vesículo-arbusculares em solos cultivados no estado de Pernambuco, Brasil. *Revista Brasileira de Botânica* 13, 89–95.
McGonigle, T.P., Evans, D.G. and Miller, M.H. (1990) Effect of degree of soil disturbance on mycorrhizal colonization and phosphorus absorption by maize in growth chamber and field experiments. *New Phytologist* 116, 629–636.
Mehta, A.P., Torma, A.E. and Murry, L.E. (1979) Effect of environmental parameters on the efficiency of biodegradation of basalt rock by fungi. *Biotechnology and Bioengineering* 21, 875–885.
Melloni, R., Siqueira, J.O. and Moreira, F.M.S. (2003) Fungos micorrízicos arbusculares em solos de área de mineração de bauxita em reabilitação. *Pesquisa Agropecuária Brasileira* 38, 267–276.
Melo, A.M.Y., Maia, L.C. and Morgado, L.B. (1997) Fungos micorrízicos arbusculares em bananeiras cultivadas no Vale do Submédio São Francisco. *Acta Botânica Brasilica* 11, 115–121.

Miller, R.M. and Jastrow, J.D. (1992) The role of mycorrhizal fungi in soil conservation. In: Bethlenfalvay, G.J. and Linderman, R.G. (eds) *Mycorrhizae in Sustainable Agriculture*. American Society Agronomy Special Publication 54, Madison, Wisconsin, pp. 29–44.

Miranda, J.C.C. and Miranda, L.N. (1997) Micorriza arbuscular. In: Vargas, M.A.T. and Hungria, M. (eds) *Biologia dos solos dos cerrados*. Planaltina: EMBRAPA-CPAC, Planaltina, Brazil, pp. 69–123.

Morton, J.B. (1986) Three new species of *Acaulospora* (Endogonaceae) from high aluminum, low pH soils in West Virginia. *Mycologia* 78, 641–648.

Morton, J.B. (2000) Evolution of endophytism in arbuscular mycorrhizal fungi of glomales. In: Bacon, C.W. and White, J.F. Jr (eds) *Microbial Endophytes*. Marcel Dekker, New York, pp. 121–140.

Morton, J.B. and Benny, G.L. (1990) Revised classification of arbuscular mycorrhizal fungi (Zygomycetes): a new order, Glomales, two new suborders, Glomineae and Gigasporineae, and two new families, Acaulosporaceae and Gigasporaceae, with an emendation of Glomaceae. *Mycotaxon* 37, 471–491.

Morton, J.B. and Bentivenga, S.P. (1994) Levels of diversity in endomycorrhizal fungi (Glomales, Zygomyctes) and their role in defining taxonomic and non-taxonomic group. *Plant and Soil* 259, 47–59.

Morton, J.B. and Redecker, D. (2001) Two new families of Glomales, Archaeosporaceae and Paraglomaceae, with two new genera *Archaeospora* and *Paraglomus*, based on concordant molecular and morphological characters. *Mycologia* 93, 181–195.

Morton, J.B., Bentivenga, S.P. and Bever, J.D. (1995) Discovery, measurement, and interpretation of diversity in arbuscular endomycorrhizal fungi (Glomales, Zygomycetes). *Canadian Journal of Botany* 73, 25–32.

Naeem, S., Thompson, L.J., Lawler, S.P., Lawton, J.H. and Woodfin, R.M. (1994) Declining biodiversity can alter the performance of ecosystems. *Nature* 368, 734–737.

Newsham, K.K., Fitter, A.H. and Watkinson, A.R. (1995a) Arbuscular mycorrhiza protect an annual grass from root pathogenic fungi in the field. *Journal of Ecology* 83, 991–1000.

Newsham, K.K., Fitter, A.H. and Watkinson, A.R. (1995b) Multi-functionality and biodiversity in arbuscular mycorrhizas. *Trends in Ecology and Evolution* 10, 407–411.

Oliveira, E., Siqueira, J.O., Lima, R.D., Colozzi-Filho, A. and Souza, P. (1990) Ocorrência de fungos micorrízicos vesículo-arbusculares em cafeeiros das regiões do alto Paranaíba e Triângulo no estado de Minas Gerais. *Hoehnea* 17, 117–125.

O'Neill, E.G., O'Neill, R.V. and Norby, R.J. (1991) Hierarchy theory as a guide to mycorrhizal research on large-scale problems. *Environmental Pollution* 73, 271–284.

Palleroni, N.J. (1994) Some reflections on bacterial diversity. *ASM News* 60, 537–540.

Paula, M.A., Siqueira, J.O., Oliveira, L.H. and Oliveira, E. (1988) Efetividade simbiótica relativa em soja de população de fungos endomicorrízicos nativos e de isolados de *Glomus macrocarpum* e *Gigaspora margarita*. *Revista Brasileira de Ciência do Solo* 12, 25–31.

Paula, M.A., Siqueira, J.O. and Döbereiner, J. (1993) Ocorrência de fungos micorrízicos vesiculo–arbusculares e de bactérias diazotróficas na cultura da batata-doce. *Revista Brasileira de Ciência do Solo* 17, 349–356.

Picone, C. (2000) Diversity and abundance of arbuscular-mycorrhizal fungus spores in tropical forest and pasture. *Biotropica* 32, 734–750.

Pirozynski, K.A. (1981) Interactions between fungi and plants through the ages. *Canadian Journal of Botany* 59, 1824–1827.

Poyu-Rojas, E. (2002) Compatibilidade simbiótica de fungos micorrízicos arbusculares com mudas de espécies arbóreas tropicais. Tese de Doutorado, Universidade Federal de Lavras, Lavras, Brazil.

Read, D.J., Leake, J.R. and Langdale, A.R. (1989) The nitrogen nutrition of mycorrhizal fungi and their host plants. In: Boddy, L., Marchant, R. and Read, D.J. (eds) *Nitrogen, Phosphorus and Sulphur Utilization by Fungi*. Cambridge University Press, Cambridge, UK, pp. 269–298.

Redecker, D., Morton, J.B. and Bruns, T.D. (2000a) Molecular phylogeny of the arbuscular mycorrhizal fungi *Glomus sinuosum* and *Sclerocystis coremioides*. *Mycologia* 92, 282–285.

Redecker, D., Kodner, R. and Graham, L.E. (2000b) Glomalean fungi from Ordovician. *Science* 289, 1920–1921.

Ricklefs, R.E. (1989) Speciation and diversity: the integration of local and regional processes. In: Otte, D. and Endler, J.A. (eds) *Speciation and Its Consequences*. Sinauer Associates, Sunderland, Massachusetts, pp. 599–622.

Saggin-Júnior, O.J. and Siqueira, J.O. (1995) Avaliação da eficiência si,biótica de fungos endomicorrízicos para o cafeeiro. *Revista Brasileira de Ciência do Solo* 19, 221–228.

Saggin-Júnior, O.J. and Siqueira, J.O. (1996) Micorrizas Arbusculares em Cafeeiro. In: Siqueira, J.O. (ed.) *Avanços em Fundamentos e Aplicação de Micorrizas*. Universidade Federal de Lavras, Lavras, Brazil, pp. 203–254.

Saggin-Júnior, O.J., Siqueira, J.O., Guimarães, P.T.G. and Oliveira, E. (1994) Interação fungos micorrízicos *versus* superfosfato e seus efeitos no crescimento a teores de nutrientes do cafeeiro em solo não fumigado. *Revista Brasileira de Ciência do Solo* 18, 27–36.

Sánchez-Diaz, M., Pardo, M., Antolín, M., Peña, J. and Aguirreola, J. (1990) Effects of water stress on photosynthetic activity in the *Medicago-Rhizobium-Glomus* symbiosis. *Plant Science* 71, 215–221.

Schenck, N.C., Siqueira, J.O. and Oliveira, E. (1989) Changes in the incidence of VA mycorrhizal fungi with changes in ecosystems. In: Vancura, V. and Kunc, F. (eds) *Interrelationships between Microorganisms and Plants in Soil.* Elsevier, Amsterdam, pp. 125–129.

Schluter, D. and Ricklefs, R.E. (1993) Species diversity: an introduction to the problem. In: Ricklefs, R.E. and Schluter, D. (eds) *Species Diversity in Ecological Communities – Historical and Geographical Perspectives.* The University of Chicago Press, Chicago, Illinois, pp. 1–10.

Schüβler, A., Schwarzott, D. and Walker, C. (2001) A new fungal phylum, the Glomeromycota: phylogeny and evolution. *Mycological Research* 105, 1413–1421.

Simon, L., Bousquet, J., Lévesque, R.C. and Lalonde, M. (1993) Origin and diversification of endomycorrhizal fungi and coincidence with vascular land plants. *Nature* 363, 67–69.

Siqueira, J.O., Sylvia, D.M., Gibson, J. and Hubbell, D.H. (1985) Spores, germination, and germ tubes of vesicular-arbuscular mycorrhizal fungi. *Canadian Journal of Microbiology* 31, 965–972.

Siqueira, J.O., Colozzi-Filho, A. and Oliveira, E. (1989) Ocorrência de micorrizas vesicular–arbusculares em agro e ecossistemas do estado de Minas Gerais. *Pesquisa Agropecuária Brasileira* 24, 1499–1506.

Siqueira, J.O., Saggin-Júnior, O.J., Flores-Aylas, W.W. and Guimarães, P.T.G. (1998) Arbuscular mycorrhizal inoculation and superphosphate application influence plant development and yield of coffee in Brazil. *Mycorrhiza* 7, 293–300.

Smith, S.E. and Read, D.J. (1997) *Mycorrhizal Symbiosis.* Academic Press, London.

Streitwolf-Engel, R., Boller, T., Wiemken, A. and Sanders, I.R. (1997) Clonal growth traits of two *Prunella* species are determined by co-occurring arbuscular mycorrhizal fungi from a calcareous grassland. *Journal of Ecology* 85, 181–191.

Stürmer, S.L. (1998) Characterization of diversity of fungi forming arbuscular endomycorrhizae in selected plant communities. PhD thesis, West Virginia University, Morgantown, Virginia.

Stürmer, S.L. (1999) Evolução, Classificação e Filogenia dos Fungos Micorrízicos Arbusculares. In: Siqueira, J.O., Moreira, F.M.S., Lopes, A.S., Guilherme, L.R.G., Faquin, V., Furtini Neto, A.E. and Carvalho, J.G. (eds) *Inter-relação Fertilidade, Biologia do Solo e Nutrição de Plantas.* Sociedade Brasileira de Ciência do Solo and Universidade Federal de Lavras, Viçosa and Lavras, Brazil, pp. 797–818.

Stürmer, S.L. and Bellei, M.M. (1994) Composition and seasonal variation of spore populations of arbuscular mycorrhizal fungi in dune soils on the island of Santa Catarina, Brazil. *Canadian Journal of Botany* 72, 359–363.

Stürmer, S.L. and Morton, J.B. (1997) Developmental patterns defining morphological characters in spores of four species in *Glomus*. *Mycologia* 89, 72–81.

Stürmer, S.L. and Morton, J.B. (1999) Taxonomic reinterpretation of morphological characters in Acaulosporaceae (Glomales) based on developmental patterns in two *Acaulospora* and one *Entrophospora* species. *Mycologia* 91, 849–857.

Stutz, J.C. and Morton, J.B. (1996) Successive pot cultures reveal high species richness of arbuscular endomycorrhizal fungi in arid ecosystems. *Canadian Journal of Botany* 74, 1883–1889.

Tilman, D., Wedin, D. and Knops, J. (1996) Productivity and sustainability influenced by biodiversity in grassland ecosystems. *Nature* 379, 718–720.

Trufem, S.F.B. and Bononi, V.L. (1985) Micorrizas vesículo-arbusculares de culturas introduzidas em áreas de cerrado. *Rickia* 12, 165–187.

Trufem, S.F.B. and Viriato, A. (1990) Fungos micorrízicos vesículo-arbusculares da Reserva biológica do Alto da Serra de Paranapiacaba, São Paulo, Brasil. *Revista Brasileira de Botânica* 13, 49–54.

Trufem, S.F.B., Otomo, H.S. and Malatinszky, S.M.M. (1989) Fungos micorrízicos vesículo-arbusculares em rizosferas de plantas em dunas do Parque Estadual da Ilha do Cardoso, São Paulo, Brasil. (1) Taxonomia. *Acta Botânica Brasilica* 3, 141–152.

Trufem, S.F.B., Malatinszky, S.M.M. and Otomo, H.S. (1994) Fungos micorrízicos arbusculares em rizosferas de plantas do litoral arenoso do Parque Estadual da Ilha do Cardoso, SP, Brasil. 2. *Acta Botânica Brasilica* 8, 219–229.

van der Heijden, M.G.A., Boller, T., Wiemken, A. and Sanders, I.R. (1998a) Different arbuscular mycorrhizal fungal species are potential determinants of plant community structure. *Ecology* 79, 2082–2091.

van der Heijden, M.G.A., Klironomos, J.N., Ursic, M., Moutoglis, P., Streitwolf-Engel, R., Boller, T., Wiemken, A. and Sanders, I.R. (1998b) Mycorrhizal fungal diversity determines plant biodiversity, ecosystem variability and productivity. *Nature* 396, 69–72.

Walker, C. (1983) Taxonomic concepts in the Endogonaceae: II. Spore wall characteristics in species descriptions. *Mycotaxon* 18, 443–455.

Wardle, D.A., Bardgett, R.D., Klironomos, J.N., Setälä, H., van der Putten, W.H. and Wall, D.H. (2004) Ecological linkages between aboveground and belowground biota. *Science* 304, 1629–1633.

Weber, O.B. and Oliveira, E. (1994) Ocorrência de fungos micorrízicos vesículo-arbusculares em citros nos estados da Bahia e Sergipe. *Pesquisa Agropecuária Brasileira* 29, 1905–1914.

Wilson, J., Ingleby, K., Mason, P.A., Ibrahim, K. and Lawson, G.J. (1992) Long-term changes in vesicular–arbuscular mycorrhizal spore populations in *Terminalia* plantations in Côte d'Ivoire. In: Read, D.H., Lewis, D.H., Fitter, A.H. and Alexander, I.J. (eds) *Mycorrhizas in Ecosystems*. CAB International, Wallingford, UK, pp. 268–275.

Wright, S.F. and Upadhyaya, A. (1998) A survey of soils for aggregate stability and glomalin, a glycoprotein produced by hyphae of arbuscular mycorrhizal fungi. *Plant and Soil* 198, 97–107.

Wright, S.F., Franke-Snyder, M., Morton, J.B. and Upadhyaya, A. (1996) Time-course study and partial characterization of a protein on hyphae of arbuscular mycorrhizal fungi during active colonization of roots. *Plant and Soil* 181, 193–203.

Zandavalli, R.B. (2001) Aspectos ecológicos e fisiológicos de micorrizas em *Araucaria angustifolia* (Bertoloni) Otto Kuntze. Dissertação de Mestrado. Universidade Federal do Rio Grande do Sul, Porto Alegre, Brazil.

11 Nitrogen-fixing Leguminosae-nodulating Bacteria

F.M.S. Moreira
Departamento de Ciência do Solo, Universidade Federal de Lavras, Caixa Postal 37, Lavras, MG, CEP 37 200-000, Brazil, e-mail: fmoreira@ufla.br

Introduction

Biological nitrogen fixation (BNF) is one of the most important soil functions for the maintenance of life on Earth. It is restricted to some highly phylogenetically diverse Prokaryote species that possess the enzyme nitrogenase, which is able to reduce N_2 to NH_3. BNF is estimated to be around 139×10^6 mg N/year in terrestrial ecosystems, while industrial N fixation that produces chemical fertilizers represents only around 49×10^6 mg N/year (Burns and Hardy, 1975). Nitrogen-fixing organisms, also called diazotrophs, can live freely in soil or in close relationships with organisms of other kingdoms such as *Protista*, *Fungi*, *Animalia* and *Plantae*. In the case of plants they are found free-living in the rhizosphere, endophytically associated with some species, or establishing a biochemically, genetically and morphologically regulated symbiosis. The establishment of mutualistic symbiosis implies that there is an exchange of benefits between partners: N_2 fixed into NH_3 (BNF) by the bacteria to the plant and CO_2 fixed into carbohydrates (photosynthesis) to the bacteria. Among other known prokaryotes establishing symbioses (cyanobacteria and actinomycetes/*Frankia*), the most important ones are those of bacteria-nodulating Leguminosae, not only because of the high economic value of many species in this family but also because of the superior efficiency of the process, which enables it to be managed for improving agricultural productivity in an environmentally sound way. This chapter provides information on Leguminosae-nodulating bacteria (LNB) characteristics and their symbioses as well as relevance in various Brazilian ecosystems.

Leguminosae-nodulating Bacteria in the Context of Prokaryote Diversity

After great technical breakthroughs, like the advances in molecular methods in the last decades (e.g. PCR, Mullis and Faloona (1987); Nobel Prize, 1992), evolutionary clocks like ribosomal genes (Kimura, 1983; Woese, 1991) have been applied intensively to Prokaryote taxonomy. This enabled the establishment of natural relationships, which was responsible for an exponential increase of described species. The GenBank site (http://www.ncbi.nlm.nih.gov/genbank/genbankstats.html) reported yearly additions to *Prokaryote* species (*Archae* and *Bacteria*) since 1993, showing how this figure had increased to 10,085 species in 2002 (Table 11.1). These figures have a strong relationship, although not linear, with the number of sequences deposited at

Table 11.1. Total number of described species as per 1993 and yearly additions thereafter within phylogenetic groups at the National Center for Biotechnology Information Database (Bethesda, Maryland).

Groups	1993	1994	1995	1996	1997	1998	1999	2000	2001	2002	All years
Archaea	66	9	19	32	23	36	17	19	12	18	443
Bacteria	1,279	367	337	417	735	459	479	458	502	498	9,642
Eukaryota	3,408	2,051	2,493	3,501	4,907	6,562	9,060	11,069	12,678	12,824	137,631
Fungi	460	191	338	453	738	905	1,245	1,440	1,417	1,689	15,022
Metazoa	1,677	988	1,054	1,409	2,356	2,653	3,943	4,655	4,717	4,690	64,699
Viridiplantae	964	668	911	1,467	1,631	2,778	3,449	4,583	6,024	6,009	51,734
Viruses	713	125	131	153	413	210	439	562	717	926	9,760

Source: available at http://www.ncbi.nlm.nih.gov/genbank/genbankstats.html

GenBank, 143,492 in 1993 and 14,976,310 in 2001, which represent almost a 100-fold increase over 8 years (Fig. 11.1). Molecular techniques are also important for revealing uncultured microbial diversity. Currently, only 26 of the approximately 52 identifiable major lineages, or phyla, within the domain Bacteria have cultivated representatives (Rappé and Giovannoni, 2003).

Prokaryote diversity can also be assessed using morphological, biochemical, physiological and genetic traits. Intraspecific (strain) variability is usually high and this is certainly the case for LNB. Another level of LNB diversity is the symbiotic relationships with plant species, which has important implications in both agricultural productivity and sustainability. Also, at that level strain variability within species is relevant. Relationships between NB strains and Leguminosae species can be divided into three main groups:

1. Non-symbiotic, i.e. the nodule structure typical of symbiosis establishment is not formed among potential partners.
2. Symbiotic non-efficient, i.e. nodules are formed mainly in roots for the majority of species and in the stems for a few species, but no nitrogen fixation occurs.
3. Symbiotic efficient, i.e. nodules are formed (in roots and/or stems) and there may be different degrees of efficiency in nitrogen fixation.

Non-symbiotic strains, i.e. strains of known LNB that have lost their ability to establish symbiosis, have been reported from soils (Segovia *et al.*, 1991; Martínez-Romero and Caballero-Mellado, 1996), but there is no published record of their occurrence in Brazilian soils.

Bacterial species/strains and plant species vary regarding both the establishment and the function of their symbiosis from highly specific, i.e. they are able to form symbiosis with just a narrow range of partner species/strains, to highly promiscuous, when they are able to establish symbiosis with a large range of partners. Thus, in treating LNB diversity, the diversity of their host partners, the Leguminosae

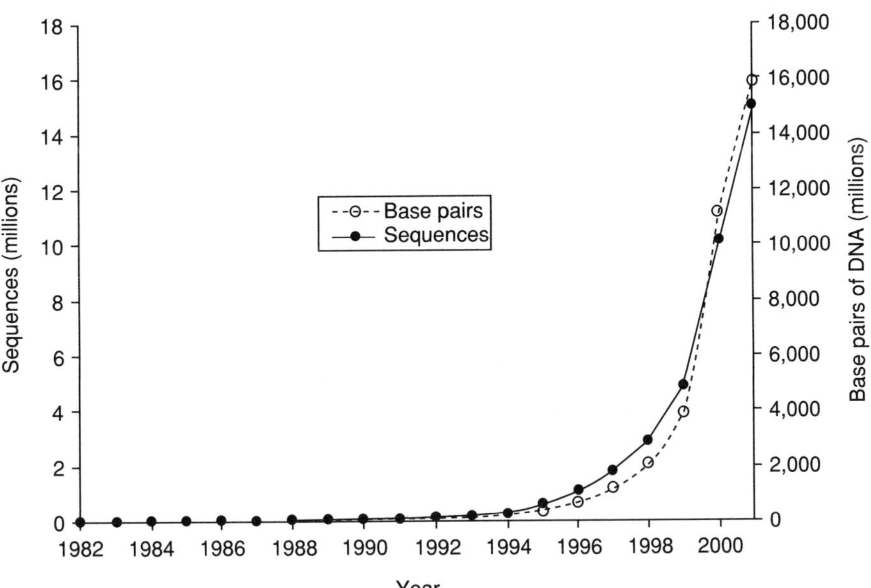

Fig. 11.1. Exponential growth of GenBank database sequences and base pairs of DNA from 1982 until 2001. Available at http://www.ncbi.nlm.nih.gov/genbank/genbankstats.html

species and the non-leguminous *Parasponia* spp. must also be considered. As there is no evidence of *Parasponia* species occurring in Brazil, only the Leguminosae species symbiosis with LNB will be considered in this chapter.

Importance of the Leguminosae in Brazil's Most Representative Ecosystems

Brazil has the highest biodiversity in the world with about 10–20% of the total extant species on the planet (Mittermeier *et al.*, 1997; Brasil MMA, 1998). Regarding the Plantae kingdom, an estimated 50,000–56,000 (20–22%) species of higher plants occur in Brazil. Among them, Leguminosae is an important family in all ecosystems in terms of both number of species and individuals. As examples, 1294 species (141 genera) were reported to occur in the Amazon (Silva *et al.*, 1989) and 555 species (71 genera) in cerrado (Kirkbride-Júnior, 1984) regions. Figures for Atlantic forests are only available from scattered areas, but certainly Leguminosae diversity is also high in this biome because it is considered one of the ecosystems with the highest diversity on the planet (Mittermeier *et al.*, 1997). Although many species occur in different regions/ecosystems, there is a high level of endemism. It can be estimated that Brazil has at least 2000 Leguminosae species, i.e. 10% of the total number (around 20,000 in 640–680 genera). From the most intact ecosystems to the most disturbed ones, frequencies of Leguminosae in relation to the total number of species are usually high (Table 11.2). Growth forms of legumes vary from herbs to trees. It is important to highlight that, in forest ecosystems, liana species are frequently found in high numbers. Although these species have not yet been explored and are scarcely mentioned, many of them have economic potential and are able to establish symbiosis with nitrogen-fixing bacteria (e.g. *Derris* spp.).

Grain legumes, most of them exotic in this country, are cultivated in large areas, playing an important role in the national economy and livelihood. One of the major export products is soybean (*Glycine max*) with about 21 million ha planted in 2003, yielding about 57 million tonnes in 2004 (http://www.ibge.net/home/estatistica/indicadores/agropecuaria/lspa03200304.shtm). The replacement of chemical fertilizers

Table 11.2. Frequency of total and nodulating Leguminosae species in relation to the total species number of other families in 32 vegetation surveys of forest ecosystems in Brazil.

State/survey date	Total species number	Diversity index H'	Leguminosae species frequency range (%)	
			Total	Nodulating[a]
Amazonas (1971–1987)	54–505	3.92–4.76	8.9–25.9[b]	4.2–11.7
Pará (1987)	101–122	4.23	13.1–20.8[b]	Nd
Maranhão (1993–1996)	104–260	4.20	13.5–18.1[b]	4.8
Rondônia (1987)	171–278	4.44–4.91	15.8–16.9[b]	9.0
Minas Gerais (1992–1994)	136–277	3.60–4.33	9.8–19.7[b]	7.4–14.6
Espírito Santo (1992)	650	–	12.9[b]	7.7
São Paulo (1992–1996)	40–176	3.0–4.13	3.6–27.5[b]	2.8–16.7
Paraná (1996)	70	3.72	20.0	15.7
Rio Grande do Sul (1992)	60–63	3.14–3.52	18.3–22.2[b]	13.3–15.9
Distrito Federal (1996)	111	–	20.7	13.5

[a]Considering potentially nodulating species.
[b]Leguminosae was the family with the highest species diversity in these surveys.
Source: modified from Pereira *et al.* (1998) based on several publications on plant surveys.

by BNF in soybean fields represented $1 billion in Brazil's agriculture in 2001 (Moreira and Siqueira, 2002). BNF by soybeans was estimated in 2001 as being 1.9 million tonnes of N. This figure is greater than the 1.7 million tonnes of N fertilizer consumed by all crops in Brazil in the same year, of which 62% was imported (www.anda.org.br). These figures stress the economic importance of BNF management in Brazilian agriculture. Other edible grain species such as *Vigna unguiculata* (cowpea) and *Phaseolus vulgaris* are the most consumed food crops along with rice. Cowpea is cultivated mainly in the north and north-east. Both species are planted by small farmers mainly as a subsistence crop, with low input technology. Attained yields and total planted area are much smaller than with soybean: 3.2 million tonnes and 4.2 million ha in 2003. *P. vulgaris* inoculation faces some problems, such as high promiscuity and, thus, competition of introduced strains with native LNB populations, sensitivity to soil acidity and temperature and a shorter period of nitrogen fixation, among others. Because cowpeas have good field responses to inoculation with nitrogen-fixing bacteria, there is a need for extension of this biotechnology among farmers.

Nodulation in Leguminosae

Since Allen and Allen's (1981) report on absence of knowledge about nodulation capability of the majority of tropical Leguminosae species, intensive surveys were made throughout the world, mainly in Brazil (Magalhães *et al.*, 1982; Faria *et al.*, 1989; Moreira *et al.*, 1992; Souza *et al.*, 1994). The total number of species analysed in the world increased from 15% to 23%, i.e. 3856 species belonging to 413 genera, from which 3397 species (88%) belonging to 317 genera (77%) are currently known as being able to nodulate (Faria *et al.*, 1999). The majority of nodulating species are within Mimosoideae (90%) and Papilionoideae (96%), but represent only 24% of Caesalpinioideae species. Unfortunately, an example of a non-nodulating species is *Caesalpinia echinata* Lam. (pau brasil), the species from which the name of our country originated. 'Brasil' (from the Portuguese côr de brasa = live-coal colour) derives from the red colour of its wood, which seems to be burning. *Eperua bijuga* Mart. Ex. Benth. (muirapiranga) is another non-nodulating species, often used for woody handicraft souvenirs, very similar to pau-brasil, because of its red colour. Besides the well-known examples of grain species (e.g. beans, peas, soybeans), most of them nodulating and belonging to Papilionoideae, an example of a nodulating woody species in the same subfamily is the famous rosewood (*Dalbergia nigra* Allem. Ex. Benth., jacarandá da Bahia). Both jacarandá da Bahia and pau-brasil are endemic to the Atlantic forest. Muirapiranga occurs in the Amazon region and in north-east Brazil.

On the basis of the figures presented above, it is clear that knowledge of symbioses of LNB with around 11,200 leguminous species is completely unknown around the world, as is the case with microsymbionts. Thus, a huge potential for research exists in this unexplored area.

Current Taxonomy of Leguminosae-nodulating Bacteria and Evolution of the Symbiosis

Taxonomy

The credit for the first isolation of nitrogen-fixing bacteria from inside leguminous nodules has often been given to Beijerinck (1888) who named it *Bacillus radicicola*. The origin of the epithet *Rhizobium leguminosarum*, retained as one of the LNB species until today, is uncertain (Young, 1999). For a long time it was ascribed to Frank (1890), but recent revision (Young, 1999) corrected it to Frank (1879), who described it but did not know that it was a nitrogen-fixing bacterium. The taxonomic revolution based on phylogenetic relationships among bacteria (Woese, 1987) by using 'evolutionary clocks' (Kimura, 1983) placed the LNB genera in the phylum α-Proteobacteria. The beginning of this century brought a new revolution to LNB taxonomy with the discovery that genera *Burkholderia* and *Ralstonia* belonging to the phylum β-Proteobacteria, as well as α-Proteobacteria belonging to the genera *Methylobacterium* and *Blastobacter* exhibit this characteristic (Chen *et al.*, 2001; Moulin *et al.*, 2001; Sy *et al.*, 2001, van Berkun and Eardly, 2002) (Table 11.3). The name 'rhizobia' was used for a long time as a collective name for the bacteria that nodulate Leguminosae originating from Rhizobiaceae, the family previously known to include all LNB (Jordan, 1984). Now, with the discovery of LNB in other phylogenetic branches of Prokaryotes this name is inappropriate.

Forty-seven species belonging to 11 genera had been described when this chapter was completed (Table 11.3). Despite the great number of legume species establishing symbioses with nitrogen-fixing bacteria already known (3397), the majority of taxonomic studies with the microsymbiont have been restricted to a few grain species, mainly Papilionoideae from temperate

Table 11.3. Genera and species of Leguminosae-nodulating bacteria with their respective host species mentioned in the original publications.

Genera/species	Hosts
Rhizobium (Frank, 1889)	
R. leguminosarum (Frank, 1879,1889), biovars *phaseoli, trifolii, viceae* (Jordan, 1984)	*P. vulgaris, P. multiflorus, P. angustifolius, Trifolium* spp., *Pisum* spp., *Vicia* spp., *Lens* spp., *M. atropurpureum, Lathyrus* spp.
R. galegae (Lindström, 1989)	*Galega orientalis, G. officinalis*
R. tropici (Martinez-Romero *et al.*, 1991)	*P. vulgaris, Leucaena sculenta, L. leucocephala*
R. etli (Segovia *et al.*, 1993)	*P. vulgaris*
R. giardinii (Amarger *et al.*, 1997) biovar *phaseoli*, Giardinii	*P. vulgaris, Phaseolus* spp., *M. atropurpureum, L. leucocephala*
R. gallicum (Amarger *et al.*, 1997)	*P. vulgaris, Phaseolus* spp., *M. atropurpureum, L. leucocephala, Onobrychis viciifolia*
R. hainanense (Chen *et al.*, 1997)	*Macroptilium lathyroides, Zornia diphila, Uraria crinita, Desmodium sinuatum, Stylosanthes guianensis, Desmodium gyroides, Acacia sinuata, Tephrosia candida, Arachis hypogea, Centrosema pubescens, Desmodium triquetrum, Desmodium heterophyllum, V. unguiculata*
R. mongolense (van Berkun *et al.*, 1998)	*Medicago ruthenica, P. vulgaris, Vicia vilosa, V. angularis*
R. huautlense (Wang *et al.*, 1998)	*Sesbania herbaceae, L. leucocephala, S. rostrata, T. repens*
R. etli (Wang *et al.*, 1999a)	*Mimosa affinis, P. vulgaris, L. leucocephala*
R. yanglingense (Tan *et al.*, 2001)	*Amphicarpaea trisperma, Coronilla varia, Gueldenstaedtia multiflora, P. vulgaris, Galega orientalis, L. leucocephala*
R. sullae (Squartini *et al.*, 2002)	*Hedysarum coronarium*
R. indigoferae (Wei *et al.*, 2002)	*Indigofera* spp.
R. loessense (Wei *et al.*, 2003)	*Astragalus* spp., *Lespedeza* spp.
Bradyrhizobium (Jordan, 1984)	
B. japonicum (Jordan, 1984)	*M. atropurpureum, Ornithopus sativus, G. max, Lupinus* spp.
B. elkanii (Kuykendall *et al.*, 1992)	*Glycine soja, G. max*
B. liaoningense (Xu *et al.*, 1995)	*G. soja, G. max, Phaseolus aureus*
B. yuanmingense (Yao *et al.*, 2002)	*Lespedeza* spp.
B. canariense (Vinuesa *et al.*, 2004)	*Chamaecytisus proliferus, Teline* spp., *Lupinus* spp., *Adenocarpus* spp., *Spartocytisus supranubius, Ornithopus* spp.

Table 11.3. Genera and species of Leguminosae-nodulating bacteria with their respective host species mentioned in the original publications. – cont'd

Genera/species	Hosts
Bradyrhizobium spp.	*M. atropurpureum*, *Aeschynomene* spp., *Crotalaria* spp., *Lotus* spp., *Vigna* spp., *Lupinus* spp., *Ornithopus* spp., *Cicer* spp., *Sesbania* spp., *Leucaena* spp., *Mimosa* spp., *Lab-lab* spp., *Acacia* spp., *Macroptilium* spp., *Glycine* spp.
Azorhizobium (Dreyfus et *al.*, 1988)[a]	
A. caulinodans (Dreyfus et *al.*, 1988)	*S. rostrata*
A. doebereinerae (Moreira et *al.*, 2005)	*S. virgata*, *S. rostrata*, *P. vulgaris*, *M. atropurpureum*
Sinorhizobium (Chen et *al.*, 1988; de Lajudie et *al.*, 1994) *Ensifer* (?)(Casida, 1982; Young, 2003)	
S. meliloti (Dangeard, 1926; Jordan et *al.*, 1984; de Lajudie et *al.*, 1994)	*Melilotus* spp., *Medicago* spp., *Trigonella* spp.
S. fredii (Scholla and Elkan, 1984; Chen et *al.*, 1988; de Lajudie et *al.*, 1994)	*G. max*, *G. soja*, *P. vulgaris*
S. xinjiangense (Chen et *al.*, 1988)	*G. max*
S. saheli (de Lajudie et *al.*, 1994)	*Sesbania* spp., *Acacia seyal*, *L. leucocephala*, *N. oleraceae*
S. teranga (de Lajudie et *al.*, 1994)	*Sesbania* spp., *Acacia* spp., *N. oleraceae*, *L. leucocephala*
S. medicae (Rome et *al.*, 1996)	*Medicago* spp.
S. arboris (Nick et *al.*, 1999)	*Acacia senegal*, *Prosopis chilensis*
S. kostiense (Nick et *al.*, 1999)	*Acacia senegal*, *Prosopis chilensis*
S. adhaerens (Willems et *al.*, 2003)	*L. leucocephala*, *P. dulce*, *Sesbania grandiflora*, *M. sativa*
S. morelense (Wang et *al.*, 2002)	*L. leucocephala*
S. kummerowiae (Wei et *al.*, 2002)	*Kummerowia stipulacea*
S. americanum (Toledo et *al.*, 2003)	*Acacia* spp., *L. leucocephala*, *P. vulgaris*
Mesorhizobium (Jarvis et *al.*, 1997)	
M. loti (Jarvis et *al.*, 1982, 1997; Jordan et *al.*, 1984)	*Wisteria frustescens*, *Caragana* spp., *Lotus* spp., *P. vulgaris*, *M. atropurpureum*, *Genista* spp., *Lupinus densiflorus*, *Anthyllis vulneraria*, *Ornithopus sativus*, *L. leucocephala*, *Cicer arietnum*, *Mimosa* spp., *Caragana arborescens*
M. huakuii (Chen et *al.*, 1991; Jarvis et *al.*, 1997)	*Astragalus sinicus*, *Vicia villosa*, *P. vulgaris*, *Sesbania* sp., *A. aliginosus*, *A. adsurgens*
M. ciceri (Nour et *al.*, 1994; Jarvis et *al.*, 1997)	*Cicer arietnum*
M. tianshanense (Chen et *al.*, 1995; Jarvis et *al.*, 1997)	*Glycyrrhiza pallidiflora*, *G.uralensis*, *G.* sp., *Halimodendron holodendron*, *Sophora alopecuroides*, *Caragana polourensis*, *G. max*, *Swainsonia salsula*
M. mediterraneum (Jarvis et *al.*, 1997; Nour et *al.*, 1995)	*Cicer arietnum*
M. plurifarium (de Lajudie et *al.*, 1998)	*Prosopis juliflora*, *Neptunia oleraceae*, *Acacia senegal*, *A. seyal*, *A. tortilis*, *L. leucocephala*, *L. pulvurulenta*, *L. diversifolia*, *C. ensiformis*, *Acacia nilotica*
M. amorphae (Wang et *al.*, 1999b)	*Amorpha fruticosa*
M. chacoense (Velazquez et *al.*, 2001)	*Prosopis alba*, *P. chilensis*, *P. flexuosa*
M. temperatum (Gao et *al.*, 2004)	*Astragalus adsurgens*
M. spetentrionale (Gao et *al.*, 2004)	*Astragalus adsurgens*
Allorhizobium (de Lajudie et *al.*, 1998)	
A. undicola (de Lajudie et *al.*, 1998)	*Neptunia natans*, *Lotus arabicus*, *Acacia seyal*, *Faidherbia albida*, *Acacia tortilis*, *M. sativa*, *Acacia senegal*
Methylobacterium	

Continued

Table 11.3. Genera and species of Leguminosae-nodulating bacteria with respective host species mentioned in the original publications. – cont'd

Genera/species	Hosts
M. nodulans (Sy *et al.*, 2001; Jourand *et al.*, 2004)	*Crotalaria* spp.
Burkholderia sp. (Moulin *et al.*, 2001)[a]	*Aspalathus* sp., *Machaerium* sp., *M. atropurpureum*
Ralstonia	
R. taiwanensis (Chen *et al.*, 2001)[b]	*Mimosa pudica*, *M. dilotricha*
Blastobacter	
B. denitrificans (van Berkun and Eardly, 2002)[c]	*Aeschynomene indica*
Devosia	
D. neptunea (Rivas *et al.*, 2002, 2003)	*Neptunia natans*

[a]*Agrobacterium* species *A. tumefaciens* (syn. *A. radiobacter*), *A. rhizogenes*, *A. rubi* and *A. vitis* and *Allorhizobium undicola* were proposed to be included in *Rhizobium* (Young *et al.*, 2001).
[b]Genus *Wautersia* (Vaneechoutte *et al.*, 2004) and later genus *Cupriavidus* (Vandamme and Conye, 2004) were proposed to accommodate this species.
[c]These authors did not describe this species but they discovered it is able to nodulate legumes.

regions (Table 11.3). Although Brazil has a relatively large number of researchers in soil microbiology, scarcity of resources is a reasonable explanation for the low number of taxonomic studies. On the other hand, Brazilian research has been stronger in field application, the best example being soybean inoculation. In spite of the great number of rhizobial species, only a few (*Rhizobium tropici*, *Azorhizobium caulinodans*, *Sinorhizobium saheli*, *Sinorhizobium teranga*, *Rhizobium hautlense*, *Mesorhizobium plurifarium*, *S. adherens*) were described based on isolates from tropical soils and/or host species. From those, only *R. tropici*, *M. plurifarium* and *S. adherens* include Brazilian isolates from *P. vulgaris*, *Chamaecrista ensiformis*, *Leucaena leucocephala*, *Leucaena* spp. and *Pithecellobium dulce*. A new species of *Azorhizobium* (*Azorhizobium johannense*) was proposed (Moreira *et al.*, 2000). This species is able to nodulate roots of both *Sesbania virgata* (a fast-growing Brazilian native shrub adapted to flooded conditions without stem nodulation) and *Sesbania rostrata* (native from Africa). However, pseudonodules are formed in the stems of the latter species, which are smaller with a whitish colour inside, indicating no nitrogen fixation. Nodulation on *S. virgata* by this new species is highly efficient. The other species of this genus, *A. caulinodans*, is the microsymbiont of *S. rostrata* (Dreyfus *et al.*, 1988), which indicates an evolutionary relationship of both symbioses although located in different continents. The epithet *A. johannense* was modified recently to *A. doebereinerae* (syn.: *A. johannae*) (Moreira *et al.*, 2005) (Fig. 11.2).

Evolution of the symbiosis

Considering data on DNA:DNA hybridization, 16S rRNA partial sequences, total protein profiles (SDS-PAGE) and also the widespread *Bradyrhizobium* symbiosis, there is no relationship between rhizobia and Leguminosae phylogenies (Young and Johnston, 1989; Moreira *et al.*, 1993; Moreira *et al.*, 1998). In Brazil most of the Leguminosae phylogenetic branches comprise LNB from diverse phylogenetic branches and vice versa (Tables 11.4A–C and 11.5). This corroborates the hypothesis of no coevolution between both partners. Young and Johnston (1989) raised three reasons for this:

1. LNB also occur free-living in soil, so, as a non-obligate symbiont they are under other selective forces.

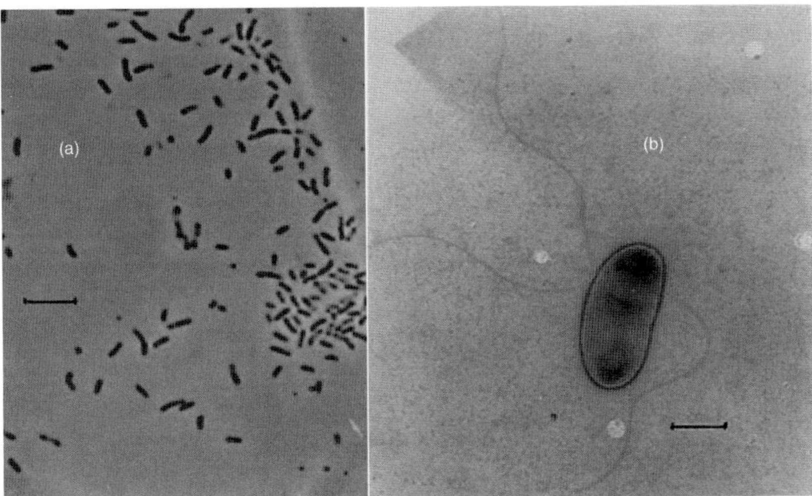

Fig. 11.2. *Azorhizobium doebereinerae* cells viewed by (a) phase-contrast microscopy (scale bar represents 5.4 μm) and (b) transmission electron microscopy (scale bar represents 0.5 μm).

Table 11.4.A. Leguminosae–Caesalpiniodeae genera from which nodulating bacteria (LNB) were isolated from Brazilian soils: extant phenotypic (*P*) or genetic characterization (*G*) and growth rate[a] of LNB in 79 medium.

Genera extant characterization	Growth rate of LNB/species identified for some strains
Campsiandra, P, G	FGs, IGs, SGs, VSGs
Chamaecrista, P, G	FGs, IGs, SGs/*M. plurifarium*
Dimorphandra, P, G	FGs, SGs, VSGs/*B. elkanii, B. japonicum, R. tropici, S. medicae*
Dycorinia	VSGs
Melanoxylum, P, G	SGs
Sclerolobium	SGs
Tachigali, P, G	IGs, SGs/*B. elkanii*
Vouacapoua, P, G	FGs, SGs

[a]Time (days) of appearance of isolated colonies: FGs = 2–3; IGs = 4–5; SGs = 6–10; VSGs ≥ 10.

Table 11.4.B. Leguminosae–Mimosoideae genera from which nodulating bacteria (LNB) were isolated from Brazilian soils: extant phenotypic (*P*) or genetic characterization (*G*) and growth rate[a] of LNB in 79 medium.

Genera extant characterization	Growth rate of LNB/species identified for some strains
Acacia, P, G	FGs, IGs, SGs, VSGs/*B. elkanii*
Abarema	FGs, IGs, VSGs
Albizia, P, G	FGs, IGs, SGs, VSGs
Anadenanthera, P	FGs, SGs, VSGs
Calliandra, P, G	FGs, SGs/*R. tropici*
Cedrelinga	VSGs
Entada, P	SGs
Enterolobium, P, G	FGs, IGs, SGs, VSGs/*B. japonicum*
Goldmania	IGs, SGs
Inga, P, G	FGs, IGs, SGs, VSGs/*Bradyrhizobium BTAi1*

Continued

Table 11.4.B. Leguminosae–Mimosoideae genera from which nodulating bacteria (LNB) were isolated from Brazilian soils: extant phenotypic (*P*) or genetic characterization (*G*) and growth rate[a] of LNB in 79 medium. – cont'd

Genera extant characterization	Growth rate of LNB/species identified for some strains
Leucaena, P, G	FGs, IGs, SGs/*R. tropici, R. leguminosarum, S. fredii, S. medicae, M. plurifarium*
Macrosamanea	FGs, VSGs
Mimosa, P, G	FGs, IGs
Mimozyganthus	FGs
Neptunia	FGs
Parapiptadenia, G	FGs
Paraserianthes, P, G	IGs, SGs/*B. elkanii*
Pentaclethra, P, G	SGs, VSGs/*B. japonicum*
Piptadenia	FGs, IGs, SGs
Pithecellobium, P, G	FGs, IGs, SGs, VSGs/*B. japonicum, S. medicae*
Plathymenia, P	FGs, IGs, SGs
Prosopis, P, G	FGs, IGs, SGs, VSGs/*Sinorhizobium* sp.
Pseudosamanea	SGs
Pseudopiptadenia	FGs
Samanea, P	IGs, SGs
Stryphnodendron, P	SGs, VSGs
Zygia	FGs

[a]Time (days) of appearance of isolated colonies: FGs = 2–3; IGs = 4–5; SGs = 6–10; VSGs ≥ 10.

Table 11.4.C. Leguminosae–Papilionoideae genera from which nodulating bacteria (LNB) were isolated from Brazilian soils: extant phenotypic (*P*) or genetic characterization (*G*) and growth rate[a] of LNB in 79 medium.

Genera extant characterization	Growth rate of LNB species identified for some strains
Abrus, P, G	SGs, *B. elkanii*
Aeschynomene	FGs
Andira	FGs, SGs
Arachis	FGs, IGs, SGs, *Bradyrhizobium* spp.
Ateleia	SGs
Bolusanthus, P	FGs
Bowdichia	IGs, SGs
Cajanus, P	FGs, SGs
Calopogonium	SGs
Canavalia	SGs
Centrolobium, P, G	FGs, IGs, *R. tropici*
Centrosema, P, G	SGs, *B. japonicum*
Chlatrotropis, P	VSGs
Cicer	IGs, FGs
Clitoria, P, G	FGs, IGs, SGs
Cyclolobium	IGs
Cratylia, P	SGs
Crotalaria	IGs, SGs
Dalbergia, P, G	FGs, SGs, VSGs, *B. japonicum, B. elkanii*
Derris, P, G	FGs, SGs, VSGs, *B. japonicum*
Desmodium	IGs, SGs
Dioclea, P	VSGs
Diplotropis	SGs, VSGs
Dolichos	FGs, SGs

Continued

Table 11.4.C. Leguminosae–Papilionoideae genera from which nodulating bacteria (LNB) were isolated from Brazilian soils: extant phenotypic (*P*) or genetic characterization (*G*) and growth rate[a] of LNB in 79 medium. – cont'd

Genera extant characterization	Growth rate of LNB/species identified for some strains
Discolobium	FGs, SGs/*Bradyrhizobium* spp.
Erythrina, P, G	FGs, IGs, SGs, VSGs
Etaballia, P	SGs
Galactia	SGs
Glycine, P, G	FGs, SGs, *B. elkanii, B. japonicum, S. fredii*
Gliricidia, P, G	FGs, *R. tropici, R. leguminosarum*
Hymenolobium	VSGs
Indigofera, G	FGs, SGs
Lathyrus	FGs
Lens	FGs
Lonchocarpus (Deguelia), P, G	FGs, SGs, *B. japonicum, R. tropici*
Lotononis	FGs
Lotus	FGs, IGs
Lupinus	FGs
Machaerium, P, G	FGs, IGs, SGs, VSGs, *B. japonicum, R. tropici*
Macroptilium, P, G	FGs, IGs, SGs, VSGs
Medicago	FGs
Monopteryx	VSGs
Mucuna, P, G	SGs
Neonotonia	IGs, SGs
Ormosia, P, G	SGs, VSGs, *B. elkanii*
Ornithopus	IGs, SGs
Pachyrhisus	FGs, SGs
Phaseolus, P, G	FGs, IGs, SGs, VSGs, *R. tropici, Burkholderia* sp., *R. leguminosarum*
Platycyamus	FGs
Platymiscium, P	SGs, VSGs
Platypodium	FGs, SGs
Poecilanthe, P	SGs
Pterocarpus, G	SGs
Pueraria, P	IGs, SGs
Sesbania, P, G	FGs, IGs, *Azorhizobium doebereinerae*
Swartzia, P, G	FGs, SGs, VSGs – *R. tropici, S. medicae*
Tephrosia, G	SGs
Tipuana	SGs
Trifolium	FGs
Vicia	FGs
Vigna, P, G	FGs, IGs, SGs, VSGs, *Bradyrhizobium* spp.

[a]Time (days) of appearance of isolated colonies: FGs = 2–3; IGs = 4–5; SGs = 6–10; VSGs ≥ 10.

2. Current Leguminosae taxonomy may not reflect the real evolutionary relationships.
3. The establishment of NB and Leguminosae symbiosis depends on some complex feature rarely evolved in the *Plantae* kingdom.

Regarding the first hypothesis much evidence emerged from the exponentially increasing number of molecular studies in the last decade. LNB as well as pathogen species share infection and symbiotic genes, indicating that lateral and horizontal transfer occurs frequently in soil (Wernegreen and Riley, 1999; Young, 2000; Thies *et al.*, 2001; Young, 2001). Corroboration of hypothesis 2 depends on extensive studies on plant molecular evolution. Although advances in molecular biology were intensively applied to Prokaryotes, they are relatively incipient for the understanding of Eucaryote phylogeny, including *Plantae Leguminosae*. Recent studies suggest that previous putative divergence of Leguminosae (Polhill, 1981; Polhill *et al.*, 1981) did not reflect true evolutionary relationships (Käss and Wink, 1996, 1997). However, even these studies were based on only 49 species of Leguminosae and 75 species of Papilionoideae, most of them from temperate regions. Advances in this area are expected to happen in the near future in order to provide a solid basis for this hypothesis, when based on a high number of taxa. Isoflavonoids and flavonoids would be the potential candidate compounds to support hypothesis 3. However, as Young and Johnston (1989) pointed out, they also can antagonize plant response in some species. Furthermore, some legume species nodulate even though no isoflavonoids are detected and many isolates respond to these compounds produced by non-nodulating plants (Shaw *et al.*, 1997).

Brazilian Culture Collections

A large number of strains have been isolated from several hosts, including woody species in Brazil. Bacterial collections have been established by some of the main Brazilian groups (Embrapa-agrobiologia, Rio de Janeiro; FEPAGRO/UFRGS, Rio Grande do Sul; IAC, São Paulo; UFLA, Minas Gerais) studying Leguminosae symbioses. There are thousands of strains isolated from many species belonging to various growth forms (trees, shrubs, herbs, lianas, etc.), the majority of them not yet fully characterized. Curators of these collections reported the following number of strains: Embrapa Agrobiologia (RJ, Prefix BR), 1300 (R. Pittard, Rio de Janeiro, 2003, personal communication); Fepagro/UFRGS (RGS, Prefix SEMIA), 1200 (E. Bagel, Rio Grande do Sul, 2003, personal communication); Instituto Agronômico de Campinas (SP, Prefix IAC), 800 (S. Freitas, Campinas, 2003, personal communication); UFLA (MG, Prefix UFLA), 2000 (F.M.S. Moreira, 2005, unpublished results). Other culture collections in Brazil are those from Embrapa Soja (PR), Embrapa Cerrados (DF), EMBRAPA Trigo (RS), Universidade Federal de Viçosa (MG), Universidade Federal do Rio de Janeiro (RJ), Centro de Energia Nuclear na Agricultura/USP (SP), Empresa Pernambucana de Pesquisa Agropecuária (PE), Escola Superior de Agricultura Luiz de Queiroz/USP (SP), Instituto Agronômico do Paraná (PR), Instituto Nacional de Pesquisas da Amazônia (AM) e Instituto de Pesquisas Tecnológicas (SP).

Phenotypic and Genotypic Approaches to Taxonomic and Diversity Studies

The international subcommittee for the taxonomy of *Rhizobium* and *Agrobacterium* proposed minimum standards for the description of new species of root- and stem-nodulating bacteria (Graham *et al.*, 1991). These included phylogenetic and phenotypic (symbiotic, cultural, morphological and physiological) traits in accordance with the polyphasic and phylogenetic taxonomy as follows: symbiotic performance with selected hosts (mainly hosts of currently recognized species), cultural and morphological characteristics, serological methods, cell lipopolysaccharide and protein-banding (SDS-PAGE) patterns, DNA:DNA relatedness (including DNA base composition % C + G), rRNA:DNA hybridization, 16S rRNA sequencing, DNA restriction fragment length polymorphism (RFLP) and multilocus enzyme electrophoresis (MEE). Since then, new techniques for characterization were described, improving not only bacterial classification but also the discrimination capability within different taxa, i.e. each of them has a specific level of resolution for bacterial classification, which might be useful for diversity studies. These techniques, also called fingerprint approaches, include digestion of genomic DNA with rare cutting-site endonucleases, followed by pulsed field gel electrophoresis (PFGE) and other RFLP-based methods; PCR-based methods such as amplification and restriction analysis of internal 16S rDNA gene regions (ARDRA), tRNA-PCR or ITS amplification and analysis of inter tRNA spacer (ITS) regions or inter 16S–23S rRNA gene regions), amplified fragment length polymorphism (AFLP) for the whole-genome analysis, random amplified polymorphic DNA (RAPD); arbitrarily primed PCR (AP-PCR), repetitive DNA elements genomic fingerprinting (REP-PCR) (Rademaker and Bruijn, 1997). Most of these techniques have been applied in the studies of nitrogen-fixing stem- and root-nodulating bacteria. Standard methods for assessment of LNB diversity from soil isolation and nodule collection at field conditions until characterization of isolates were recently reviewed (Moreira and Pereira, 2001).

Diversity of LNB in Brazil

At a given soil condition in time, Leguminosae species can be found either nodulating or non-nodulating with native LNB populations. If nodules are found, LNB can be isolated and characterized, providing information on the diversity of microsymbionts to that particular plant species in that particular environmental condition. To isolate and enumerate LNB from a diverse microbial community, such as occurs in the soil, a method that clearly separates LNB from other species is required. The plant infection technique (by using trap species) makes use of the nodulation process itself to estimate LNB populations in soil. LNB diversity in Brazil was searched in both ways, i.e. nodules collected both from field and from trap species.

Diversity of LNB from nodules collected in field/nursery

Although LNB from different genera can be found in Brazilian soils, the genus *Bradyrhizobium* seems the most important one as it is found to be naturally predominating among the microsymbionts of forest species, and also among forage and green manure genera/species (Table 11.4A–C and 11.6). *Bradyrhizobium* strains were isolated from all studied genera in Caesalpinioideae and from 84% and 80% of the studied genera in Mimosoideae and Papilionoideae forest species, respectively. *Bradyrhizobium* spp. are typically slow growers (SGs) or very slow growers (VSGs) (less frequently) that alkalinize 79 medium (Fred and Waksman, 1928). Many SG and VSG alkalinizing strains, isolated from nodules collected from field or nursery conditions, analysed by phenotypic and genotypic techniques, such as SDS-PAGE of total proteins, PCR-RADP, MEE, REP-PCR and 16S rRNA sequence, were identified as belonging to the genus *Bradyrhizobium* (Moreira *et al.*, 1993; Coutinho *et al.*, 1995; Moreira *et al.*, 1995; F.M.S. Moreira, 2005, unpublished results). Partial 16S rRNA sequences of 22 SG strains were highly similar to *Bradyrhizobium elkanii* or *Bradyrhizobium japonicum* (Table 11.5).

The wide occurrence of *Bradyrhizobium* symbiosis with native species conflicts with previous reports that *B. japonicum* was introduced in Brazil as a bacterial inoculant for soybean (Martínez-Romero and Caballero-Mellado, 1996; Santos *et al.*, 1999, based on several references; Hungria *et al.*, 2000). It must be considered that the majority of these strains were isolated from nodules collected in native forests. Thus, an introduced strain should have not only a high competitive ability but also an exceptional performance to colonize diverse soil conditions, and a promiscuous behaviour in relation to a wide genetic diversity of hosts. All these should be achieved in less than 50 years as the first inoculant strains for soybeans were introduced in Brazil by 1950. This is not in accordance with evolutionary and adaptation times normally required. Instead, the wide distribution of symbioses with *Bradyrhizobium* within almost all nodulating phylogenetic branches of Leguminosae in tropical regions and higher frequency of SGs in the most primitive groups of Caesalpinioideae, Mimosoideae and Papilionoideae (Table 11.4A–C) corroborate

Table 11.5. *Bradyrhizobium* 16S rRNA sequences (GenBank) from Brazilian and introduced strains with respective origins (host, region/state/country).

Sequence/most similar sequence (%) or number of nucleotides different of 230	Species	Origin: host genus/ species (subfamily)[a]	Reference origin: region/state/country
L20781 (SEMIA587)	*B. elkanii* USDA 76ᵀ (100% similar 188 bp)	*G. max* (*P*)	Rumjanek *et al.*, 1993
L208867 (BR29W = SEMIA5019)			Rio de Janeiro, BR
AJ003235 (L41530-ATCC10324ᵀ – 2 bp)	*B. japonicum*	*Dalbergia* (*P*) *Dimorphandra* (*C*) *Pentaclethra* (*M*) *Derris* (*P*) *Machaerium* (*P*) *Enterolobium* (*M*) *Lonchocarpus* (*P*) *Acacia* (*M*)	Moreira *et al.*, 1998 Amazon and Atlantic forests, BR
M55490 (0 bp)	*B. elkanii* USDA 76ᵀ	*Dalbergia* (*P*) *Ormosia* (*P*) *Tachigali* (*C*) *Clitoria* (*P*) *Acacia* (*M*) *Paraserianthes* (*M*) *Abrus* (*P*)	Moreira *et al.*, 1998 Amazon and Atlantic forests, BR
L41530 (0 bp)	*B. japonicum* ATCC10324ᵀ	*Inga* (*M*)	Moreira *et al.*, 1998 Amazon forest, BR
M55492 (0 bp)	*Bradyrhizobium* BTAi1	*Erythrina* (*P*)	Moreira *et al.*, 1998 Atlantic forest, BR
AJ003236 (L41530-ATCC10324ᵀ – 3 bp)	*B. japonicum*	*Pithecellobium* (*M*)	Moreira *et al.*, 1998 Atlantic forest, BR
AF237422 (SEMIA5019) (99% M55487)	*B. elkanii*	*G. max* (*P*)	Chueire *et al.*, 2000, south, BR
AF234890 (SEMIA587) (100% X70402)	*B. elkanii*		South, BR
AF234889 (SEMIA5080 = CPAC7) (99% L23330)	*B. japonicum*		Cerrado, BR
AF234888 (SEMIA5079 = CPAC15) (99% B.jᵀ)	*B. japonicum*		Cerrado, BR
AF236087 (SEMIA586 = CB1809) (100% SEMIA5080)	*B. japonicum*		Australia inoculant?
AF236086 (SEMIA566) (100% SEMIA5079)	*B. japonicum*		US inoculant
AY117676 (SEMIA5039)	*B. japonicum*	*G. max* (*P*)	Ferreira and Hungria, 2002, south, BR
Y117673 (R17)	*B. elkanii*		Fernandes *et al.*, unpublished, south, BR
AY117674 (R35)	*B. japonicum*		
AY117675 (R45)	*B. japonicum*	*G. max* (*P*)	
AY117676 (SEMIA5085)	*B. japonicum*	*G. max* (*P*)	Chueire *et al.*, unpublished, south, BR

[a]*C* = Caesalpinioideae, *M* = Mimosoideae, *P* = Papilionoideae.

the evolutionary hypothesis that symbiosis with *Bradyrhizobium* was established earlier than those with fast growers (e.g. *Rhizobium*), which was postulated by Norris (1965).

According to Freire and Vernetti (1999) 27 *Bradyrhizobium* strains have been used for soybean inoculant production in Brazil since 1950. They mentioned that most of them (except five from the University of Wisconsin, USA; two from USDA, USA; and one from CSIRO, Australia) were from Brazilian soils from south (14), south-east (3) and central (2) regions, respectively. Soybean inoculant strains were introduced in all five regions of Brazil. In Amazonas state, from where most of the forest strains studied by Moreira (1991) were isolated, there are records of soybean inoculant (Turfal) application both in flooded areas near the Solimões River (varzea) and in 'terra firme' soils in Manaus since 1975 (Rahman, 1977; Yuyama and Oliveira, 1997). No yield response to inoculation was observed (Rahman, 1977; Yuyama and Oliveira, 1997) at sites from both ecosystems, indicating existence of native populations able to nodulate soybeans. Strains used as inoculants at that time were SEMIA543, SEMIA587, SEMIA527, SEMIA532, SEMIA586 and SEMIA566. Strains isolated from Amazon forest species, almost 10 years later, were genetically different from these strains (Table 11.5). Strain 566, identified as *B. japonicum* (AF236086) (Chueire *et al.*, 2000) and isolated from a soil inoculated previously with a US commercial inoculant in Rio Grande do Sul (Freire and Vernetti, 1999) had nodule occupancy much lower than BR29 and SEMIA587 (inoculant strains since 1968 and 1979, respectively) when soybeans were introduced in cerrado soils in the 1970s: 2% and 90%, respectively (Vargas *et al.*, 1981). However, it has been observed that SEMIA566 serogroup nodule occupancy is becoming more expressive with time, reaching 60% after more than 10 years (Vargas *et al.*, 1994). Silva *et al.* (1998) reported significant differences in protein profiles within serogroup SEMIA566 strains with the parental strain, which could be related to the adaptive response in competitiveness of the introduced populations. Adaptation of other introduced *Bradyrhizobium* strains (SEMIA5039, SEMIA5020, 532c, SEMIA586/CB1809) to Brazilian soils also resulted in a great diversity in both genetic and symbiotic characteristics (Santos *et al.*, 1999). Although some papers reported high similarity among indigenous soil isolates and strains introduced as inoculants (Ferreira and Hungria, 2002), high variability of native populations of *Bradyrhizobium*-nodulating soybeans has been observed (Santos *et al.*, 1999; Ferreira *et al.*, 2000).

Bradyrhizobium native strains nodulating cowpea were shown to also nodulate soybeans (Martins *et al.*, 1997). Strain INPA03-11B, highly efficient in cowpea (Fig. 11.3) and soybean symbiosis (Miguel and Moreira, 2001), was isolated from a *Centrosema* sp. growing in a field previously cultivated with inoculated soybeans in Manaus, but its 16S rRNA sequence is not similar to the inoculant strains used at that time (F.M.S. Moreira, 2005, unpublished results). These results indicate that native populations could also be a source of LNB for soybeans, including efficient strains.

The recently described *Methylobacterium* has culture characteristics similar to *Bradyrhizobium*, i.e. alkalinization and slow growth in 79 medium. In our laboratory a *Methylobacterium* was isolated from Amazon region soil by using cowpea as trap species, so probably other isolates may be confirmed as belonging to this genus after further characterization of UFLA's collection.

Fast-growing strains belonging to *Sinorhizobium fredii* and known to nodulate the Asian soybean genotype were isolated from forest species (Moreira *et al.*, 1993, 1998); however, they were not able to nodulate soybean cultivar 'Vencedora' under controlled conditions where BR29 was used as a positive control. Fast-growing strains with phenotypic characteristics different from *S. fredii* were isolated from about 50% of Brazilian soil samples (collected from Paraná, Rio Grande do Sul, Amazonas, Distrito Federal, Minas Gerais and Goiás) by using Asian genotypes and

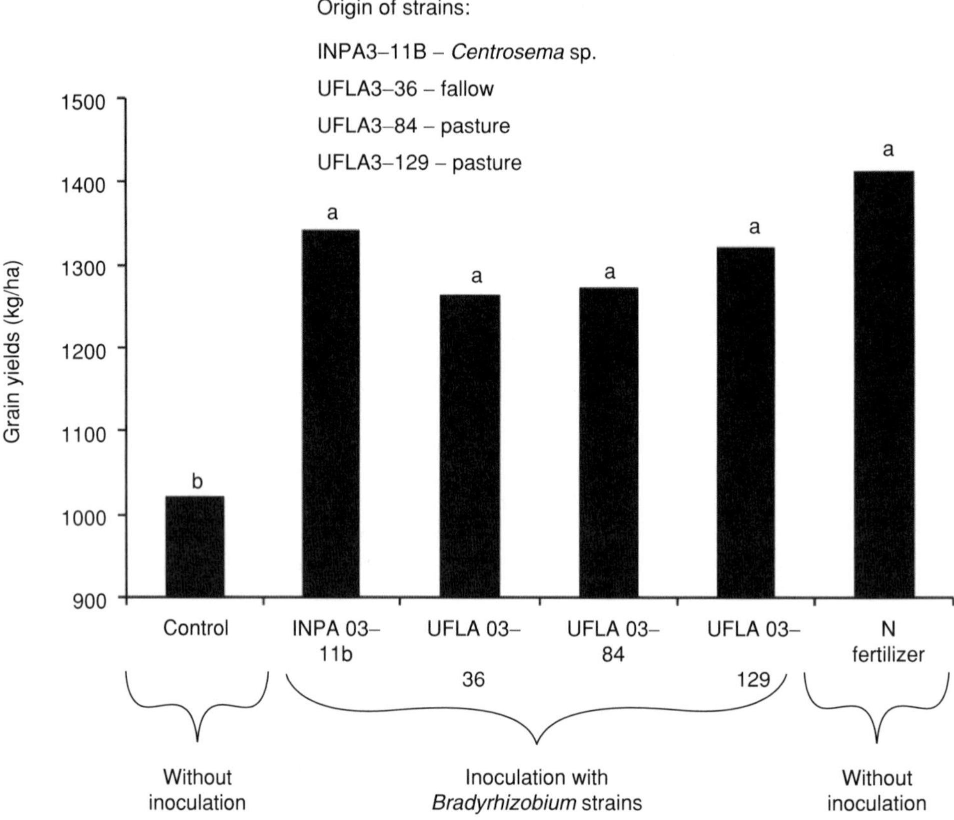

Fig. 11.3. Cowpea inoculated with selected *Bradyrhizobium* strains isolated from Amazon region land use systems. (Source: extracted from Lacerda *et al.*, 2004.)

one modern genotype (Hungria *et al.*, 2001a). Although some of them fixed N_2 as much as *Bradyrhizobium* inoculant strains, only a few of them were able to compete with these inoculant strains (Hungria *et al.*, 2001b).

Strains with intermediary growth rate were also isolated, but from a small percentage of native forest genera (14% Caesalpinioideae, 26% Mimosoideae and 13% Papilionoideae) (Table 11.6). Some of them were assigned to the genus *Mesorhizobium* (Jarvis *et al.*, 1997; de Lajudie *et al.*, 1998). Fast growers (*Rhizobium*, *Sinorhizobium* and *Mesorhizobium*) occur in symbiosis with nearly 40% of both Caesalpinioideae and Papilionoideae genera, but they are highly frequent among Mimosoideae (81%). Species of *Rhizobium*, *Azorhizobium*, *Sinorhizobium* and *Mesorhizobium* (Moreira *et al.*, 1998) (Tables 11.4A–C) and *Burkholderia* (Moreira *et al.*, 2002) were identified among these isolates.

Fast growers are much more diverse than SGs, which is indicated not only by the much higher number of fast-growing species described in relation to SGs (Table 11.3) but also by genotypic and phenotypic characterization from soil populations. When 171 strains were characterized by SDS-PAGE in terms of total proteins, the 51 fast and intermediary growers could be differentiated in 12 clusters and 20 strains had isolated positions, while the major group comprising 120 slow or very SGs could be differentiated in a lower number of 8 clusters and 10 strains having isolated positions (Moreira *et al.*, 1993). The same was observed with other tech-

Table 11.6. Culture characteristics (on 79 medium) of 705 Leguminosae-nodulating bacteria strains isolated from Amazon and Atlantic forest species belonging to 49 genera.

		Culture characteristics of 79 medium[a]			
Origin subfamily	Total number of strains	FGs (% of total number)	IGs (% of total number)	SGs (% of total number)	VSGs (% of total number)
Caesalpiniodeae	64	14	4	67	14
Mimosoideae	444	40	15	36	10
Papilionoideae	197	20	10	53	17
Total	705				

[a]*Rhizobium, Sinorhizobium, Mesorhizobium, Burkholderia*: FGs = fast growers, usually acidifiers; IGs = intermediate growers, usually acidifiers. *Bradyrhizobium*: SGs = slow growers, alkalinizers; VSGs = very slow growers, alkalinizers.
Source: modified from Moreira (1991).

niques like 16S rRNA sequencing (Moreira *et al.*, 1998). The higher diversity in fast growers may be due to a higher frequency of plasmids, which are subject to lateral and horizontal transfer.

Nod and *nif* genes of fast growers are located in the so-called *pSym* plasmids, making these characteristics more unstable. In SGs *nod* and *nif* genes are located in the chromosome, so their expressed characteristics are more stable. This is one of the reasons why agricultural management is more successful in *Bradyrhizobium* symbiosis, as with soybeans. *P. vulgaris*, a promiscuous host nodulating with at least ten fast-growing LNB species (Table 11.3), has still several other constraints for establishment and stability of strains introduced by inoculation as they compete with native populations for infection sites and can also lose nitrogen fixation characteristics under adverse climatic conditions (e.g. high temperature).

Diversity of LNB by using trap plant species

When referring to the diversity of interactions between LNB and Leguminosae species, the concept of promiscuity (or low specificity) is quite ambiguous as it is highly dependent on the number of species tested. For instance, the behaviour of the well-known promiscuous species *P. vulgaris*, *L. leucocephala*, *Macroptilium atropurpureum* and *V. unguiculata*, which have been used widely for assessment of LNB diversity in soils, can vary depending on various features. Lewin *et al.* (1987) found *V. unguiculata* as the host with the lowest specificity for symbiosis with 35 fast-growing tropical rhizobia tested, as compared with *L. leucocephala* and *M. atropurpureum*. However, considering other results listed in Table 11.7, *L. leucocephala* has the broadest host range when compared with others, including *V. unguiculata*. Thus, this classification can be changed as more strains and new species are tested, and other promiscuous hosts are discovered. Anyway, promiscuity of these species is quite evident and cannot be refuted. They also have the advantage of producing large amounts of small seeds that can be easily manipulated under axenic conditions. This is the reason why they have been used in many studies as trap species for assessing LNB diversity.

Recent studies have shown *L. leucocephala* trapped mainly *R. tropici* IIA and IIB from soil samples of the cerrado region (90% of the total number of isolates) while *P. vulgaris* trapped *R. leguminosarum* bv. *phaseoli*/*Rhizobium etli* besides *R. tropici* IIA and IIB. *P. vulgaris* exhibited the highest promiscuity, 50%, compared to only 10% in *L. leucocephala* isolates (Mercante *et al.*, 1998). Methods used for identification of LNB species in this study were growth in

Table 11.7. Rhizobia species reported to establish symbiosis with promiscuous hosts (based on bacterial species description papers listed in Table 11.3; Lewin *et al.*, 1987;[a] Martins *et al.*, 1997; Moreira *et al.*, 1998; Moulin *et al.*, 2001; Moreira *et al.*, 2002).

Promiscuous host species	Nodulating bacteria species
M. atropurpureum	*R. leguminosarum; R. giardini; R. galiccum; Mesorhizobium loti; B. japonicum; B.* spp.; fast-growing rhizobia;[a] *Burkholderia* spp.
L. leucocephala	*Rhizobium tropici; R. giardini; R. galiccum; R. hautlense; R. leguminosaum; M. plurifarium; S. saheli; S. meliloti; S. fredii; S. medicae; S. teranga; Bradyrhizobium* spp.; fast-growing rhizobia[a]
V. unguiculata	*Bradyrhizobium* spp.; fast-growing rhizobia[a]
P. vulgaris	*R. leguminosarum; Rhizobium tropici; R. etli; R. giardini; R. galiccum; R. mongolense; M. loti; M. huakuii; S. fredii; Burkholderia* sp.*; A. doebereinerae*

[a]Lewin *et al.* found *V. unguiculata* as the host with lowest specificity/highest promiscuity among 14 plant species in relation to fast-growing rhizobia, even compared with *L. leucocephala* and *M. atropurpureum*.

Luria broth medium; hybridization with probes to nodABC (*Sinorhizobium meliloti*), nodSU (*Sinorhizobium* sp.), hupSL (*R. leguminosarum* bv viceae) and ORF3 (*R. leguminosarum* bv *phaseoli*) and colony morphology in 79 medium, which is not completely reliable as other rhizobia species present the same characteristics, which was pointed out by these authors.

LNB population densities and efficiency of LNB trapped by *M. atropurpureum* from diverse Land Use Systems (LUS) (disturbed forests, pastures, crops, fallows and agroforestry systems) in Amazonia were quite variable. The highest density (log of most probable number of cells/g soil) and efficiency (shoot dry matter weight) were found in pastures, while the lowest density and efficiency were found in agroforestry systems and disturbed forests (Pereira, 2000), respectively. Populations within and among LUS exhibited high diversity in culture characteristics in 79 medium. SGs alkalinizing this medium (*Bradyrhizobium* spp.) predominate in fallows but not in other LUS (Fig. 11.4). For instance, crops like *P. vulgaris*, which establish symbiosis predominantly with fast growers, also stimulate fast grower populations in soils (Fig. 11.4). In another study in our laboratory, it was found that *V. unguiculata* trapped a higher diversity of LNB cultural types than *P. vulgaris* from a bauxite mine spoil submitted to diverse rehabilitation strategies (Melloni, 2001).

Fast- and slow-growing strains were isolated and phenotypically characterized by using *Crotalaria juncea*, *P. vulgaris* and *Cajanus cajan* as trap hosts from pastures and cerrado forest in São Paulo state. Soil types and vegetation influenced the size of rhizobia populations and specificity decrease in the following order: *P. vulgaris* < *C. juncea* < *C. cajan* (Lombardi, 1995).

Leguminosae native species belonging to genera nodulating with diverse culture types of LNB are possible candidates as native trap species (Tables 11.4A–C). One potential genus is *Inga*. It has a large number of species and together with *Mimosa* and *Acacia* comprises two-thirds of the total number of Mimosoideae species. These genera have the tropical humid region of South America as their centre of origin (Irwin, 1981), so they are well adapted to these soil conditions. Silva *et al.* (1989) reported 83 *Inga* species occurring in Amazonia. Species of this genus are fast growers, producing a large quantity of seeds and are ubiquitous components of home gardens in the Amazon region because of the widespread utilization of their edible fruits. However, their seeds have short viability. Under diverse soil conditions and vegetation types, these species frequently nodulate even under conditions where nodules are absent in other nodulating species, probably indicating promiscuity regarding native LNB populations.

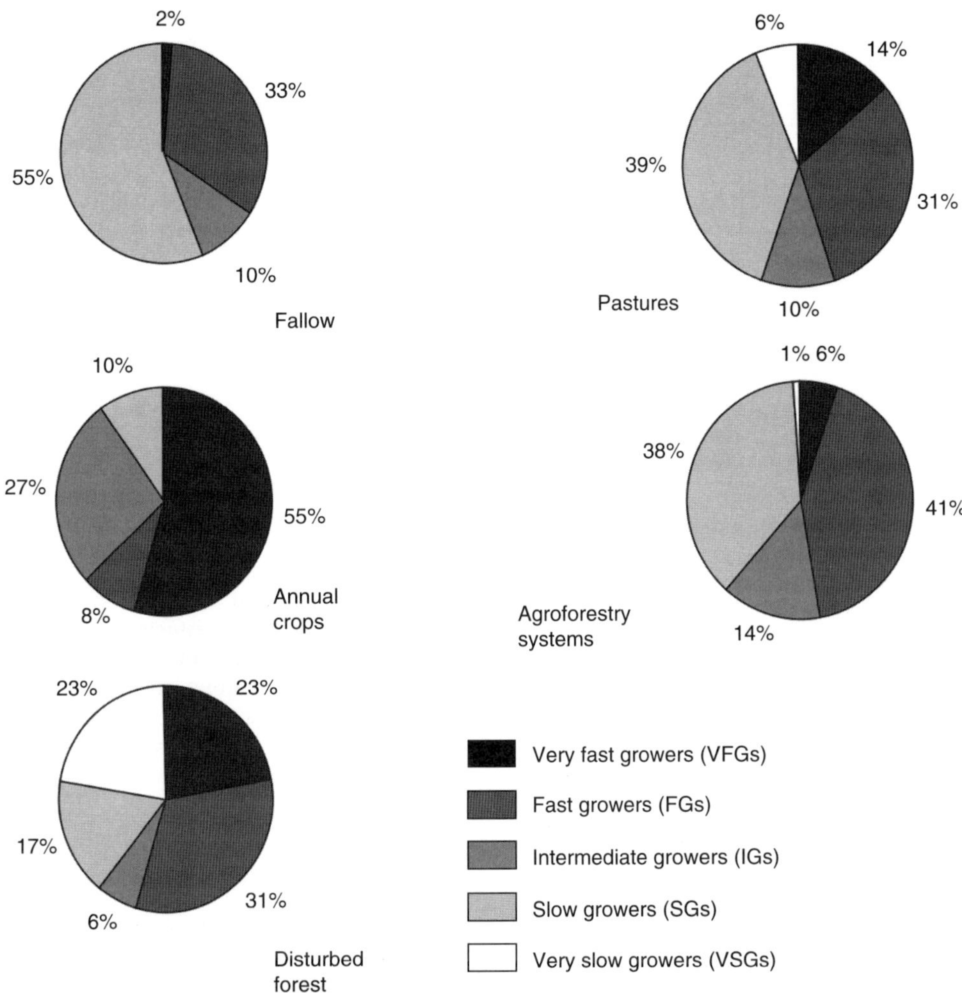

Fig. 11.4. Frequency of culture types trapped by siratro from diverse land use systems in the Amazon region as characterized by the time of appearance (days) of isolated colonies in 79 medium (Fred and Waksman, 1928). VFGs = 1; FGs = 2–3; IGs = 4–5; SGs = 6–10; VSGs ≥ 10. Frequency from a total in: fallow = 126; pasture = 261; agroforestry = 106; forest = 111; annual crops = 110 isolates. (Source: extracted from Pereira, 2000.)

Stem-nodulating LNB

Most LNB establish symbiosis on the roots, but a few species are reported to nodulate plant stems. In Brazil, *Aeschynomene* and *Discolobium* species were found to be stem-nodulated (Loureiro *et al.*, 1994, 1995). Microsymbionts seem to belong to *Bradyrhizobium* spp. and were found to be photosynthetic (Loureiro *et al.*, 1993).

Nitrogen-fixation efficiency

Diversity must also be considered regarding nitrogen-fixation efficiency. This feature can vary even among strains within the same LNB genus nodulating a given host species. For instance, *Arachis pintoi*, a perennial forage crop also used as a cover crop in agroforestry systems due to its shade adaptation, whose diversity centre is South

America, exhibited a wide range of growth and N content in response to inoculation with 230 bradyrhizobia strains. These responses also varied depending on the two plant genotypes tested. Tolerance to low pH of all strains tested *in vitro* varied, but most of them were able to grow well when the pH ranged from 4.5 to 6.8. Also growth and nodulation of both plant genotypes were not affected when cultivated in N-free nutrient solution with pH within the same range (Purcino *et al.*, 2000). By screening a large group of strains regarding their efficiency with a host species as well as their adaptation to soil and environmental conditions, inoculant strains are selected.

Strain Selection and Inoculant Production

Efficient strains have been selected and recommended for inoculant production for 109 leguminous species, including those used for green manure, grains, pasture and multi-purpose trees (Table 11.8). In 2001 according to Rede de laboratórios para recomendação, padronização e difusão da tecnologia de inoculantes microbiológicos de interesse agrícola (RELARE – Laboratory network for recommendation, standardization and diffusion of inoculant technology relevant to agriculture), 14 million doses of inoculant were produced and commercialized (one dose contains 125 g and is applied at 1 ha), the vast majority, i.e. 99%, being for soybeans. Thus, the remaining 1% was for the other 108 species.

Due to the huge bacterial diversity in the tropics and adaptation of these to diverse conditions, potential inoculant strains are being or can be selected for many host species, replacing or adding to those already recommended. For instance, among several Amazonian isolates, *Bradyrhizobium* strains were found to be more efficient than the currently recommended strain (BR2001) for cowpea (Lacerda *et al.*, 2004). Even in soil from very disturbed ecosystems like rehabilitated bauxite mining areas, efficient and diverse LNB populations were found by using both cowpeas (*V. unguiculata*) and beans (*P. vulgaris*) as trap hosts (Melloni, 2001). Among *Bradyrhizobium* strains iso-

Table 11.8. Leguminosae species with strains selected for inoculant production and used in Brazil.

Utilization: Leguminosae species

Grains: *C. max, P. vulgaris, V. unguiculata, Arachis hypogeae, Cicer arietnum, Lens sculenta, Pisum sativum*

Pasture (temperate species): *Lathyrus odoratus, Lotus corniculatus, L. pedunculatus, L. subflorus, L. tenuis, Medicago polymorpha, M. sativa, Ornithopus sativus, Trifolium pratense, T. repens, T. semipilosum, T. subterraneum, T. vesiculosum, Vicia faba*

Pasture (tropical species): *C. cajan, Centrosema* spp., *Desmodium canum, D. intortum, D. ovalifolium, Galactia striata, Indigofera hirsuta, Lablab purpureus, Lotononis balsesli, M. atropurpureum, Macrotyloma axilares, Neonotonia wightii, Stylosanthes* spp.

Green manure: *Calopogonio* spp., *C. ensiformis, C. juncea, C. spectabilis, Cyamopolis tetragonoloba, Lupinus* spp., *Pueraria phaseoloides, Stizolobium aterrinum*

Multiple use (trees): *Acacia angustissima, A. auriculiformis, A. crassicarpa, A. decurrens, A. farnesiana, A. holosericea, A. mangium, A. melanoxylon, A. podalyriaefolia, A. salicina, A. saligna, Acosmium nitens, Aeschynomene sensitiva, Albizia lebbeck, A. procera, Ateleia glazioviana, Balizia pedicellaris, Bowdichia virgiloides, Calliandra surinamensis, C. ensiformis, Clitoria farchildiana, D. nigra, Dimorphandra exaltata, Diphysa robinoides, E. contortsiliquum, E. cyclocarpum, E. timbouva, Erythrina falcata, E. fusca, E. speciosa, E. variegata, E. verna, Falcataria molucana, Gliricidia sepium, Goldmania paraguenses, Hydrochorea corymbosa, Inga marginata, I. thibaudiana, Leucaena diversifolia, L. leucocephala, Lonchocarpus costatus, Melanoxylon brauna, Mimosa acutistipula, M. artemiziana, M. bimucronata, M. caesalpiniifolia, M. camporum, M. flocculosa, M. pellita, M. scabrella, M. tenuiflora, Parapiptadenia pterosperma, P. rigida, Piptadenia gonoacantha, Pithecelobium tortum, Poecilanthe parviflora, Prosopis chilensis, P. juliflora, Pseudosamanea guachapele, Samanea saman, Sclerolobium paniculatum, Sesbania exasperata, S. virgata, Stryphnodendron guianensis, Tephrosia sinapou*

Sources: from checklist of IPAGRO/UFRGS and Faria *et al.* (1999).

lated from these sites by using cowpea as a trap species, a great many were shown to have high efficiency in both Leonard jars (Motta, 2002) and in soil in symbiosis with this species (Lacerda *et al.*, 2004). However, other results from Leonard jar experiments showed Amazonian isolates as having the same or lower efficiency in improving cowpea growth (shoot dry matter weight) and N content (Neves *et al.*, 1992).

New efficient and competitive strains were selected for *P. vulgaris* among isolates from bean areas in Paraná state, which had never been inoculated (Hungria *et al.*, 2000) and also from cerrado areas (Mostasso *et al.*, 2002). These *P. vulgaris* strains show mixed characteristics of *R. tropici* IIa and IIb and are adapted to high acidity and high temperature. Pereira *et al.* (2000) also found efficient strains (Leonard jars) to this host species isolated from field nodules collected in the Amazon region. Some of these strains were identified as *R. leguminosarum* (F.M.S. Moreira, 2001, unpublished results) and were efficient in field experiments (Soares, 2004). As these strains were isolated from acid soil conditions in a high-temperature environment they have a high potential as inoculants in tropical soils.

Institutions responsible for culture collection and distribution of these strains to inoculant industries are FEPAGRO/UFRGS, Rio Grande do Sul state (strains with prefix SEMIA) and Embrapa-Agrobiologia, Rio de Janeiro state (strains with prefix BR). An annual or biannual meeting of RELARE has as the main objective the recommendation of inoculant strains. It was created in accordance with regulatory requirements of Decree no. 75583 (9 April 1975) article 23 that stated: 'Inoculants produced with recommended strains must be registered only if they are recommended by public institutions'. The RELARE participants are researchers from diverse Brazilian institutions, representatives of the Ministry of Agriculture as well as those from private industries that produce commercial inoculants. Considering diversity in nodulating hosts, bacterial strains, environmental conditions and crop management, there is a great potential to expand commercial BNF in this country.

Factors Affecting LNB Populations

Presence and size of indigenous LNB populations is a function of climate, soil, crop history and management (Singleton *et al.*, 1992). Numbers of LNB in soil are reported to vary from 0 to 10^6 cells/g of soil (Weaver and Frederick, 1974).

Evaluation methods

Methodological aspects such as trap host species or culture medium can affect the assessment of LNB populations. Woomer *et al.* (1988), testing 5 hosts (*Leucaena leucocephala*, *M. atropurpureum* (siratro) *Vicia sativa*, *Medicago sativa*, *Trifolium repens*) as rhizobia trap species from diverse sites, found densities ranging from 1.1 to 4.8 $\log_{10}$ cells/g of soil, siratro being the species that usually trapped the highest numbers in the majority of the sites. Bonetti *et al.* (1984) found in plantations of forest tree species near Manaus, densities up to 2 $\times 10^2$ cells/g of soil by using siratro as a trap species. Lower densities were found when nutrient solution was used in plastic pouches than when it was agarized (Fahreus tubes). Pereira (2000) found densities ranging from 15×10^0 to 2×10^4 cells/g of soil in five land use systems in western Amazon region by inoculating soil suspensions on siratro cultivated in nutrient solution (plastic pouches).

Agronomic practices

These may also affect LNB diversity. No tillage (versus conventional tillage) and crop rotation was shown to increase bradyrhizobia–soybean diversity detected by RADP in Paraná, south Brazil (Ferreira *et al.*, 2000). However, no differences in LNB diversity trapped by *C. cajan*, detected also by RADP profiles, were found among no-till and conventional tillage soybean in south-east São Paulo, Brazil (Coutinho *et al.*, 1999). This diversity was even lower than prior to the soybean planting under pasture.

Soil pH

Soil acidity is one of the major characteristics of tropical soils. Hence, it is expected that these bacteria are adapted to high acidity. Indeed, legume forest species nodulate abundantly at natural conditions with pH values around 4.0 in Amazonia (Magalhães and Blum, 1984; Moreira *et al.*, 1992). Furthermore, some LNB isolations from very acid soils were possible only in acidified medium (pH 5.0) (Souza *et al.*, 1984). Araújo (1994) found the highest incidence of *P. vulgaris* spontaneous nodulation in soils in Goiás state with pH around 5.7–5.9. Sharp decreases were found at pH values below 5.5 and mainly above 6.3.

Soil and plant species origins seem to affect LNB tolerance to acidity. In general, strains isolated from acid soils grow better in acid culture medium. Silva and Franco (1984) tested 211 strains from various origins and found 95 strains able to grow at pH 4.6 plus 25 mM $Al_2(SO_4)_3$. The frequency of acid-tolerant strains as related to their plant species origin decreased in the order: Caesalpinioideae (86%), Mimosoideae (49%) and Papilionoideae (29%). Also for *P. vulgaris* native isolates in culture medium a high incidence was found for pH 4.5 tolerance (111 out of 155) in São Paulo state (Vargas and Denardin, 1992).

Liming and antibiotic tolerance

Liming is a common practice to raise pH and decrease available Al content for better growth and yield of most crops, which has been shown to change microbial community composition (Siqueira and Moreira, 1997). Several studies indicate that liming of acidic soils increased the relative incidence of actinomycetes in soil and also the proportion of bacteria, including rhizobia, resistant to antibiotics (e.g. streptomycin) (Baldani *et al.*, 1982; Scotti *et al.*, 1982; Pereira, 1995). Vargas *et al.* (1992) found native rhizobia isolated from *P. vulgaris* cultivated in acid soils (4.5–5.5) in Paraná state, less tolerant to various antibiotics than those from soil, limed or not, with pH values above this range up to 6.6. Ramos *et al.* (1987) also found that the antibiotic-sensitive isolates from *P. vulgaris* mainly originated from unlimed soils. A large number of cowpea rhizobia isolates from diverse origins (LUS, edaphic–climatic conditions) showed a large variation in intrinsic antibiotic resistance (IAR). An increase in IAR was associated with an increase in soil pH and P and Al content (Xavier *et al.*, 1998). This seems a strategy to overcome antagonism problems as soybean and cowpea nodulation were shown to increase after liming of very acid Amazonian soils, especially after slashing and burning of fallows (Alfaia *et al.*, 1988; Neves *et al.*, 1992).

Andrade *et al.* (2002a) showed that the structure of native rhizobia populations from *P. vulgaris* in south Brazil was affected by liming. Rhizobia species richness and diversity (Shannon index) as well as abundance increased along a soil-liming gradient (Andrade *et al.*, 2002a,b). Subpopulations of *R. tropici* IIB indicated by pattern types (PCR-mediated RFLP of IGS 16S–23S) in soils that received liming were the most diverse. Meanwhile, there was a higher number of *R. leguminosarum* types in soils with low pH than in soil with a high pH. However, *R. leguminosarum* strains were found among the native populations nodulating *P. vulgaris* in Amazonia (Pereira, 2000) at pH 6.7 and no strains similar to *R. tropici* were found. Thus, soil pH also drastically affects composition of LNB populations and liming practice can have beneficial effects on diversity, probably not only due to the effect of pH but also due to the supply of Ca and Mg, which are important nutrients for Leguminosae symbiosis establishment, development and functioning.

Despite the acidic characteristic of soils and adaptation of strains to this, inoculants are produced in Brazil with a strain culture pH around 7.0. It was demonstrated that recommended strains for soybean (BR29, SEMIA587), *V. unguiculata* (INPA3-11B) and *Enterolobium contortsiliquum* (BR4406) grew and survived better on peat-based inoculants at pH 6.0, thus indicating that inoculants should have this pH value as a means of 'acid habituation' to tropical soils (Miguel and Moreira, 2001).

Nutrients

Phosphorus is the main limiting factor in tropical soils (see other chapters of this book). Its deficiency drastically affects plant growth and severely constrains nodulation of Leguminosae species (Fig. 11.5). At very low soil nutrient contents, as in the case of the experiment presented in Fig. 11.5, which was 1 mg/kg, mycorrhizas do not supply amounts adequate for plant growth. However when P is supplied, significant responses are observed in rhizobia and mycorrhiza inoculation as well as for liming and micronutrient application, but these effects are variable according to plant species (Fig. 11.5). Micronutrient deficiencies especially Mo can occur in acid soils. Responses of *P. vulgaris* to this micronutrient have been observed (Franco and Day, 1980; Amane

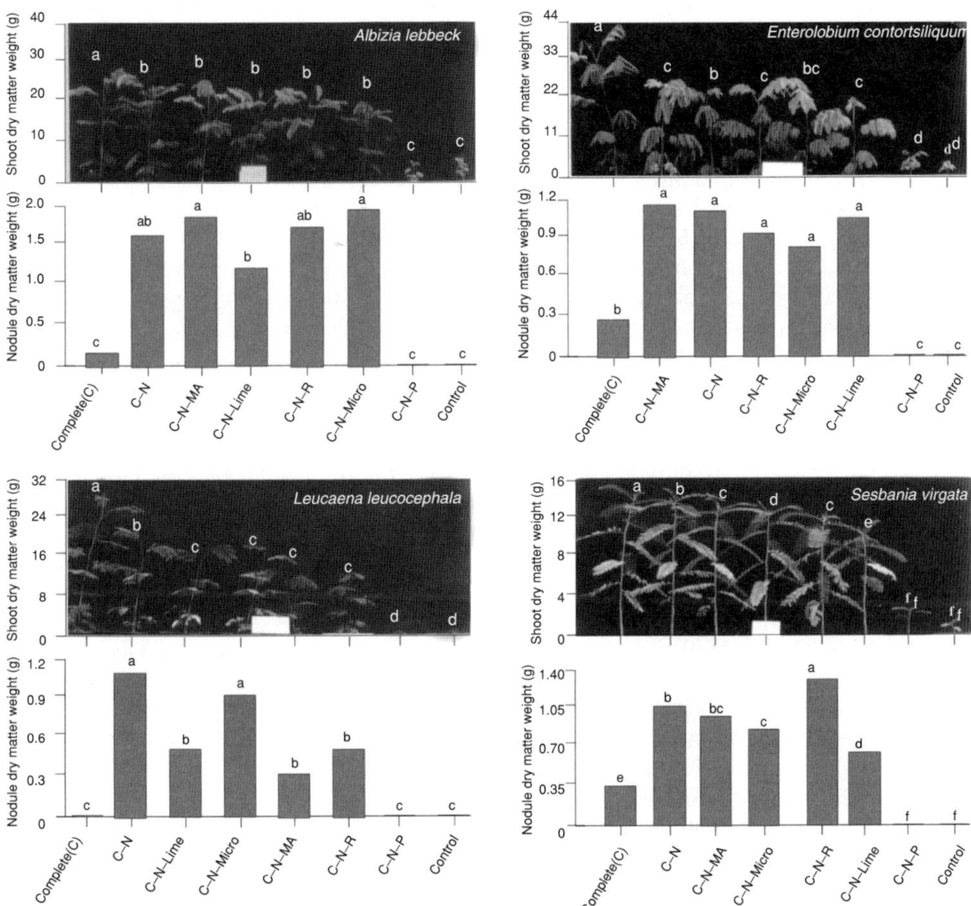

Fig. 11.5. Shoot and nodule dry weight of four tree legume species under limiting factors of a clayey latosol. Four-month-old plants in greenhouse pot experiments. Treatments: complete (C) = fertilization with K + P + micronutrients (micro) + liming (lime) + mineral nitrogen (N) and rhizobia (R) and arbuscular mycorrhizal fungi (MA) inoculations; C–N = complete minus mineral nitrogen; C–N–micro = complete minus mineral nitrogen and minus micronutrients; C–N–R = complete minus mineral nitrogen and minus rhizobia inoculation; C–N–MA = complete minus mineral nitrogen and minus arbuscular mycorrhizal fungi inoculation; C–N–lime = complete minus mineral nitrogen and minus liming; C–N–P = complete minus mineral nitrogen and phosphorus; control = no amendments or inoculation. (Source: adapted from Moreira and Siqueira, 1995.)

et al., 1999; Fullin *et al.*, 1999). However, they are strongly dependent on soil pH and seed content.

Mining

Mining is an important economic activity in Brazil. However, it is responsible for extensive soil degradation. Mining, industrial activity and the disposal of heavy metals have been the cause of high pollution levels, especially in tropical soils due to low organic matter contents. High organic matter content has been shown to decrease availability of these elements to plant nutrition in temperate soils (Hayes and Traina, 1998). Sixty tropical strains from various origins (Leguminosae subfamilies, plant species, Amazon region, Atlantic forests, crops, contaminated soil) exhibited a wide range of tolerance to heavy metals in culture media (Matsuda *et al.*, 2002a). Similar to what has been found for temperate isolates (Angle *et al.*, 1993), toxicity increases in the order Zn < Cd < Cu; however, maximum tolerated concentrations were higher for tropical isolates: 800, 60 and 60 mg/l, respectively (Trannin *et al.*, 2001a; Matsuda *et al.*, 2002a). *Bradyrhizobium* is by far the most tolerant genus followed by *Sinorhizobium/Mesorhizobium/Rhizobium* and then by *Azorhizobium* (Matsuda *et al.*, 2002a). Two strains, later identified as *Burkholderia* (INPA353B, BR3460) (F.M.S. Moreira, 2004, unpublished results), exhibited the same tolerance as *Sino/Meso/Rhizobium*. *B. elkanii* strain BR29 (SEMIA5019), one of the inoculant strains recommended for soybeans was relatively tolerant, with a maximum tolerated concentration (MTC) to Zn, Cd and Cu of 600, 40 and 20 mg/l, respectively. Tolerance in culture media was correlated with survival in contaminated soil. The most tolerant strains BR4406 and UFLA 01–457, both isolated from *E. contortsiliquum*, had about 20% cell survival (log CFU 9.7/log CFU 10.4) after 28 days in soil with 1250, 206 and 67 mg/dm^3, respectively, to DTPA-extracted Zn, Cd and Cu (Matsuda *et al.*, 2002b). Nodulating Leguminosae species also vary in their capability to tolerate heavy metals. *Acacia mangium* and *E. contortsiliquum* were the most tolerant of the species tested up to now, even considering herbaceous legumes or non-leguminous species such as *Eucalyptus* spp. (Table 11.9). However, plants with N nutrition by mineral nitrogen were in general less sensitive than those with N supplied by symbiosis with nitrogen-fixing bacteria (Mostasso, 1997; Trannin *et al.*, 2001b). From the legume symbiosis studied up to now *E. contortisiliquum* and BR4406 (Fig. 11.6) is the most promising for revegetation of contaminated sites. *E. contortsiliquum* is a fast-growing woody species producing large quantities of seeds with long-term viability. BR4406 is the recommended strain for this species because of its superior efficiency in fixing nitrogen and high tolerance to heavy metals as shown above.

Effects of rehabilitation of bauxite mine soils with diverse practices, including the use of legume species such as *Mimosa scabrella* and *C. cajan*, grasses and *Eucalyptus*, on LNB diversity was assessed by using *V. unguiculata* and *P. vulgaris* as trap species (Melloni, 2001; Melloni *et al.*, 2005). In most cases rehabilitation strategies increased NB efficiency and diversity (based on culture characteristics) in relation to natural reference areas. This corroborates the fact that BNF in natural ecosystems in equilibrium is lower than in agroecosystems. Moreira *et al.* (1992) and Moreira and Franco (1994) reported that legumes nodulate abundantly in those ecosystems in which N is limiting such as soils flooded by the Amazon River. On the other hand, nodulation is quite scarce in tropical forests.

Concluding Remarks

Nitrogen fixation in a given environment is a function of LNB diversity (epecially at strain level), Leguminosae species diversity and environmental conditions (physical, biological and chemical factors), as many soil and climatic factors may affect the establishment and function of the

Table 11.9. Leguminosae-nodulating species[a] studied for heavy metal tolerance in Brazil.

Reference	Studied host species	Most tolerant (ascending order)
Mostasso (1997)	*E. contortsiliquum*	*E. contortsiliquum*
	Enterolobium timbouva	*Mucuna aterrina*
	S. virgata	
	Leucaena sp.	
	Macroptilium atropurpureum	
	Vigna radiata	
	V. unguiculata	
	Canavalia brasiliensis	
	Mucuna aterrina	
	P. vulgaris	
Siqueira *et al.* (1999)	*E. contortsiliquum*	*E. contortsiliquum*[b]
	Albizia lebbeck	
Grazziotti (1999)	*A. mangium*	*A. mangium*[c]
Marques *et al.* (2000)	*A. mangium*	*A. mangium*[d]
	Mimosa caesalpinifolia	*Mimosa caesalpinifolia*[d]
	Anadenanthera peregrina	*Platypodium gonoacantha*[d]
	Machaerium nictidans	
	Pipitadenia gonoacantha	
	Platypodium gonoacantha	
Trannin *et al.* (2001b)	*E. contortsiliquum*	*E. contortsiliquum*
	A. mangium	
	S. virgata	
Carneiro *et al.* (2002)	*Arachis pintoi, T. repens*[e]	*Pffafia* sp. (non-legume), *Trifolium repens*

[a]Inoculated with recommended selected rhizobia strains.
[b]Among five species, including non-nodulating legumes and non-legumes.
[c]Among 13 other *Eucalyptus* spp. and three *Pinus* spp.
[d]Among 20 species, including non-nodulating legumes and non-legumes.
[e]Tested with 29 other species, most grasses and Compositae (1), Malvaceae (1), Solanaceae (1), Cyperaceae (1) and Amaranthaceae (1).

Fig. 11.6. *B. japonicum* BR4406 root nodules in *Enterolobium schomburkii.*

symbiosis. The overall functionality of LNB symbiosis in a given ecosystem is almost impossible to predict as it results from an infinite combination of these multiple factors, most of which are unknown and their relationships quite complex. The most accurate known *in situ* techniques are those applying the ^{15}N isotope (Vose, 1980). However, to find adequate non-fixing reference species is often a problem and results are species-based and semi-quantitative or qualitative (Högberg, 1986; Yoneyama *et al.*, 1993). Nodulation is a relatively easy parameter to estimate N_2 fixation comparatively across land use systems. For instance, under field conditions in the Amazon region, the percentage of plants found with nodules decreased in the order primary forest > fallow > disturbed areas > plantations (Moreira and Franco, 1994). Nodulation in seasonally flooded areas is far greater than in non-flooded ones. The same is true for sandy soils. High mineral N availability inhibits N_2 fixation, whereas depletion stimulates this function, including nodulation. In the case of pristine forest, efficient mycorrhiza-mediated cycling is sufficient to sustain the low ecosystem demands for metabolic activity and the very slow growth of plants. Thus, legume species nodulation capability surveys in pristine forest are limited because almost no nodulation occurs there. Then, harvesting seeds of potentially N_2-fixing species is necessary to grow plants under nursery conditions where high N demands stimulate nodulation (Moreira, 1995, 1997). Native soils can be used as a substrate to trap native rhizobia species/strains; however, fertilization with nutrients other than N must be recommended to avoid other limitations. The percentage of species nodulating increased from 65% to 73% in a terra firme soil when it was fertilized (Moreira *et al.*, 1992).

In field and nursery surveys, about one-third of all species and half of all total genera of Leguminosae have been analysed for nodulation and phenotypic characteristics of their microsymbionts in Brazil (Magalhães *et al.*, 1982; Faria *et al.*, 1989; Moreira *et al.*, 1992; Souza *et al.*, 1994). However, only very few strains were characterized genetically (Table 11.4A–C). Therefore, native LNB biodiversity, as well as the effects of land use systems and agronomic practices on LNB ecology, is only poorly understood in Brazil. Besides that, it must be stressed that host species of known LNB species listed in Table 11.3 are only those from which LNB were isolated or were tested with. Thus, an infinite number of symbiotic relationships (i.e. nodulation capability and efficiency) among Leguminosae species as a whole and known LNB species is still to be explored.

Although management of BNF is very successful for the introduced species *G. max*, lack of both knowledge and extension of this biotechnology limits its application to the large number of ecologically and economically important native species in Brazil. This biotechnology could have a great contribution for sustainability in tropical agriculture and deserves more attention. Actions necessary to explore and manage the potential of BNF in Brazil include continuity of surveys on Leguminosae nodulation capability, characterization/identification/selection of microsymbionts, studies on the symbiotic relationships of the diverse legume species with LNB strains and knowledge of LNB diversity and ecology under diverse soil conditions and land use systems.

Acknowledgements

We are thankful to Rosa Maria Pittard, curator of Rhizobia collection at EMBRAPA-Agrobiologia, for information on culture characteristics of some strains; to Geraldo B. Cruz, Marlene A. Souza and Rafaela Nóbrega for *A. doebereinerae* photomicrographs and to CNPq for research fellowship.

References

Alfaia, S.S., Magalhães, F.M.M., Yuyama, K. and Muraoka, T. (1988) Efeito da aplicação de calcário e micronutrientes em latossolo amarelo da Amazônia Central. *Acta Amazônica* 18, 13–25.

Allen, O.N. and Allen, E. (1981) *The Leguminosae: A Source Book of Characteristics, Uses and Nodulation.* The University of Wisconsin Press, Madison, Wisconsin.

Amane, M.I.V., Vieira, C., Novais, R.F. and Araújo, G.A.A. (1999) Adubação nitrogenada e molíbdica da cultura do feijão na zona da mata de Minas Gerais. *Revista Brasileira de Ciência do Solo* 23, 643–650.

Amarger, N., Machret, V. and Laguerre, G. (1997) *Rhizobium gallicum* sp. nov. and *Rhizobium giardinii* sp. nov., from *Phaseolus vulgaris* nodules. *International Journal of Systematic Bacteriology* 47, 996–1006.

Andrade, D.S., Murphy, P.J. and Giller, K.E. (2002a) The diversity of *Phaseolus*-nodulating rhizobial populations is altered by liming of acid soils planted with *Phaseolus vulgaris* L. in Brazil. *Applied and Environmental Microbiology* 68, 4025–4034.

Andrade, D.S., Murphy, P.J. and Giller, K.E. (2002b) Effects of liming and legume/cereal cropping on populations of indigenous rhizobia in an acid Brazilian Oxisol. *Soil Biology and Biochemistry* 34, 447–485.

Angle, J.S., McGrath, S.P., Chaudri, A.M., Chaney, R.L. and Giller, K.E. (1993) Inoculation effects on legumes grown in soil previously treated with sewage sludge. *Soil Biology and Biochemistry* 25, 575–580.

Araújo, R.S. (1994) Caracterização morfológica, fisiológica e bioquímica do rizóbio. In: Hungria, M. and Araujo, R.S. (eds) *Manual de métodos empregados em estudos de microbiologia agrícola.* EMBRAPA – CNPAF/CNPSo, Brasília, Brazil, pp. 157–170.

Baldani, J.I., Baldani, V.L.D., Xavier, D.F., Boddey, R.M. and Dobereiner, J. (1982) Effect of liming on the number of actinomycetes and on the percentage of streptomycin resistant bacteria in the rhizosphere. *Revista de Microbiologia* 13, 250–263.

Beijerinck, M.W. (1888) Die Bacterien der Papilionaceen-knöllchen. *Botanik Zeitung* 46, 725–735, 741–750, 757–771, 781–790, 797–804.

Bonetti, R., Oliveira, L.A. and Magalhães, F.M.M. (1984) *Rhizobium* spp. populations and mycorrhizal associations in some plantations of forest tree species. *Pesquisa Agropecuária Brasileira* 19, 137–142.

Brasil MMA. Ministério do Meio Ambiente, dos Recursos Hídricos e da Amazônia Legal (1998) *Apresentação da diversidade biológica brasileira: Primeiro Relatório Nacional para a Convenção sobre Diversidade Biológica.* Ministério do Meio Ambiente, Brasília, Brazil.

Burns, R.C. and Hardy, R.W.F. (1975) *Nitrogen Fixation in Bacteria and Higher Plants.* Springer-Verlag, New York.

Carneiro, M.A.C., Siqueira, J.O. and Moreira, F.M.S. (2002) Behaviour of herbaceous species in soil mixes with different degrees of contamination with heavy metals. *Pesquisa Agropecuária Brasileira* 37, 1629–1638.

Casida, L.E. Jr (1982) *Ensifer adhaerens* gen. nov., sp. nov.: a bacterial predator of bacteria in soil. *International Journal of Systematic Bacteriology* 32, 339–345.

Chen, W.X., Yan, G.H. and Li, J.L. (1988) Numerical taxonomic study of fast-growing soybean rhizobia and a proposal that *Rhizobium fredii* be assigned to *Sinorhizobium* gen. nov. *International Journal of Systematic Bacteriology* 38, 392–397.

Chen, W.X., Li, G.S., Qi, Y.L., Wang, E.T., Yuan, H.L. and Li, J.L. (1991) *Rhizobium huakuii* sp. nov. isolated from the root nodules of *Astragalus sinicus. International Journal of Systematic Bacteriology* 41, 275–280.

Chen, W.X., Wang, E., Wang, S., Li, Y., Chen, X. and Li, Y. (1995) Characteristics of *Rhizobium tianshanense* sp. nov., a moderately and slowly growing root nodule bacterium isolated from an arid saline environment in Xinjiang, People's Republic of China. *International Journal of Systematic Bacteriology* 45, 153–159.

Chen, W.X., Tan, Z.Y., Gao, J.L., Li, Y. and Wang, E.T. (1997) *Rhizobium hainanense* sp. nov., isolated from tropical legumes. *International Journal of Systematic Bacteriology* 47, 870–873.

Chen, W.M., Laevens, S., Lee, T.M., Coenye, T., Vos, P., Mergeay, M. and Vandamme, P. (2001) *Ralstonia taiwanensis* sp. nov. isolated from nodules of *Mimosa* species and sputum of a cystic fibrosis patient. *International Journal of Systematic and Evolutionary Microbiology* 51, 1729–1735.

Chueire, L.M.O., Bagel, E., Ferreira, M.C., Grange, L., Campo, R.J., Mostasso, F.L., Andrade, D.S., Pedrosa, F.O. and Hungria, M. (2000) *Classificação taxonômica, baseada na caracterização molecular, das estirpes de rizóbio recomendadas para as culturas da soja e do feijoeiro.* Embrapa Soja, Londrina, PR, Brazil.

Coutinho, H.L.C., Oliveira, V.M., Hollanda, L.M., Moreira, F.M.S. and Franco, A.A. (1995) Diversity of rhizobia isolated from nodules of legumes occurring in the Atlantic and Amazonian rainforests.

In: Abstracts of the 7th International Symposium on Microbial Ecology. International Committee on Microbial Ecology/Brazilian Society for Microbiology, Santos, São Paulo, Brazil, 162 pp.

Coutinho, H.L.C., Oliveira, V.M., Lovato, A., Maia, A.H.N. and Manfio, G.P. (1999) Evaluation of the diversity of rhizobia in Brazilian agricultural soils cultivated with soybeans. *Applied Soil Ecology* 391, 1–9.

Dangeard, P.A. (1926) *Botaniste, Paris* 16, 1–275.

de Lajudie, P., Willems, A., Pot, B., Dewettinck, D., Maestrojuan, G., Neyra, M., Collins, M.D., Dreyfus, B., Kersters, K. and Gillis, M. (1994) Polyphasic taxonomy of rhizobia: emendation of the genus *Sinorhizobium* and description of *Sinorhizobium meliloti* com. nov.; *Sinorhizobium saheli* sp. nov.; and *Sinorhizobium teranga* sp. nov. *International Journal of Systematic Bacteriology* 44, 715–733.

de Lajudie, P., Willems, A., Nick, G., Moreira, F., Molouba, F., Hoste, B., Torck, U., Neyra, M., Collins, M.D., Lindstrom, K., Dreyfus, B. and Gillis, M. (1998) Characterization of tropical tree rhizobia and description of *Mesorhizobium plurifarium* sp. nov. *International Journal of Systematic Bacteriology* 48, 369–382.

Dreyfus, B., Garcia, J.L. and Gillis, M. (1988) Characterization of *Azorhizobium caulinodans* gen. nov. sp. nov., a stem-nodulating nitrogen-fixing bacterium isolated from *Sesbania rostrata*. *International Journal of Systematic Bacteriology* 38, 89–98.

Faria, S.M., Lewis, G.P., Sprent, J.I. and Sutherland, J.M. (1989) Occurrence of nodulation in the *Leguminosae*. *New Phytologist* 111, 607–619.

Faria, S.M., Lima, H.C., Olivares, F.L., Melo, R.B. and Xavier, R.P. (1999) Nodulação em espécies florestais: especificidade hospedeira e implicações na sistemática de *Leguminosae*. In: Siqueira, J.O., Moreira, F.M.S., Lopes, A.S., Guilherme, L.R.G., Faquin, V., Furtini Neto, A.E. and Carvalho, J.G. (eds) *Soil Fertility, Soil Biology and Plant Nutrition Interrelationships*. SBCS/UFLA/DCS, Lavras, Brazil, pp. 667–686.

Ferreira, M.C. and Hungria, M. (2002) Recovery of soybean inoculant strains from uncropped soils in Brazil. *Field Crops Research* 79, 139–152.

Ferreira, M.C., Andrade, D.S., Chueire, L.M.O., Takemura, S.M. and Hungria, M. (2000) Tillage method and crop rotation effects on the population sizes and diversity of bradyrhizobia nodulating soybean. *Soil Biology and Biochemistry* 32, 627–637.

Franco, A.A. and Day, J.M. (1980) Effects of lime and molybdenum on nodulation and nitrogen fixation of *Phaseolus vulgaris* L. in acid soils of Brazil. *Turrialba* 30, 99–105.

Frank, B. (1879) Ueber die Parasiten in den Wurzelan-schwillungen der Papilionaceen. *Botanik Zeitung* 37, 376–387, 394–399.

Frank, B. (1889) Ueber die Pilzsymbiose der Leguminosen. *Berichte der Deutschen Botanischen Gesellschaft* 7, 332–346.

Frank, B. (1890) *Landwirtschaftliche Jahrbucher* 19, 563.

Fred, E.B. and Waksman, S.A. (1928) *Laboratory Manual of General Microbiology*. McGraw-Hill, New York/London.

Freire, J.R.J. and Vernetti, F.J. (1999) The research on soybeans, selection of rhizobia and production of inoculants in Brazil. *Pesquisa Agropecuária Gaúcha* 5, 117–126.

Fullin, C.A., Zangrande, M.B., Sani, J.A., Mendonça, L.F. and Filho, N.D. (1999) Nitrogen and molybdenum fertilization in dry bean under irrigated conditions. *Pesquisa Agropecuária Brasileira* 34, 1145–1149.

Gao, J.L., Turner, S.L., Kan, F.L., Wang, E.T., Tan, Z.Y., Qiu, Y.H., Terefework, Z., Young, J.P.W., Lindstrom, K. and Chen, W.X. (2004) *Mesorhizobium septentrionale* sp. nov. and *Mesorhizobium temperatum* sp. nov. isolated from *Astragalus adsurgens* growing in the northern regions of China. *International Journal of Systematic and Evolutionary Microbiology* 5, 2003–2012.

Graham, P.H., Sadowsky, M.J., Keyser, H.H., Barnet, Y.M., Bradley, R.S., Cooper, J.E., De Ley, J., Jarvis, B.D.W., Roslycky, E.B., Strijdom, B.W. and Young, J.P.W. (1991) Proposed minimum standards for the description of new genera and species of root- and stem-nodulating bacteria. *International Journal of Systematic Bacteriology* 41, 582–587.

Grazziotti, P.H. (1999) Behavior of ectomycorrhizal fungi, *Acacia mangium* and species of *Pinus* and *Eucalyptus* isolated in soil contamined with heavy metal. PhD thesis, Universidade Federal de Lavras, Lavras, Brazil.

Hayes, K.F. and Traina, S.J. (1998) Metal speciation and its significance in ecosystem health. In: Huang, P.M. (ed.) *Soil Chemistry and Ecosystem Health*. Soil Science Society of America, Madison, Wisconsin, pp. 45–84 (SSSA Special Publication No. 52).

Högberg, P. (1986) Nitrogen-fixation and nutrient relations in savanna woodland trees (Tanzania). *Journal of Applied Ecology* 23, 675–688.

Hungria, M., Andrade, D.S., Chueire, L.M.O., Probanza, A., Guttierrez-Mañero, F.J. and Megías, M. (2000) Isolation and characterization of new efficient and competitive beans (*Phaseolus vulgaris* L.) rhizobia from Brazil. *Soil Biology and Biochemistry* 32, 1515–1528.

Hungria, M., Campo, R.J., Chueire, L.M.O., Grange, L. and Megías, M. (2001a) Symbiotic effectiveness of fast-growing rhizobial strains isolated from soybean nodules in Brazil. *Biology and Fertility of Soils* 33, 387–394.

Hungria, M., Chueire, L.M.O., Coca, R.G. and Megías, M. (2001b) Preliminary characterization of fast growing rhizobial strains isolated from soyabean nodules in Brazil. *Soil Biology and Biochemistry* 33, 349–1361.

Irwin, H.S. (1981) Preface. In: Polhill, R.M. and Raven, P.H. (eds) *Advances in Legume Systematics*. Part I. Royal Botanic Gardens, Kew, UK.

Jarvis, B.D.W., Pankhurst, C.E. and Patel, J.J. (1982) *Rhizobium loti*, a new species of legume root nodule bacteria. *International Journal of Systematic Bacteriology* 32, 378–380.

Jarvis, B.D.W., van Berkum, P., Chen, W.X., Nour, S.M., Fernandez, M.P., Cleyet-Marel, J.C. and Gillis, M. (1997) Transfer of *Rhizobium loti, Rhizobuim huakuii, Rhizobium ciceri, Rhizobium mediterraneum* and *Rhizobium tianshanense* to *Mesorhizobium* gen. nov. *International Journal of Systematic Bacteriology* 47, 895–898.

Jordan, D.C. (1984) *Rhizobiaceae* Conn 1938. In: Krieg, N.R. and Holt, J.D. (eds) *Bergey's Manual of Systematic Bacteriology*. Williams and Wilkins, London, pp. 234–244.

Jourand, P., Giraud, E., Bena, G., Sy, A., Willems, A., Gillis, M., Dreyfus, B. and de Lajudie, P. (2004) *Methylobacterium nodulans* sp. nov., for a group of aerobic facultatively methylotrophic, legume root-nodule-forming and nitrogen-fixing bacteria. *International Journal of Systematic and Evolutionary Microbiology* 54, 2269–2273.

Käss, E. and Wink, M. (1996) Molecular evolution of the *Leguminosae*: three sub-families based on rbcL sequences. *Biochemical Systematics and Ecology* 24, 365–378.

Käss, E. and Wink, M. (1997) Phylogenetic relationships in the *Papilionoideae* (family *Leguminosae*) based on nucleotide sequences of cpDNA (rbcL) and ncDNA (ITS 1 and 2). *Molecular Phylogenetics and Evolution* 8, 65–88.

Kimura, M. (1983) *The Neutral Theory of Molecular Evolution*. Cambridge University Press, Cambridge, UK.

Kirkbride-Júnior, J.H. (1984) Legumes of cerrado. *Pesquisa Agropecuária Brasileira* 19, 23–46.

Kuykendall, L.D., Saxena, B., Devine, T.E. and Udell, S. (1992) Genetic diversity in *Bradyrhizobium japonicum* and a proposal for *Bradyrhizobium elkanii* sp. nov. *Canadian Journal of Microbiology* 38, 501–505.

Lacerda, A.M., Moreira, F.M.S., Andrade, M.J.B. and Soares, A.L.L. (2004) Yield and nodulation of cowpea inoculated with selected strains. *Revista Ceres* 51, 67–82.

Lewin, A., Rosenberg, C., Meyer, H., Wong, C.H., Nelson, L., Manen, J.F., Stanley, J., Dowling, D.N., Dénarie, J. and Broughton, W.J. (1987) Multiple host-specificity loci of the broad host-range Rhizobium sp. NGR234 selected using the widely compatible legume *Vigna unguiculata*. *Plant Molecular Biology* 8, 447–459.

Lindström, K. (1989) *Rhizobium galegae*, a new species of legume root nodule bacteria. *International Journal of Systematic Bacteriology* 39, 365–367.

Lombardi, M.L.C.O. (1995) Diversity of wild rhizobia from soils of São Paulo state, Brasil. PhD thesis, Escola Superior de Agricultura Luiz de Queiroz, Piracicaba, São Paulo, Brazil.

Loureiro, M.F., Hungria, M., Sampaio, M.J.A.M., Franco, A.A. and Baldani, J.I. (1993) Photosynthetic characteristics of strains of rhizobia isolated from stem nodules of *Aeschynomene fluminensis* grown in the Pantanal region Brazil. In: Palacios, R., Mora, J. and Newton, W.E. (eds) *New Horizons in Nitrogen Fixation*. Kluwer, Dordrecht, The Netherlands.

Loureiro, M.F., De Faria, S.M., James, E.K., Pott, A. and Franco, A.A. (1994) Nitrogen-fixing stem nodules of the legume, *Discolobium pulchellum* Benth. *New Phytologist* 128, 283–295.

Loureiro, M.F., James, E.K., Sprent, J.I. and Franco, A.A. (1995) Stem and root nodules on the tropical wetland legume *Aeschynomene fluminensis*. *New Phytologist* 130, 531–544.

Magalhães, L.M.S. and Blum, W.E.H. (1984) Nodulation and growth of *Cedrelinga catenaeformis* Ducke at experimental stands in the Manaus region–Amazonas. *Pesquisa Agropecuária Brasileira* 19, 59–164.

Magalhães, F.M.M., Magalhães, L.M.S., Oliveira, L.A. and Döbereiner, J. (1982) Ocorrência de nodulação em leguminosas florestais nativas de terra firme da região de Manaus–AM. *Acta Amazônica* 12, 509–514.

Marques, T.C.L.L.S.M., Moreira, F.M.S. and Siqueira, J.O. (2000) Growth and metal concentration of woody species in a heavy metal contaminated soil. *Pesquisa Agropecuária Brasileira* 35, 121–132.

Martínez-Romero, E. and Caballero-Mellado, J. (1996) *Rhizobium* phylogenies and bacterial genetic diversity. *Critical Reviews in Plant Sciences* 12, 113–140.
Martínez-Romero, E., Segovia, L., Mercante, F.B., Franco, A.A., Graham, P. and Pardo, M.A. (1991) *Rhizobium tropici,* a new species nodulating *Phaseolus vulgaris* L. Beans and *Leucaena* trees. *International Journal of Systematic Bacteriology* 41, 417–426.
Martins, L.M.V., Neves, M.C.P. and Rumjanek, N.G. (1997) Growth characteristics and symbiotic efficiency of rhizobia isolated from cowpea nodules of the north-east region of Brazil. *Soil Biology and Biochemistry* 29, 1005–1010.
Matsuda, A., Moreira, F.M.S. and Siqueira, J.O. (2002a) Tolerance of rhizobia genera from different origins to zinc, copper and cadmium. *Pesquisa Agropecuária Brasileira* 37, 343–355.
Matsuda, A., Moreira, F.M.S. and Siqueira, J.O. (2002b) Survival of *Bradyrhizobium* and *Azorhizobium* in heavy metal contaminated soil. *Revista Brasileira Ciência Solo* 26, 249–256.
Melloni, R. (2001) Density and diversity of fixing nitrogen bacteria and arbuscular mycorrhizal fungi in bauxite mined soils. PhD thesis, Universidade Federal de Lavras, Lavras, MG, Brazil.
Melloni, R., Moreira, F.M.S., Nóbrega, R.S.A. and Siqueira, J.O. (2005) Efficiency and phenotypic diversity among nitrogen fixing bacteria nodulating cowpea [*Vigna unguiculata* (L.) Walp.] and beans (*Phaseolus vulgaris* L.) in bauxite mined soils under rehabilitation. *Revista Brasileira de Ciência Solo* 29 (in press).
Mercante, F.M., Cunha, C.O., Straliotto, R., Ribeiro Júnior, S.Q., Vanderleyden, J. and Franco, A.A. (1998) *Leucaena leucocephala* as a trap-host for *Rhizobium tropici* strains from the Brazilian 'cerrado' region. *Revista de Microbiologia* 29, 49–58.
Miguel, D.L. and Moreira, F.M.S. (2001) Influence of medium and peat pH on the behaviour of Bradyrhizobium strains. *Revista Brasileira de Ciência Solo* 25, 873–883.
Mittermeier, R.A., Myers, N., Gil, P.R. and Mittermeier, C.G. (1997) *Hotspots – Earth's Biologically Richest and Most Endangered Terrestrial Ecoregions,* Cemex Conservation International, Mexico City.
Moreira, F.M.S. (1991) Characterization of rhizobia strains isolated from forest species belonging to diverse divergence groups of *Leguminosae* introduced or native to Amazon and Atlantic forests. PhD thesis, Universidade Federal Rural do Rio de Janeiro, Rio de Janeiro, Brazil.
Moreira, F.M.S. (1995) Nodulation and growth of leguminous species in two Amazonia soils, Brazil. *Revista Brasileira de Ciência do Solo* 19, 197–204.
Moreira, F.M.S. (1997) Nursey growth and nodulation of forty-nine woody legume species native from Amazonia. *Revista Brasileira de Ciência do Solo* 21, 581–590.
Moreira, F.M.S. and Franco, A.A. (1994) Rhizobia–host interactions in tropical ecosystems in Brazil. In: Sprent, J.I. and Mckey, D. (eds) *Advances in Legume Systematics 5: The Nitrogen Factor,* Royal Botanic Gardens, Kew, UK, pp. 63–74.
Moreira, F.M.S.M. and Pereira, E.G. (2001) Microsymbionts: rhizobia. In: Swift, M. and Bignell, D. (eds) *Standard Methods for Assessment of Soil Biodiversity and Land Use Practice.* International Centre for Research in Agroforestry, Bogor, Indonesia, pp. 19–24.
Moreira, F.M.S. and Siqueira, J.O. (1995) Growth, nodulation and arbuscular mycorrhizal colonization of four woody legumes in a low fertility soil. In: Boddey, R.M. and Resende, A.S. (eds) *Programme and Abstracts International Symposium on Sustainable Agriculture for the Tropics – The Role of Biological Nitrogen Fixation,* Angra dos Reis, RJ, Brazil, pp. 164–165.
Moreira, F.M.S. and Siqueira, J.O. (2002) *Microbiologia e bioquímica do solo.* Editora UFLA, Lavras, MG, Brazil.
Moreira, F.M.S., Silva, M.F. and Faria, S.M. (1992) Occurrence of nodulation in legume species in the Amazon region of Brazil. *New Phytologist* 121, 563–570.
Moreira, F.M.S., Gillis, M., Pot, B., Kersters, K. and Franco, A.A. (1993) Characterization of rhizobia isolated from different divergence groups of tropical *Leguminosae* by comparative polyacrylamide gel electrophoresis of their total proteins. *Systematic Applied Microbiology* 16, 135–146.
Moreira, F.M.S., Martinez-Romero, E., Segovia, L. and Franco, A.A. (1995) Genetic diversity of rhizobia and bradyrhizobia from native tropical species characterized by multilocus enzyme eletrophoresis. In: *Abstracts of the 7th International Symposium on Microbial Ecology.* Santos, São Paulo, Brazil, 88 pp.
Moreira, F.M.S., Haukka, K. and Young, J.P.W. (1998) Biodiversity of rhizobia isolated from a wide range of forest legumes in Brazil. *Molecular Ecology* 7, 889–895.
Moreira, F.M.S., Carvalho, Y., Gonçalves, M., Haukka, K., Young, P.J.W., Faria, S.M., Franco, A.A., Cruz, L.M. and Pedrosa, F.O. (2000) *Azorhizobium johanense* sp. nov. and *Sesbania virgata* (Caz.) Pers.: a highly specific symbiosis. In: Pedrosa, F.O., Hungria, M., Yates, G.Y. and Newton, W.E. (eds) *Nitrogen Fixation: From Molecules to Crop Productivity.* Kluwer, Dordrecht, The Netherlands, 197 pp.

Moreira, F.M.S., Tiedje, J. and Marsh, T.L. (2002) *Burkholderia* spp. are among fast growing symbiotic diazotrophs isolated from diverse land use systems in Amazônia and from Brazilian *Leguminosae* forest species. In: *Memorias da XXI Reunión Latinoamericana de rhizobiologia*, Cocoyoc, Mexico, pp. 45–46.

Moreira, F.M.S., Cruz, L.M., Faria, S.M., Marsh, T., Martinez-Romero, E., Pedrosa, F.O., Pitard, R. and Young, P.J.W. (2005) *Azorhizobium doeberinerae* sp. nov. microsymbiont of *Sesbania virgata* (Caz.) Pers. *Systematic and Applied Microbiology* (in press).

Mostasso, F.L. (1997) Growth and nodulation of legume species in soil contaminated soil with heavy metals. MSc thesis, Universidade Federal de Lavras, Lavras, MG, Brazil.

Mostasso, L., Mostasso, F.L., Dias, B.G., Vargas, M.A.T. and Hungria, M. (2002) Selection of bean (*Phaseolus vulgaris* L.) rhizobial strains for the Brazilian cerrados. *Field Crop Research* 73, 121–132.

Motta, J.S. (2002) Phenotypic diversity and symbiotic efficiency of *Bradyrhizobium* strains isolated from rehabilitated bauxite mining areas. 2002. MSc thesis, Universidade Federal de Lavras, Lavras, MG, Brazil.

Moulin, L., Munive, A., Dreyfus, B. and Bolvin-Masson, C. (2001) Nodulation of legumes by members of the β sub-class of *Proteobacteria. Nature* 411, 948–950.

Mullis, K.B. and Faloona, F.A. (1987) Specific synthesis of DNA *in vitro* via polymerase-catalyzed chain reaction. *Methods in Enzymology* 155, 335–351.

Neves, M.C.P., Ramos, M.L.G., Martinazzo, A.F., Botelho, G.R. and Döbereiner, J. (1992) Adaptation of more efficient soybean and cowpea strains to replace established populations. In: Mungoloy, K., Gueye, M. and Spencer, D.S.C. (eds) *Biological Nitrogen Fixation and Sustainability of Tropical Agriculture*. John Wiley & Sons, New York, pp. 219–223.

Nick, G., de Lajudie, P., Eardly, B.D., Suomalainen, S., Paulin, L., Zhang, X., Gillis, M. and Lidstrom, K. (1999) *Sinorhizobium arboris* sp. nov. and *Sinorhizobium kostiense* sp.nov., isolated from leguminous trees in Sudan and Kenya. *International Journal of Systematic Microbiology* 49, 1359–1368.

Norris, D.O. (1965) Acid production by Rhizobium: a unifying concept. *Plant and Soil* 22, 143–166.

Nour, S.M., Cleyet-Marel, J.C., Beck, D., Effosse, A. and Fernandes, M.P. (1994) Genotypic and phenotypic diversity of *Rhizobium* isolated from chickpea (*Cicer arietnum* L.). *Canadian Journal of Microbiology* 40, 345–354.

Nour, S.M., Cleyet-Marel, J.C., Normand, P. and Fernandes, M.P. (1995) Genomic heterogeneity of strains nodulating chickpea (*Cicer arietnum* L.) and description of *Rhizobium mediterraneum* sp. nov. *International Journal of Systematic Bacteriology* 45, 640–648.

Pereira, J.C. (1995) Ecologia da comunidade bacteriana em solos de cerrados. PhD thesis, Universidade Federal Rural do Rio de Janeiro, RJ, Brazil.

Pereira, E.G. (2000) Diversity of rhizobia isolated from different land use systems in Amazon region. PhD thesis, Universidade Federal de Lavras, Lavras, Brazil.

Pereira, E.G., Trannin, I.C.B., Moreira, F.M.S. and Siqueira, J.O. (1998) Ocorrência de leguminosas e de nodulação em relação a biodiversidade vegetal em ecossistemas florestais brasileiros. In: *Abstracts FertBIO 98*, Caxambu, SP, Brazil, 218 pp.

Pereira, E.G., Lacerda, A.M., Lima, A.S., Moreira, F.M.S., Carvalho, D. and Siqueira, J.O. (2000) Genotypic, phenotypic and symbiotic diversity amongst rhizobia isolates from *Phaseolus vulgaris* L. growing in the Amazon region. In: *The Biology and Fertility of Tropical Soils*, TSBF report 1997–1998. TSBF, c/o UNESCO, Nairobi.

Polhill, R.M. (1981) *Papilionoideae*. In: Polhill, R.M. and Raven, P.H. (eds) *Advances in Legume Systematics: Part 1*. Royal Botanic Gardens, Kew, UK.

Polhill, R.M., Raven, P.H. and Stirton, C.H. (1981) Evolution and systematics of the *Leguminosae*. In: Polhill, R.M. and Raven, P.H. (eds) *Advances in Legume Systematics: Part 1*. Royal Botanic Gardens, Kew, UK.

Purcino, H.M.A., Festin, P.M. and Elkan, G.H. (2000) Identification of effective strains of *Bradyrhizobium* for *Arachis pintoi. Tropical Agriculture* 77, 226–231.

Rademaker, J.L.W. and Bruijn, F.D. (1997) Characterization and classification of microbes by rep-PCR genomic fingerprinting and computer-assisted pattern analysis. In: Caetano-Anollés, G. and Gresshoff, P.M. (eds) *DNA Markers: Protocols, Applications, and Overviews*. John Wiley & Sons, New York.

Rahman, F. (1977) Introdução e melhoramento de soja na várzea do rio Solimões (Caldeirão, Cacau Pirera), no período de 1975 a 1976. *Acta Amazônica* 7, 449–454.

Ramos, M.L.G., Magalhães, N.F.M. and Boddey, R.M. (1987) Native and inoculated rhizobia isolated from field grown *Phaseolus vulgaris*: effects of liming an acid soil on antibiotic resistance. *Soil Biology and Biochemistry* 19, 179–185.

Rappé, S.J. and Giovannoni, S. (2003) The uncultured microbial majority. *Annual Review of Microbiology* 57, 369–394.

Rivas, R., Velázquez, E., Willems, A., Vizcaíno, N., Subba-Rao, N., Mateos, P.F., Gillis, M., Dazzo, F.D. and Martinez-Molina, E. (2002) A new species of *Devosia* that forms a unique nitrogen-fixing root-nodule symbiosis with the aquatic legume *Neptunia natans* (L.f.) Druce. *Applied and Environmental Microbiology* 68, 5217–5222.

Rivas, R., Willems, A., Subba-Rao, N., Mateos, P.F., Dazzo, F.D., Kroppenstedt, R.M., Martinez-Molina, E. Gillis, M. and Vizcaíno, N. (2003) Description of *Devosia neptuniae* sp.nov. that nodulates and fix nitrogen in symbiosis with *Neptunia natans*, an aquatic legume from India. *Systematic and Applied Microbiology* 26, 47–53.

Rome, S., Fernadez, M.P., Brunel, B., Normand, P. and Cleyet-Marel, J.C. (1996) *Sinorhizobium medicae* sp. nov., isolated from annual *Medicago* spp. *International Journal of Systematic Bacteriology* 46, 972–980.

Rumjanek, N.G., Dobert, R.C., van Berkum, P. and Triplett, E.W. (1993) Common soybean inoculant strains in Brazil are members of *Bradyrhizobium elkanii*. *Applied and Environmental Microbiology* 59, 4371–4373.

Santos, M.A., Vargas, M.A.T. and Hungria, M. (1999) Characterization of soybean bradyrhizobium strains adapted to the Brazilian savannas. *FEMS Microbiology Ecology* 30, 261–272.

Scholla, M.H. and Elkan, G.H. (1984) *Rhizobium fredii* sp. nov., a fast-growing species that effectively nodulates soybeans. *International Journal of Systematic Bacteriology* 34, 484–486.

Scotti, M.R.M.M.L., Sá, N.M.H., Vargas, M.A.T. and Dobereiner, J. (1982) Streptomycin resistance of *Rhizobium* isolates from Brazilian cerrados. *Anais de Academia Brasileira de Ciências* 54, 733–738.

Segovia, L., Piñero, D., Palacios, R. and Martinez-Romeiro, E. (1991) Genetic structure of a soil population of nonsymbiotic *Rhizobium leguminosarum*. *Applied Environmental Microbiology* 57, 426–433.

Segovia, L., Young, J.P.W. and Martinez-Romero, E. (1993) Reclassification of American *Rhizobium leguminosarum* biovar phaseoli type I strains as *Rhizobium etli* sp. nov. *International Journal of Systematic Bacteriology* 43, 374–377.

Shaw, J.E., Reynolds, T. and Sprent, J. (1997) A study of the symbiotic importance and location of *nod* gene inducing compounds in two widely nodulating and two non-nodulating tropical tree species. *Plant and Soil* 188, 77–82.

Silva, G.G. and Franco, A.A. (1984) Selection of *Rhizobium* spp. strains in culture medium for acid soils. *Pesquisa Agropecuária Brasileira* 19, 169–173.

Silva, D.R.C., Leitão, M.R.S.M.M. and Vargas, M.A.T. (1998) Changes in protein profiles (SDS-PAGE) of competitive *Bradyrhizobium* sp. strains during soybean pre-infection stage. *Pesquisa Agropecuaria Brasileira* 33, 1375–1388.

Silva, M.F., Carreira, L.M.M., Tavares, A.S., Ribeiro, I.C., Jardim, M.A.G., Lobo, M.G.A. and Oliveira, J. (1989) Leguminosas da Amazônia brasileira: Lista prévia. In: *Anais Congresso Nacional De Botânica*, Sociedade Brasileira de Botância, Belém, PA, Brazil, pp. 193–237.

Singleton, P.W., Bohlool, B.B. and Nakao, P.L. (1992) Legume response to rhizobial inoculation in the tropics: myths and realities. In: Lal, R. and Sanchez, P.A. (eds) *Myths and Science of Soil of the Tropics*. Soil Science Society of America, Madison, Wisconsin, pp. 135–155.

Siqueira, J.O. and Moreira, F.M.S. (1997) Microbial populations and activities in highly-weathered acidic soils: highlights of the Brazilian research. In: Moniz, A.C., Furlani, A.M.C., Schaffert, R.E., Fageria, N.K., Rosolem, C.A. and Cantarella, H. (eds) *Plant–Soil Interactions at Low pH*. Brazilian Soil Science Society, Campinas, SP, Brazil, pp. 139–156.

Siqueira, J.O., Pouyú, E. and Moreira, F.M.S. (1999) Arbuscular mycorrhizae on post-transplant growth of woody outplants in soil with excess of heavy metals. *Revista Brasileira de Ciência do Solo* 23, 569–580.

Soares, A.L.L. (2004) Agronomic efficiency of selected rhizobia strains and diversity of native soil populations able to nodulate cowpea and beans in Perdões. MSc thesis, Universidade Federal de Lavras, Lavras, MG, Brazil.

Souza, L.A.G., Magalhães, F.M.M. and Oliveira, L.A. (1984) Evaluation of the growth of *Rhizobium* from leguminous forest trees on different culture media. *Pesquisa Agropecuária Brasileira* 19, 165–168.

Souza, L.A.G., Silva, M.F. and Moreira, F.W. (1994) Capacity of nodulation of 100 *Leguminosae* in Amazônia. *Acta Amazônica* 24, 9–18.

Squartini, A., Struffi, P., Döring, H., Selnska-Pobell, S., Tola, E., Giacomini, A., Vendramin, E., Velázquez, E., Mateos, P.F., Matínez-Molina, E., Dazzo, F.B., Casella, S. and Nuti, M.P. (2002) *Rhizobium sullae* sp. nov. (formerly '*Rhizobium hedysari*'), the root-nodule microsymbiont of *Hedysarum coronarium* L. *International Journal of Systematic and Evolutionary Microbiology* 52, 1267–1276.

Sy, A., Giraud, E., Jourand, P., Garcia, N., Willems, A., de Lajudie, P., Prin, Y., Neyra, M., Gillis, M., Boivin-Masson, C. and Dreyfus, B. (2001) Methylotrophic *Methylobacterium* bacteria nodulate and fix nitrogen in symbiosis with legumes. *Journal of Bacteriology* 183, 214–220.

Tan, Z.Y., Kan, G.X., Wang, E.T., Reinhold-Hurek, B. and Chen, W.X. (2001) *Rhizobium yanglingense* sp.nov., isolated from arid and semi-arid regions in China. *International Journal of Systematic and Evolution Microbiology* 51, 901–914.

Thies, J.E., Holmes, E.M. and Vachot, A. (2001) Application of molecular techniques to studies in *Rhizobium* ecology: a review. *Australian Journal of Experimental Agriculture* 41, 299–319.

Toledo, I., Lloret, L. and Martinez-Romero, E. (2003) *Sinorhizobium americanum* sp.nov., a new Sinorhizobium species nodulating Acacia spp. in Mexico. *Systematic and Applied Microbiology* 26, 54–64.

Trannin, I.C.B., Moreira, F.M.S. and Siqueira, J.O. (2001a) Tolerance of *Bradyrhizobium* and *Azorhizobium* strains and isolates to copper, cadmium and zinc *'in vitro'*. *Revista Brasileira de Ciência do Solo* 25, 305–316.

Trannin, I.C.B., Moreira, F.M.S. and Siqueira, J.O. (2001b) Growth and nodulation of *Enterolobium contortsiliquum, Acacia mangium* and *Sesbania virgata* in heavy metal contaminated soil. *Revista Brasileira de Ciência do Solo* 25, 753–763.

van Berkun, P. and Eardly, B.D. (2002) The aquatic budding bacterium *Blastobacter denitrificans* is a nitrogen-fixing symbiont of *Aeschinomene indica*. *Applied and Environmental Microbiology* 68, 1132–1136.

van Berkun, P., Beyene, D., Bao, G., Campbell, T.A. and Eardly, B. (1998) *Rhizobium mongolense* sp. nov. is one of three rhizobial genotypes identified which nodulate and form nitrogen-fixing symbioses with *Medicago ruthenica* [(L.) Ledebour]. *International Journal of Systematic Bacteriology* 48, 13–22.

Vandamme, P. and Conye, T. (2004) Taxonomy of the genus *Cupriavidus*: a tale of lost and found. *International Journal of Systematic Bacteriology* 54, 2285–2289.

Vaneechoutte, M., Kamfer, P., De Baere, T., Falsen, E. and Verschraegen, G. (2004) *Wautersia* gen.nov. sp. nov., a new genus accommodating the phylogenetic lineage including *Ralstonia eutropha* and related species, and proposal of *Ralstonia* [*Pseudomonas*] *syzygii* (Roberts *et al.* 1990) comb. nov. *International Journal of Systematic Bacteriology* 54, 317–327.

Vargas, A.A.T. and Denardin, N.D. (1992) Tolerance to acidity and soil aluminium by rhizobia strains of beans isolated in São Paulo state, Brazil. *Revista Brasileira e Ciência do Solo* 16, 337–342.

Vargas, M.A.T., Peres, J.R.R. and Suhet, A.R. (1981) Reinoculação de soja em função dos sorogrupos de *Rhizobium japonicum* predominantes em solos do Cerrados. In: *Anais Seminario Nacional De Pesquisa De Soja*, Embrapa-CNPSo, Londrina, PR, Brazil, pp. 715–723.

Vargas, A.A.T., Denardin, N.D. and van Berkun, P. (1992) Tolerance of indigenous bean rhizobia to antibiotics and possible relationship with soil acidity factors. *Revista Brasileira de Ciência do Solo* 16, 331–336.

Vargas, M.A.T., Mendes, I.C., Suhet, A.R. and Peres, J.R.R. (1994) Inoculation of soybean in cerrado soils with established populations of *Bradyrhizobium japonicum*. *Revista de Microbiologia* 25, 245–250.

Velazquez, E., Igual, J.M., Willems, A., Fernandez, M.P., Munoz, E., Mateos, P.F., Abril, A., Toro, N., Normand, P., Cervantes, M., Gillis, M. and Martinez-Molina, E. (2001) *Mesorhizobium chacoense* sp. nov., a novel species that nodulates *Prosopis alba* in the Chaco Arido region (Argentina). *International Journal of Systematic and Evolution Microbiology* 51, 1011–1021.

Vinuesa, P., Léon-Barrios, M., Silva, C., Willems, A., Jarabo-Lorenzo, A., Pérez-Galdona, R., Werner, D. and Martinez-Romero, E. (2005) *Bradyrhizobium canariense* sp. nov., an acid-tolerant endosymbiont that nodulates genistoid legumes (*Papilionoideae*:Genisteae) from Canary Islands, along with *Bradyrhizobium japonicum* bv. genistearum, *Bradyrhizobium* genospecies α and *Bradyrhizobium* genospecies β. *International Journal of Systematic and Evolutionary Microbiology* 55, 569–575.

Vose, P.B. (1980) Introduction to nuclear techniques in agronomy and plant biology. Pergamon Press, Oxford, UK.

Wang, E.T., van Berkum, P., Beyene, D., Sui, X.H., Dorado, O., Chen, W.X. and Martinez-Romero, E. (1998) *Rhizobium huautlense* sp. nov., a symbiont of *Sesbania herbacea* that has a close phylogenetic relationship with *Rhizobium galegae*. *International Journal of Systematic Bacteriology* 48, 687–699.

Wang, E.T., Rogel-Hernández, A., Santos, A.G., Martinez-Romero, J., Cevallos, M.A. and Martinez-Romero, E. (1999a) *Rhizobium etli* bv. *mimosae*, a novel biovar isolated from *Mimosa affinis*. *International Journal of Systematic Bacteriology* 49, 1479–1491.

Wang, E.T., van Berkum, P., Sui, X.H., Beyene, D., Chen, W.X. and Martinez-Romero, E. (1999b) Diversity of rhizobia associated with *Amorpha fruticosa* isolated from Chinese soils and description of *Mesorhizobium amorphae* sp.nov. *International Journal of Systematic Bacteriology* 49, 51–65.

Wang, E.T., Tan, Z.Y., Willems, A., Fernández-López, M., Rinhold-Hurek, B. and Martínez-Romero, E. (2002) *Sinorhizobium morelense* sp. nov., a *Leucena leucocephala*-associated bacterium that is highly resistant to multiple antibiotics. *International Journal of Systematic Microbiology* 52, 1687–1693.

Weaver, R.W. and Frederick, L.R. (1974) Effect of inoculum rate on competitive nodulation of *Glycine max* L. Merrill. II. Field studies. *Agronomy Journal* 66, 233–236.

Wei, G.H., Wang, E.T., Tan, M.E., Zhu, M.E. and Chen, W.X. (2002) *Rhizobium indigoferae* sp. nov. and *Sinorhizobium kummerowieae* sp. nov., respectively isolated from *Indigofera* spp. and *Kummerowia stipulacea*. *International Journal of Systematic and Evolutionary Microbiology* 52, 2231–2239.

Wei, G.H., Wang, E.T., Zhu, M.E., Wang, E.T., Han, S.Z. and Chen, W.X. (2003) Characterization of rhizobia isolated from legume species within the genera *Astragalus* and *Lespedeza* grown in Loess plateau region of China and description of *Rhizobium loessense* sp. nov. *International Journal of Systematic and Evolutionary Microbiology* 53, 1575–1583.

Wernegreen, J.J. and Riley, M.A. (1999) Comparison of the evolutionary dynamics of symbiotic and housekeeping loci: a case for the genetic coherence of rhizobial lineages. *Molecular Biology and Evolution* 16, 98–113.

Willems, A., Fernández-Lopez, M., Muñoz-Adelantado, E., Goris, J., Vos, P., Martínez-Romero, E., Toro, N. and Gillis, M. (2003) Description of new *Ensifer* strains from nodules and proposal transfer *Ensifer adhaerens* Cassida 1982 to *Sinorhizobium* as *Sinorhizobium adhaerens* comb. nov. Request for an opinion. *International Journal of Systematic and Evolutionary Microbiology* 53, 1207–1217.

Woese, C.R. (1987) Bacterial evolution. *Microbiological Reviews* 51, 221–271.

Woese, C.R. (1991) Prokaryote systematics: The evolution of a science. In: Ballows, A., Trüper, H.G., Dworkin, M., Harder, W. and Schleifer, K. (eds) *The Prokaryotes: A Handbook on the Biology of Bacteria, Ecophysiology, Isolation, Identification, Applications,* 2nd edn. Springer-Verlag, New York, pp. 3–18.

Woomer, P., Singleton, P.W. and Bohlool, B.B. (1988) Ecological indicators of rhizobia in tropical soils. *Applied and Environmental Microbiology* 54, 1112–1116.

Xavier, G.R., Martins, L.M.V., Neves, M.C.P. and Rumjanek, N.G. (1998) Edaphic factors as determinants for the distribution of intrinsic antibiotic resistance in a cowpea rhizobia population. *Biology and Fertility of Soils* 27, 386–392.

Xu, L.M., Ge, C., Cui, Z., Li, J. and Fan, H. (1995) *Bradyrhizobium liaoningense* sp. nov., isolated from the root nodules of soybeans. *International Journal of Systematic Bacteriology* 45, 706–711.

Yao, Z.Y., Kan, F.L., Wang, E.T. and Chen, W.X. (2002) Characterization of rhizobia that nodulate legume species within the genus *Lespedeza* and description of *Bradyrhizobium yuanmigense* sp. nov. *International Journal of Systematic Bacteriology* 52, 2219–2230.

Yoneyama, T., Muraoka, T., Murakami, T. and Bookerd, N. (1993) Natural abundance of ^{15}N in tropical plants with emphasis on tree legumes. *Plant and Soil* 153, 295–304.

Young, J.M. (1999) Correction to the authority of *Rhizobium leguminosarum*. *International Journal of Systematic Bacteriology* 49, 1943.

Young, J.P.W. (2000) Molecular evolution in diazotrophs: do the genes agree? In: Pedrosa, F.O., Hungria, M., Yates, G.Y. and Newton, W.E. (eds) *Nitrogen Fixation: From Molecules to Crop Productivity.* Kluwer, Dordrecht, The Netherlands, pp. 161–164.

Young, J.M. (2001) Implications of alternative classifications and horizontal gene transfer for bacterial taxonomy. *International Journal of Systematic and Evolutionary Microbiology* 51, 945–953.

Young, J.M. (2003) The genus name *Ensifer* Casida 1982 takes priority over *Sinorhizobium* Chen *et al.* 1988, and *Sinorhizobium morelense* Wang *et al.* 2002 is a junior synonym of *Ensifer adhaerens* Casida 1982. Is the combination '*Sinorhizobium adhaerens*' Casida 1982; Willems *et al.* 2002 legitimate? Request for an opinion. *International Journal of Systematic and Evolutionary Microbiology* 53, 2107–2110.

Young, J.M., Kuykendall, L.D., Martinez-Romero, E., Kerr, A. and Sawada, H. (2001) A revision of *Rhizobium* Frank 1889, with an emended description of the genus, and the inclusion of all species of *Agrobacterium* Conn 1942 and *Allorhizobium undicola* de Lajudie *et al.* 1998 as new combinations: *Rhizobium radiobacter, R. rhizogenes, R. rubi, R. undicola* and *R. vitis. International Journal of Systematic Evolution Microbiology* 51, 89–103.

Young, J.P.W. and Johston, A.W.B. (1989) The evolution of specificity in the legume-*Rhizobium* symbiosis. *Tree* 4, 341–349.

Yuyama, K. and Oliveira, L.A. (1997) Pesquisas com culturas anuais para a produção de grãos. In: Noda, H., Souza, L.A.G. and Fonseca, O.J.M. (eds) *Duas décadas de contribuições do INPA à pesquisa agronômica no trópico úmido.* Instituto Nacional de Pesquisas da Amazônia, pp. 89–109.

Index